Linux系列丛书

Linux自动化运维实战指南

吴光科　朱轩溢　杜　臣 ◎ 编著

北京理工大学出版社
BEIJING INSTITUTE OF TECHNOLOGY PRESS

内 容 简 介

本书系统地论述了 Linux 运维领域的各种技术，全书共 11 章，主要内容包括 Shell 企业编程基础、Shell 编程高级企业实战、自动化运维发展、Puppet 自动运维企业实战、Ansible 自动运维企业实战、SaltStack 自动运维企业实战、企业邮件服务器实战、Jenkins 持续集成企业实战、SVN 版本管理实战、Git 版本管理企业实战及 ELK 日志平台企业实战。

本书可作为系统管理员、网络管理员、在校大学生、Linux 运维工程师、Linux 系统管理人员及从事云计算、网站开发、测试、设计等人员的参考用书。

版权专有 侵权必究

图书在版编目（CIP）数据

Linux 自动化运维实战指南 / 吴光科，朱轩溢，杜臣编著. -- 北京：北京理工大学出版社，2025.1.
ISBN 978-7-5763-4996-2

Ⅰ. TP316.85-62

中国国家版本馆CIP数据核字第2025Z0D923号

责任编辑：江 立		文案编辑：江 立	
责任校对：周瑞红		责任印制：施胜娟	

出版发行 / 北京理工大学出版社有限责任公司
社　　址 / 北京市丰台区四合庄路 6 号
邮　　编 / 100070
电　　话 /（010）68944451（大众售后服务热线）
　　　　　（010）68912824（大众售后服务热线）
网　　址 / http://www.bitpress.com.cn

版 印 次 / 2025年1月第1版第1次印刷
印　　刷 / 三河市中晟雅豪印务有限公司
开　　本 / 787 mm × 1020 mm　1/16
印　　张 / 21.5
字　　数 / 477千字
定　　价 / 99.00 元

图书出现印装质量问题，请拨打售后服务热线，负责调换

前言 PREFACE

Linux 是当今三大操作系统（Windows、macOS、Linux）之一。Linux 系统创始人 Linus Torvalds（林纳斯·托瓦兹）21 岁的时候，用 4 个月的时间创建了 Linux 内核的第一个版本，并于 1991 年 10 月 5 日正式对外发布。该版本继承了 UNIX 操作系统以网络为核心的思想，是一个性能稳定的多用户网络操作系统。

随着互联网的飞速发展，IT 技术引领时代潮流，而 Linux 技术又是一切 IT 技术的基石，应用领域包括个人计算机、服务器、嵌入式应用、智能手机、云计算、大数据、人工智能、数字货币、区块链等。

我出生在贵州省一个贫困的小山村，从小经历了山里砍柴、放牛、挑水、做饭，日出而作、日落而归的朴素生活，看到父母一辈子都在小山村里，没有见过大城市，所以我从小立志要走出大山，让父母过上幸福的生活。正是这样一个信念让我不断地努力，大学毕业至今，我在"北漂"的 IT 运维路上走了 10 多年，从初创小公司到国企、机关单位，再到图吧、研修网、京东商城等一线 IT 企业，分别担任过 Linux 运维工程师、Linux 运维架构师、运维经理，直到今天创办了京峰教育培训机构。

一路走来，感谢生命中遇到的每一个人，是大家的帮助，让我不断地进步和成长，也让我明白了一个人活着不应该只为自己和自己的家人，也应该为这个社会贡献出哪怕只是一点点的价值。

为了帮助更多的人通过技术改变自己的命运，我决定编写《Linux 自动化运维实战指南》。虽然市面上有很多 Linux 书籍，但是很难找到一本关于 Linux 运维 Nginx Web、集群、分布式、动静分离、负载均衡、反向代理、Location、Rewrite、Tomcat、JVM 详解、参数优化、LVS、Keepalived、负载均衡算法、转发方式、DR、NAT、TUN 等技术，同时还包含 Linux 运维安全、安全加固、DDoS、黑客暴力破解、防御策略和方法等详细、全面的主流技术的书籍，这是我编写本书的初衷！

本书读者对象包括系统管理员、网络管理员、在校大学生、Linux 运维工程师、Linux 系统管理人员及从事云计算、网站开发、测试、设计等人员。

尽管我花费了大量的时间和精力核对书中的代码和语法，但其中难免还会存在一些纰漏，恳请读者批评指正。

吴光科

致 谢
THANKS

感谢 Linux 之父 Linus Torvalds，他不仅创造了 Linux 操作系统，还影响了整个开源世界，同时也影响了我的一生。

感谢我亲爱的父母，含辛茹苦地把我们兄弟三人抚养长大，是他们对我无微不至的照顾，让我有更多的精力和动力工作，并帮助更多的人。

感谢潘彦伊、周飞、何红敏、周孝坤、杨政平、王帅、李强、刘继刚、常青帅、孙娜、花杨梅、吴俊、李芬伦、陈洪刚、黄宗兴、代敏、杨永琴、姚钗、王志军、谭陈诚、王振、杨浩鹏、张德、刘建波、洛远、谭庆松、李涛、张强、刘峰、周育佳、谢彩珍、王奇、李建堂、张建潮、佘仕星、潘志付、薛洪波、王中、朱愉、左堰鑫、齐磊、韩刚、舒畅、何新华、朱军鹏、孟希东、黄鑫、陈权志、胡智超、焦伟、曾地长、孙峰、黄超、陈宽、罗正峰、潘禹之、揭长华、姚仑、高玲、陈培元、秦业华、沙伟青、戴永涛、唐秀伦、金鑫、石耀文、梁凯、彭浩、唐彪、郭大德、田文杰、柴宗虎、张馨、赵武星、王永明、何庆强、张镇卿、周聪、周玉海、周泊江、吴啸烈、卫云龙、刘祥胜等挚友多年来对我的信任和支持。

感谢腾讯公司腾讯课堂所有课程经理及平台老师，感谢 51CTO 学院院长一休和全体工作人员对我及京峰教育培训机构的大力支持。

感谢京峰教育培训机构的每位学员对我的支持和鼓励，希望他们都学有所成，成为社会的中流砥柱。感谢京峰教育培训机构 COO 蔡正雄，感谢京峰教育培训机构的全体老师和助教，是他们的大力支持，让京峰教育能够帮助更多的学员。

最后感谢我的爱人黄小红，是她一直在背后默默地支持我、鼓励我，让我有更多的精力和时间去完成这本书。

目 录
CONTENTS

第 1 章　Shell 企业编程基础 ·············· 1
　1.1　Shell 编程入门 ·············· 1
　1.2　Shell 脚本及编写 Hello World 程序 ·············· 2
　1.3　Shell 编程 ·············· 3
　　　1.3.1　变量详解 ·············· 3
　　　1.3.2　系统变量 ·············· 4
　　　1.3.3　环境变量 ·············· 4
　　　1.3.4　用户变量 ·············· 4
　1.4　if 条件语句实战 ·············· 5
　1.5　Shell 编程括号和符号详解 ·············· 7
　　　1.5.1　括号详解 ·············· 7
　　　1.5.2　符号详解 ·············· 7
　1.6　MySQL 数据库备份脚本 ·············· 8
　1.7　LNMP 一键自动化安装脚本 ·············· 8
　1.8　for 循环语句实战 ·············· 11
　1.9　while 循环语句实战 ·············· 13
　1.10　case 选择语句实战 ·············· 16
　1.11　select 选择语句实战 ·············· 17
　1.12　Shell 编程函数实战 ·············· 18
　1.13　Shell 编程"四剑客" ·············· 19
　　　1.13.1　find ·············· 19
　　　1.13.2　sed ·············· 21
　　　1.13.3　awk ·············· 24
　　　1.13.4　grep ·············· 27
　1.14　Shell 数组编程 ·············· 29

第 2 章　Shell 编程高级企业实战 ·············· 32
　2.1　Shell 编程 Linux 操作系统备份脚本 ·············· 32

2.2 Shell 编程收集服务器信息脚本 ·· 34
2.3 Shell 编程拒绝恶意 IP 地址登录脚本 ·· 36
2.4 Shell 编程 LAMP 部署脚本 ·· 37
2.5 Shell 编程 LNMP 部署脚本 ·· 40
2.6 Shell 编程 MySQL 主从复制脚本 ·· 43
2.7 Shell 编程修改 IP 地址及主机名脚本 ·· 45
2.8 Shell 编程 Zabbix 安装配置脚本 ··· 48
2.9 Shell 编程 Nginx 虚拟主机脚本 ·· 50
2.10 Shell 编程 Nginx、Tomcat 脚本 ·· 52
2.11 Shell 编程管理 Linux 操作系统的系统用户和系统组脚本 ····································· 55
2.12 Shell 编程 Vsftpd 虚拟用户管理脚本 ··· 58
2.13 Shell 编程 Apache 多版本软件安装脚本 ·· 60
2.14 Shell 编程局域网 IP 地址探活脚本 ·· 62
2.15 Shell 编程 Apache 虚拟主机管理脚本 ··· 65
2.16 Shell 编程实现 Apache 高可用脚本 ·· 67
2.17 Shell 编程拒绝黑客攻击 Linux 脚本 ·· 68
2.18 Shell 编程 mysqldump 数据库自动备份脚本 ·· 69
2.19 Shell 编程 MySQL 主从自动配置脚本 ··· 71
2.20 Shell 编程部署 Tomcat 多实例脚本 ··· 74
2.21 Shell 编程 Nginx 日志切割脚本 ··· 76
2.22 Shell 编程 Tomcat 实例和 Nginx 均衡脚本 ·· 76
2.23 Shell 编程密码远程执行命令脚本 ·· 87
2.24 Shell 编程密码远程复制文件脚本 ·· 88
2.25 Shell 编程 Bind DNS 管理脚本 ··· 89
2.26 Shell 编程 Docker 虚拟化管理脚本 ··· 94
2.27 Shell 编程脚本 ·· 99
    2.27.1 Shell 编程采集服务器硬件信息脚本 ·· 99
    2.27.2 Shell 编程 Linux 操作系统初始化脚本 ·· 99
    2.27.3 Shell 编程 Xtrabackup 数据库自动备份脚本 ··· 99
    2.27.4 Shell 编程 Linux 服务器免密钥分发脚本 ··· 99
    2.27.5 Shell 编程 Nginx 多版本软件安装脚本 ·· 100
    2.27.6 Shell 编程自动收集软件、端口、进程脚本 ·· 100
    2.27.7 Shell 编程 LVS 负载均衡管理脚本 ·· 100

2.27.8 Shell 编程 Keepalived 管理脚本 100
2.27.9 Shell 编程 Discuz 门户网站自动部署脚本 101
2.27.10 Shell 编程监控 Linux 磁盘分区容量脚本 101

第 3 章 自动化运维发展 102
3.1 传统运维方式简介 102
3.2 自动化运维简介 103
3.3 自动化运维的具体内容 103
3.4 建立高效的 IT 自动化运维管理 103
3.5 IT 自动化运维工具 104
3.6 IT 自动化运维体系 104

第 4 章 Puppet 自动运维企业实战 106
4.1 Puppet 入门 106
4.2 Puppet 工作原理 107
4.3 Puppet 安装配置 108
4.4 Puppet 企业案例演示 111
4.5 Puppet 常见资源及模块 113
4.6 Puppet file 资源案例 114
4.7 Puppet package 资源案例 117
4.8 Puppet service 资源案例 119
4.9 Puppet exec 资源案例 121
4.10 Puppet cron 资源案例 124
4.11 Puppet 日常管理与配置 125
4.11.1 Puppet 自动认证 125
4.11.2 Puppet 客户端自动同步 127
4.11.3 Puppet 服务器主动推送 128
4.12 Puppet 批量部署案例 129
4.12.1 Puppet 批量修改静态 IP 地址案例 129
4.12.2 Puppet 批量配置 NTP 同步服务器 131
4.12.3 Puppet 自动部署及同步网站 132

第 5 章 Ansible 自动运维企业实战 135
5.1 Ansible 工具特点 135
5.2 Ansible 运维工具原理 135
5.3 Ansible 管理工具安装配置 136

5.4　Ansible 工具参数详解 ... 138
5.5　Ansible ping 模块实战 ... 139
5.6　Ansible command 模块实战 ... 139
5.7　Ansible copy 模块实战 ... 141
5.8　Ansible YUM 模块实战 ... 143
5.9　Ansible file 模块实战 ... 144
5.10　Ansible user 模块实战 ... 145
5.11　Ansible cron 模块实战 ... 147
5.12　Ansible synchronize 模块实战 ... 149
5.13　Ansible Shell 模块实战 ... 151
5.14　Ansible service 模块实战 ... 152
5.15　Ansible Playbook 应用 ... 154
5.16　Ansible 配置文件详解 ... 160
5.17　Ansible 性能调优 ... 161

第 6 章　SaltStack 自动运维企业实战 ... 164

6.1　SaltStack 运维工具特点 ... 164
6.2　SaltStack 运维工具简介 ... 164
6.3　SaltStack 运维工具原理 ... 165
6.4　SaltStack 平台配置实战 ... 166
6.5　SaltStack 节点 Hosts 及防火墙设置 ... 166
6.6　SaltStack 管理工具安装配置 ... 167
6.7　SaltStack 工具参数详解 ... 168
6.8　SaltStack ping 模块实战 ... 169
6.9　SaltStack cmd 模块实战 ... 170
6.10　SaltStack copy 模块实战 ... 171
6.11　SaltStack pkg 模块实战 ... 172
6.12　SaltStack service 模块实战 ... 172
6.13　SaltStack 配置文件详解 ... 173
6.14　SaltStack State 自动化实战 ... 174
  6.14.1　SLS 文件企业实战案例一 ... 176
  6.14.2　SLS 文件企业实战案例二 ... 176
  6.14.3　SLS 文件企业实战案例三 ... 177
  6.14.4　SLS 文件企业实战案例四 ... 177

目录

  6.14.5 SLS 文件企业实战案例五 178
  6.14.6 SLS 文件企业实战案例六 178
  6.14.7 SLS 文件企业实战案例七 179
  6.14.8 SLS 文件企业实战案例八 179

第 7 章 企业邮件服务器实战 180

 7.1 邮件服务器简介 180
 7.2 Sendmail 安装配置 182
 7.3 Dovecot 服务配置 184
 7.4 Sendmail 别名配置 185
 7.5 测试邮件收发 185
 7.6 配置 Open WebMail 186
 7.7 Postfix 入门简介 189
 7.8 Postfix 服务安装 190
 7.9 Postfix 服务器配置 190
 7.10 Foxmail 本地邮箱配置 192
 7.11 PostfixAdmin 配置 194
 7.12 Roundcube GUI Web 配置 198
 7.13 Postfix 虚拟用户配置 204
 7.14 Postfix+ExtMail 配置实战 210
 7.15 Postfix+ExtMan 配置实战 215
 7.16 MailGraph_ext 安装配置 217
 7.17 Postfix+ExtMan 虚拟用户注册 219
 7.18 基于 ExtMan 自动注册并登录 220

第 8 章 Jenkins 持续集成企业实战 225

 8.1 传统部署网站的流程 225
 8.2 目前主流部署网站的流程 226
 8.3 Jenkins 持续集成简介 227
 8.4 Jenkins 持续集成组件 228
 8.5 Jenkins 平台实战部署 228
 8.6 Jenkins 相关概念 229
 8.7 Jenkins 平台设置 231
 8.8 Jenkins 构建 job 工程 234
 8.9 Jenkins 自动部署 237

8.10 Jenkins 插件安装 239

8.11 Jenkins 邮件配置 243

8.12 Jenkins 多实例配置 247

8.13 Jenkins+Ansible 高并发构建 253

第 9 章 SVN 版本管理实战 256

9.1 SVN 服务器简介 256

9.2 SVN 的功能特性 256

9.3 SVN 的架构剖析 257

9.4 SVN 的组件模块 259

9.5 SVN 分支概念剖析 259

9.6 基于 YUM 构建 SVN 服务器 260

9.7 SVN 二进制+Apache 整合实战 261

9.8 基于 MAKE 构建 SVN 服务器 263

9.9 SVN 源码+Apache 整合实战 265

9.10 SVN 客户端命令实战 266

9.11 Svnserve.conf 文件配置参数剖析 269

9.12 Passwd 文件参数剖析 270

9.13 Authz 文件参数剖析 270

第 10 章 Git 版本管理企业实战 272

10.1 版本控制的概念 272

10.2 本地版本控制系统 272

10.3 集中化版本控制系统 273

10.4 分布式版本控制系统 274

10.5 Git 版本控制系统简介 275

10.6 Git 和 SVN 的区别 275

10.7 Git 版本控制系统实战 279

10.8 配置 Git 版本仓库 280

10.9 Git 获取帮助 283

第 11 章 ELK 日志平台企业实战 284

11.1 ELK 架构原理深入剖析 285

11.2 ElasticSearch 配置实战 287

11.3 ElasticSearch 配置故障演练 289

11.4 ElasticSearch 插件部署实战 290

11.5	Kibana Web 安装配置	292
11.6	Logstash 客户端配置实战	294
11.7	ELK 收集系统标准日志	294
11.8	ELK-Web 日志数据图表	295
11.9	ELK-Web 中文汉化支持	297
11.10	Logstash 配置详解	299
11.11	Logstash 自定义索引实战	302
11.12	Grok 语法格式剖析	304
11.13	Redis 高性能加速实战	305
11.14	ELK 收集 MySQL 日志实战	305
11.15	ELK 收集 Kernel 日志实战	306
11.16	ELK 收集 Nginx 日志实战	308
11.17	ELK 收集 Tomcat 日志实战	310
11.18	ELK 批量日志集群实战	311
11.19	ELK 报表统计 IP 地域访问量	313
11.20	ELK 报表统计 Nginx 访问量	315
11.21	Filebeat 日志收集实战	317
11.22	Filebeat 案例实战	319
11.23	Filebeat 收集 Nginx 日志	319
11.24	Filebeat 自定义索引	321
11.25	Filebeat 收集多个日志	323
11.26	Kibana Web 安全认证	325
11.27	ELK 增加 X-pack 插件	328

第 1 章 Shell 企业编程基础

说到 Shell 编程,很多从事 Linux 运维工作的朋友都不陌生,都对 Shell 有基本的了解。初学者刚开始接触 Shell 的时候,可能感觉编程非常困难,但慢慢就会发现,Shell 是所有编程语言中最容易上手、最容易学习的编程脚本语言之一。

本章将介绍 Shell 编程入门、Shell 编程变量、if、for、while、case、select 等基本语句的案例演练及 Shell 编程"四剑客"(find、sed、awk、grep)的深度剖析等内容。

1.1 Shell 编程入门

Shell 是操作系统的最外层,可以合并编程语言以控制进程和文件,以及启动和控制其他程序。

Shell 通过提示符让用户输入,并向操作系统解释该输入,然后处理来自操作系统的任何结果输出。简单来说,Shell 就是用户和操作系统之间的一个命令解释器。

Shell 语言是用户与 Linux 操作系统之间沟通的桥梁,用户既可以输入并执行命令,也可以利用 Shell 脚本编程运行,如图 1-1 所示。

Shell 语言的种类非常多,常见的有以下几种。

(1) Bourne Shell (/usr/bin/sh 或 /bin/sh)。

(2) Bourne Again Shell (/bin/bash)。

(3) C Shell (/usr/bin/csh)。

(4) K Shell (/usr/bin/ksh)。

(5) Shell for Root (/sbin/sh)。

不同的 Shell 语言的语法有所不同,一般不能交换使用。最常用的 Shell 语言是 Bash,也就是 Bourne Again Shell。Bash 由于易用和免费,在日常工作中被广泛使用,也是大多数 Linux 操作系统默认的 Shell 环境。

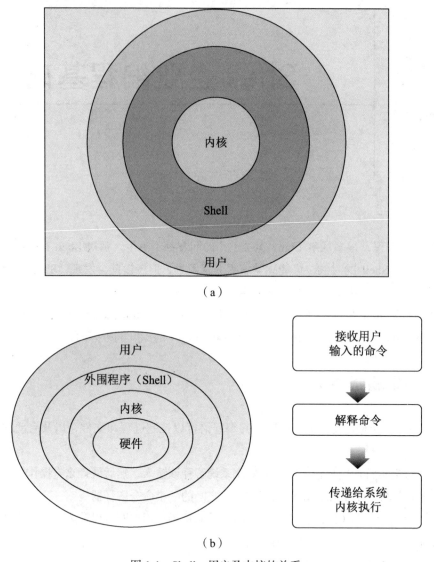

图 1-1 Shell、用户及内核的关系

（a）用户和内核关系；（b）用户、内核和硬件关系

Shell、Shell 编程、Shell 脚本、Shell 命令之间有什么区别呢？简单来说，Shell 是一个整体的概念，Shell 脚本内置 Shell 命令，Shell 命令则是 Shell 编程底层具体的语句和实现方法。

1.2 Shell 脚本及编写 Hello World 程序

想要熟练掌握 Shell 编程语言，需要大量的练习，初学者可以用 Shell 打印"Hello World"

字符串，寓意开始新的里程。

Shell 脚本编程需要注意以下事项。

（1）Shell 脚本名称中的英文区分大小写。

（2）不能使用特殊符号、空格来命名。

（3）Shell 脚本以.sh 结尾。

（4）不建议命名为纯数字，一般以脚本功能命名。

（5）Shell 脚本内容首行需以"#!/bin/bash"开头。

（6）Shell 脚本中变量名称尽量使用大写字母，字母间不能使用"-"，可以使用"_"。

（7）Shell 脚本变量名称不能以数字、特殊符号开头。

Shell 脚本编写完，如需运行该脚本，可以使用 chmod o+x first_shell.sh 命令赋予可执行权限，然后通过命令./first_shell.sh 执行。还可以使用命令/bin/sh first_shell.sh 直接运行脚本，不需要执行权限，脚本执行最终显示的效果是一样的。

初学者学习 Shell 编程时，可以将在 Shell 终端运行的各种命令依次写入脚本，把 Shell 脚本当成 Shell 命令的堆积。

1.3 Shell 编程

1.3.1 变量详解

Shell 是非类型的解释型语言，不像 C++、Java 语言编程时需要事先声明变量，Shell 给一个变量赋值，实际上就是定义了变量，在 Linux 操作系统支持的所有 Shell 中，都可以用赋值符号（=）为变量赋值。Shell 变量为弱类型，定义变量不需要声明类型，但在使用时需要明确变量的类型。

Shell 编程中的变量分为 3 种，分别是系统变量、环境变量和用户变量。Shell 变量名在定义时，首字符必须为字母（a~z、A~Z），不能以数字开头，中间不能有空格，可以使用下画线（_），不能使用半字线（-），也不能使用标点符号。

例如，定义变量 A=jfedu.net，其中 A 为变量名，jfedu.net 是变量的值。变量名有格式规范，变量的值可以随意指定。变量定义完成后，如需引用变量，可以使用$A。

Shell 脚本 var.sh 内容如下。

```
#!/bin/bash
#By author jfedu.net 2021
A=123
echo "Printf variables is $A."
```

执行该 Shell 脚本，结果将显示 Printf variables is jfedu.net。

1.3.2　系统变量

系统变量主要用于对参数和命令返回值的判断。系统变量详解如下。

```
$0                          #当前脚本的名称
$n                          #当前脚本的第 n 个参数, n=1,2,…,9
$*                          #当前脚本的所有参数(不包括程序本身)
$#                          #当前脚本的参数个数(不包括程序本身)
$?                          #命令或程序执行完成后的状态,返回 0 表示执行成功
$$                          #程序本身的 PID 号
```

1.3.3　环境变量

环境变量主要是在程序运行时设置。环境变量详解如下。

```
PATH                        #命令所示路径,以冒号分隔
HOME                        #打印用户主目录
SHELL                       #显示当前 Shell 类型
USER                        #打印当前用户名
ID                          #打印当前用户 ID 信息
PWD                         #显示当前所在路径
TERM                        #打印当前终端类型
HOSTNAME                    #显示当前主机名
```

1.3.4　用户变量

用户变量又称局部变量，主要在 Shell 脚本内部或者临时局部使用。用户变量详解如下。

```
SITE=jfedu.net              #自定义变量 SITE
IP1=192.168.1.11            #自定义变量 IP1
IP2=192.168.1.12            #自定义变量 IP2
WEB=www.jd.com              #自定义变量 WEB
SQL_DB=jfedu001             #自定义变量 SQL_DB
SQL_PWD=1qaz@WSX            #自定义变量 SQL_PWD
NGX_SOFT=nginx-1.16.0.tar.gz #自定义变量 NGX_SOFT
BAK_DIR=/data/backup/       #自定义变量 BAK_DIR
GATEWAY=192.168.0.1         #自定义变量 GATEWAY
```

创建 Echo 打印菜单 Shell 脚本，脚本代码如下。

```
#!/bin/bash
#auto install httpd
#By author jfedu.net 2021
```

```
echo -e '\033[32m-----------------------------\033[0m'
FILE=httpd-2.2.31.tar.bz2
URL=http://mirrors.cnnic.cn/apache/httpd/
PREFIX=/usr/local/apache2/
echo -e "\033[36mPlease Select Install Menu:\033[0m"
echo
echo "1)官方下载Httpd文件包."
echo "2)解压apache源码包."
echo "3)编译安装Httpd服务器."
echo "4)启动HTTPD服务器."
echo -e '\033[32m-----------------------------\033[0m'
sleep 20
```

运行脚本，结果如图 1-2 所示。

图 1-2　Echo 打印菜单脚本

1.4　if 条件语句实战

Shell 编程中，if、for、while、case 等条件语句用得非常多，熟练掌握以上条件语句及语法，对编写 Shell 脚本有非常大的益处。

if 条件语句通常以 if 开头，以 fi 结尾，也可加入 else 或 elif 进行多条件的判断。if 条件语句的表达方式如下。

```
if   (表达式)
    语句1
else
    语句2
fi
```

Shell 脚本编程中，if 条件语句案例如下。

（1）比较两个整数的大小。

```
#!/bin/bash
#By author jfedu.net 2021
```

```
NUM=100
if (( $NUM > 4 )) ;then
    echo "The Num $NUM more than 4."
else
    echo "The Num $NUM less  than 4."
fi
```

(2)判断系统目录是否存在。

```
#!/bin/bash
#judge DIR or Files
#By author jfedu.net 2021
if [ ! -d /data/20210515 -a ! -d /tmp/2021/ ];then
    mkdir -p /data/20210515
fi
```

(3)多个条件测试分数判断。

```
#!/bin/bash
#By author jfedu.net 2021
scores=$1
if [[ $scores -eq 100 ]]; then
    echo "very good!";
elif [[ $scores -gt 85 ]]; then
    echo "good!";
elif [[ $scores -gt 60 ]]; then
    echo "pass!";
elif [[ $scores -lt 60 ]]; then
    echo "no pass!"
fi
```

if 常见判断逻辑运算符详解如下。

```
-f                      #判断文件是否存在 eg: if [ -f filename ]
-d                      #判断目录是否存在 eg: if [ -d dir     ]
-eq                     #等于,应用于整型比较 equal
-ne                     #不等于,应用于整型比较 not equal
-lt                     #小于,应用于整型比较 letter
-gt                     #大于,应用于整型比较 greater
-le                     #小于或等于,应用于整型比较
-ge                     #大于或等于,应用于整型比较
-a                      #双方都成立(and)eg:逻辑表达式 -a 逻辑表达式
-o                      #单方成立(or)eg:逻辑表达式 -o 逻辑表达式
-z                      #空字符串
||                      #单方成立
&&                      #双方都成立表达式
```

1.5　Shell 编程括号和符号详解

1.5.1　括号详解

Shell 编程中，尤其是使用 if 条件语句时，经常会使用到( )、(( ))、[ ]、[[ ]]、{ }等括号，以下为几种括号使用方法的简单对比。

```
( )
#用于多个命令组、命令替换及初始化数组,多用于Shell命令组,例如,JF=(jf1 jf2 jf3),其
#中括号左右不保留空格
(( ))
#整数扩展、运算符、重新定义变量值,算术运算比较,例如,((i++))、((i<=100)),其中括号左
#右不保留空格
[ ]
#Bash 内部命令,[ ]与test是等同的,正则字符范围、引用数组元素编号,不支持+-*/数学运
#算符,逻辑测试使用-a、-o,通常用于字符串比较、整数比较以及数组索引,其中括号左右要保留
#空格
[[ ]]
#Bash 程序语言的关键字,不是一个命令,[[ ]]结构比[ ]结构更加通用,不支持+-*/数学运算
#符,逻辑测试使用&&、||,通常用于字符串比较、逻辑运算符等,其中括号左右要保留空格
{}
#主要用于命令集合或者范围,例如,mkdir -p /data/201{7,8}/,其中括号左右不保留空格
```

1.5.2　符号详解

Shell 编程中，使用变量时，经常会使用变量前导符、反斜杠、单引号、双引号、反引号等符号，以下为几种符号使用方法的简单对比。

（1）变量前导符（$）主要用于引用 Shell 编程中的变量，例如，定义变量 JF=www.jfedu.net，引用值需要用$JF。

（2）反斜杠（\）主要用于对特定的字符实现转义，保留原有意义，例如，echo "\$JF"结果会打印$JF，而不会打印 www.jfedu.net。

（3）单引号（' '）又称强引，不具有变量置换的功能，使任意字符还原为字面意义，可实现屏蔽 Shell 元字符的功能。

（4）双引号（" "）又称弱引，具有变量置换的功能，保留变量前导符、转义符、反引号原字符的功能。

（5）反引号（` `）位于键盘 Tab 键上面一行的键，用作命令替换（相当于$(…)）。

1.6 MySQL 数据库备份脚本

MySQL 数据库备份是运维工程师的工作内容之一，以下为自动备份 MySQL 数据库脚本。

```
#!/bin/bash
#auto backup mysql
#By author jfedu.net 2021
#Define PATH 定义变量
BAK_DIR=/data/backup/mysql/`date +%Y-%m-%d`
MYSQLDB=webapp
MYSQLPW=backup
MYSQLUSR=backup
#must use root user run scripts 必须使用root用户运行,$UID 为系统变量
if
   [ $UID -ne 0 ];then
   echo This script must use the root user!!!
   sleep 2
   exit 0
fi
#Define DIR and mkdir DIR 判断目录是否存在,不存在则新建
if
   [ ! -d $BAKDIR ];then
   mkdir -p $BAKDIR
fi
#Use mysqldump backup Databases
/usr/bin/mysqldump -u$MYSQLUSR -p$MYSQLPW -d $MYSQLDB >$BAKDIR/webapp_db.sql
echo "The mysql backup successfully"
```

1.7 LNMP 一键自动化安装脚本

现基于 if 条件语句和变量的知识，介绍一键源码安装 LNMP 脚本。

编写脚本应养成先分解脚本各个功能的习惯，以便于快速写出更高效的脚本。

一键源码安装 LNMP 脚本，可以拆分为以下步骤。

（1）LNMP 打印菜单。

① 安装 Nginx Web 服务。

② 安装 MySQL DB 服务。

③ 安装 PHP Web 服务。

④ 整合 LNMP 架构。

⑤ 启动 LNMP 服务。

（2）编写 LNMP 一键部署脚本，要先熟练掌握手动方式部署 LNMP 架构。auto_install_lnmp_v1.sh 脚本代码如下。

```bash
#!/bin/bash
#auto install LNMP Web
#by author www.jfedu.net
#######################
#Install Nginx Web
yum install -y wget gzip tar make gcc
yum install -y pcre pcre-devel zlib-devel
wget -c http://nginx.org/download/nginx-1.16.0.tar.gz
tar zxf nginx-1.16.0.tar.gz
cd nginx-1.16.0
useradd -s /sbin/nologin www -M
./configure --prefix=/usr/local/nginx --user=www --group=www --with-http_stub_status_module
make && make install
/usr/local/nginx/sbin/nginx
setenforce 0
systemctl stop firewalld.service

#Install MySQL database
cd ../
yum install -y gcc-c++ ncurses-devel cmake make perl gcc autoconf
yum install -y automake zlib libxml2 libxml2-devel libgcrypt libtool bison
wget -c http://mirrors.163.com/mysql/Downloads/MySQL-5.6/mysql-5.6.51.tar.gz
tar -xzf mysql-5.6.51.tar.gz
cd mysql-5.6.51
cmake . -DCMAKE_INSTALL_PREFIX=/usr/local/mysql56/ \
-DMYSQL_UNIX_ADDR=/tmp/mysql.sock \
-DMYSQL_DATADIR=/data/mysql \
-DSYSCONFDIR=/etc \
-DMYSQL_USER=mysql \
-DMYSQL_TCP_PORT=3306 \
-DWITH_XTRADB_STORAGE_ENGINE=1 \
-DWITH_INNOBASE_STORAGE_ENGINE=1 \
-DWITH_PARTITION_STORAGE_ENGINE=1 \
-DWITH_BLACKHOLE_STORAGE_ENGINE=1 \
-DWITH_MYISAM_STORAGE_ENGINE=1 \
-DWITH_READLINE=1 \
-DENABLED_LOCAL_INFILE=1 \
-DWITH_EXTRA_CHARSETS=1 \
-DDEFAULT_CHARSET=utf8 \
-DDEFAULT_COLLATION=utf8_general_ci \
-DEXTRA_CHARSETS=all \
-DWITH_BIG_TABLES=1 \
```

```
-DWITH_DEBUG=0
make
make install
#Config MySQL Set System Service
cd /usr/local/mysql56/
\cp support-files/my-large.cnf /etc/my.cnf
\cp support-files/mysql.server /etc/init.d/mysqld
chkconfig --add mysqld
chkconfig --level 35 mysqld on
mkdir -p /data/mysql
useradd mysql
/usr/local/mysql56/scripts/mysql_install_db --user=mysql --datadir=/data/mysql/ --basedir=/usr/local/mysql56/
ln -s /usr/local/mysql56/bin/* /usr/bin/
service mysqld restart
#Install PHP Web 2021
cd ../
yum install libxml2 libxml2-devel -y
wget http://mirrors.sohu.com/php/php-5.6.28.tar.bz2
tar jxf php-5.6.28.tar.bz2
cd php-5.6.28
./configure --prefix=/usr/local/php5 --with-config-file-path=/usr/local/php5/etc --with-mysql=/usr/local/mysql56/ --enable-fpm
make
make install

#Config LNMP Web and Start Server
cp php.ini-development   /usr/local/php5/etc/php.ini
cp /usr/local/php5/etc/php-fpm.conf.default /usr/local/php5/etc/php-fpm.conf
cp sapi/fpm/init.d.php-fpm /etc/init.d/php-fpm
chmod o+x /etc/init.d/php-fpm
/etc/init.d/php-fpm start

cat>/usr/local/nginx/conf/nginx.conf<<EOF
worker_processes  1;
events {
    worker_connections  1024;
}
http {
    include       mime.types;
    default_type  application/octet-stream;
    sendfile        on;
    keepalive_timeout  65;
```

```
    server {
        listen       80;
        server_name  localhost;
        location / {
           fastcgi_pass   127.0.0.1:9000;
           fastcgi_index  index.php;
           fastcgi_param  SCRIPT_FILENAME $document_root$fastcgi_script_name
           include        fastcgi_params;
        }

    location ~ .*\.(php|jsp|cgi)$
        {
        fastcgi_pass   127.0.0.1:9000;
           fastcgi_index  index.php;
           fastcgi_param  SCRIPT_FILENAME $document_root$fastcgi_script_name;
           include        fastcgi_params;
        }

    location ~ .*\.(htm|html|png|jpeg|gif|txt|js|css|doc)$
        {
        root html;
        expires 30d;
        }
       }
}
EOF
echo "
<?php
phpinfo();
?>">/usr/local/nginx/html/index.php
/usr/local/nginx/sbin/nginx -s reload
```

1.8 for 循环语句实战

for 循环语句主要用于对某个数据域进行循环读取或对文件进行遍历，通常用于循环某个文件或者列表。其语法格式以 for…do 开头，以 done 结尾。for 语法格式如下。

```
for var in （表达式）
do
    语句1
done
```

Shell 脚本编程中，for 循环语句的案例如下。

（1）打印百度、淘宝和腾讯的企业官网网址。

```
#!/bin/bash
#By author jfedu.net 2021
for website in www.baidu.com www.taobao.com www.qq.com
do
    echo $website
done
```

（2）打印 1~100 的数字。

```
#!/bin/bash
#By author jfedu.net 2021
for  i  in  'seq 1 100'          #seq 表示列出数据范围
do
    echo "NUM is $i"
done
```

（3）求数字 1~100 的总和。

```
#!/bin/bash
#By author jfedu.net 2021
#auto sum 1 100
j=0
for ((i=1;i<=100;i++))
do
    j='expr $i + $j'
done
echo $j
```

（4）对系统日志文件进行分组打包。

```
#!/bin/bash
#By author jfedu.net 2021
for  i  in  'find /var/log -name "*.log"'
do
    tar -czf 2021_log$i.tgz $i
done
```

（5）批量远程主机文件传输。

```
#!/bin/bash
#auto scp files for client
#By author jfedu.net 2021
for i in 'seq 100 200'
do
    scp -r /tmp/jfedu.txt root@192.168.1.$i:/data/webapps/www
done
```

（6）批量在远程机器上执行命令。

```
#!/bin/bash
#auto scp files for client
#By author jfedu.net 2021
for i in `seq 100 200`
do
     ssh -l root 192.168.1.$i 'ls /tmp'
done
```

（7）打印 10 s 的等待提示。

```
for ((j=0;j<=10;j++))
do
     echo -ne "\033[32m-\033[0m"
     sleep 1
done
echo
```

1.9 while 循环语句实战

while 循环语句与 for 循环语句功能类似，主要用于对某个数据域进行循环读取、对文件进行遍历，通常用于循环某个文件或者列表，满足循环条件会一直循环，不满足则退出循环。其语法格式以 while…do 开头，以 done 结尾。while 语法格式如下。

```
while  (表达式)
do
     语句 1
done
```

Shell 脚本编程中 while 循环语句的案例如下。

（1）打印百度、淘宝、腾讯的企业官网网址。

```
#!/bin/bash
#By author jfedu.net 2021
while read line                              #read 指令用于读取行或读取变量
do
    echo $line
done <jfedu.txt
```

其中，jfedu.txt 文件的内容为：

```
www.baidu.com
www.taobao.com
www.qq.com
```

（2）每秒输出 Hello World。

```
#!/bin/bash
#By author jfedu.net 2021
while sleep 1
do
     echo -e "\033[32mHello World.\033[0m"
done
```

其中，jfedu.txt 文件的内容为：

```
www.baidu.com
www.taobao.com
www.qq.com
```

（3）打印 1～100 的数字。

```
#!/bin/bash
#By author jfedu.net 2021
i=0
while ((i<=100))              #此处只打印1~100,当i>100时结束
do
     echo $i
     i=`expr $i + 1`          #expr用于逻辑运算
done
```

（4）求数字 1～100 的总和。

```
#!/bin/bash
#By author jfedu.net 2021
#auto sum 1 100
j=0
i=1
while ((i<=100))
do
     j=`expr $i + $j`
     ((i++))
done
echo $j
```

（5）逐行读取文件。

```
#!/bin/bash
#By author jfedu.net 2021
while read line
do
   echo $line;
done < /etc/hosts
```

（6）判断输入的 IP 地址格式是否正确。

```
#!/bin/bash
#By author jfedu.net 2021
#Check IP Address
read -p "Please enter ip Address,example 192.168.0.11 ip": IPADDR
echo $IPADDR|grep -v "[Aa-Zz]"|grep --color -E "([0-9]{1,3}\.){3}[0-9]{1,3}"
while [ $? -ne 0 ]
do
        read -p "Please enter ip Address,example 192.168.0.11 ip": IPADDR
        echo $IPADDR|grep -v "[Aa-Zz]"|grep --color -E "([0-9]{1,3}\.){3}[0-9]{1,3}"
done
```

（7）每 5 s 循环一次，判断/etc/passwd 路径下的文件是否被非法修改。

```
#!/bin/bash
#Check File to change
#By author jfedu.net 2021
FILES="/etc/passwd"
while true
do
        echo "The Time is 'date +%F-%T'"
        OLD=`md5sum $FILES|cut -d" " -f 1`
        sleep 5
        NEW=`md5sum $FILES|cut -d" " -f 1`
        if [[ $OLD != $NEW ]];then
                echo "The $FILES has been modified."
        fi
done
```

（8）每 10 s 循环一次，判断 jfedu 用户是否登录系统。

```
#!/bin/bash
#Check File to change
#By author jfedu.net 2021
USERS="jfedu"
while true
do
        echo "The Time is 'date +%F-%T'"
        sleep 10
        NUM=`who|grep "$USERS"|wc -l`
        if [[ $NUM -ge 1 ]];then
                echo "The $USERS is login in system."
        fi
done
```

1.10　case 选择语句实战

case 选择语句主要用于对多个选择条件进行匹配输出，与 if…elif 语句结构类似，通常用于脚本传递输入参数、打印出输出结果及内容，其语法格式以 case…in 开头，以 esac 结尾。case 语法格式如下。

```bash
#!/bin/bash
#By author jfedu.net 2021
case $1 in
    Pattern1)
    语句 1
    ;;
    Pattern2)
    语句 2
    ;;
    Pattern3)
    语句 3
    ;;
esac
```

Shell 脚本编程中 case 选择语句的案例如下。

（1）打印 monitor 及 archive 选择菜单。

```bash
#!/bin/bash
#By author jfedu.net 2021
case $1 in
      monitor)
      monitor_log
      ;;
      archive)
      archive_log
      ;;
      help)
      echo -e "\033[32mUsage:{$0 monitor | archive |help }\033[0m"
      ;;
      *)
      echo -e "\033[32mUsage:{$0 monitor | archive |help }\033[0m "
esac
```

（2）自动修改 IP 地址脚本菜单。

```bash
#!/bin/bash
#By author jfedu.net 2021
case $i in
        modify_ip)
```

```
                change_ip
                ;;
                modify_hosts)
                change_hosts
                ;;
                exit)
                exit
                ;;
                *)
                echo -e "1) modify_ip\n2) modify_ip\n3)exit"
esac
```

1.11 select 选择语句实战

select 选择语句主要用于选择，通常用于选择菜单的创建，可以配合 PS3 做打印菜单的输出信息，其语法格式以 select…in do 开头，以 done 结尾。select 语法格式如下。

```
select i in （表达式）
do
        语句
done
```

Shell 脚本编程中 select 选择语句案例如下。

（1）打印开源操作系统选择。

```
#!/bin/bash
#By author jfedu.net 2021
PS3="What you like most of the open source system?"
select i in CentOS RedHat Ubuntu
do
        echo "Your Select System: "$i
done
```

（2）打印 LAMP 选择菜单。

```
#!/bin/bash
#By author jfedu.net 2021
PS3="Please enter you select install menu:"
select i in http php mysql quit
do
case $i in
        http)
        echo Test Httpd.
        ;;
        php)
        echo Test PHP.
```

```
                ;;
                mysql)
                echo Test MySQL.
                ;;
                quit)
                echo The System exit.
                exit
        esac
done
```

1.12 Shell 编程函数实战

Shell 允许将一组命令集或语句形成一个可用块,这些块称为 Shell 函数。Shell 函数的优点在于只需定义一次,后期便可随时使用,无须在 Shell 脚本中重复添加语句块,其语法格式以"function name(){"开头,以"}"结尾。

Shell 编程函数默认不能将参数传入"()"内部,Shell 函数参数跟随函数名称传递,例如,name args1 args2。

```
function name (){
        command1
        command2
        ...
}
name args1 args2
```

(1)创建 Apache 软件安装函数,给 Apache_install 函数传递参数 1。

```
#!/bin/bash
#auto install LAMP
#By author jfedu.net 2021
#Httpd define path variable
H_FILES=httpd-2.2.31.tar.bz2
H_FILES_DIR=httpd-2.2.31
H_URL=http://mirrors.cnnic.cn/apache/httpd/
H_PREFIX=/usr/local/apache2/
function Apache_install()
{
#Install httpd Web Server
if [[ "$1" -eq "1" ]];then
    wget -c $H_URL/$H_FILES && tar -jxvf $H_FILES && cd $H_FILES_DIR &&./configure --prefix=$H_PREFIX
    if [ $? -eq 0 ];then
        make && make install
        echo -e "\n\033[32m-------------------------------------------------\033[0m"
```

```
            echo -e "\033[32mThe $H_FILES_DIR Server Install Success!\033[0m"
        else
            echo -e "\033[32mThe $H_FILES_DIR Make or Make install ERROR,Please
Check......"
            exit 0
        fi
    fi
}
Apache_install 1
```

（2）创建 judge_ip 函数判断 IP 地址的格式。

```
#!/bin/bash
#By author jfedu.net 2021
judge_ip(){
        read -p "Please enter ip Address,example 192.168.0.11 ip": IPADDR
        echo $IPADDR|grep -v "[Aa-Zz]"|grep --color -E "([0-9]{1,3}\.)
{3}[0-9]{1,3}"
}
judge_ip
```

1.13 Shell 编程"四剑客"

1.13.1 find

通过对以上基础语法知识的学习，读者对 Shell 编程有了更进一步的理解，即 Shell 编程不再是简单命令的堆积，而是各种特殊的语句、语法、编程工具及命令的集合。

在 Shell 编程工具中，"四剑客"工具的使用最为广泛。Shell 编程"四剑客"包括 find、sed、grep 和 awk。熟练掌握"四剑客"工具会使 Shell 编程能力得到极大的提升。

find 工具主要用于操作系统文件和目录的查找，其语法参数格式如下。

```
find    path    -option    [ -print ]    [ -exec    -ok    command ]    { } \;
```

其中，option 常用参数详解如下。

```
-name      filename           #查找名为 filename 的文件
-type      b/d/c/p/l/f        #查找块设备文件、目录文件、字符设备文件、管道文件、符号链接
                              #文件、普通文件
-size      n[c]               #查找长度为 n 块[或 n 字节]的文件
-perm                         #按执行权限查找
-user      username           #按文件属主查找
-group     groupname          #按组查找
-mtime     -n +n              #按文件更改时间查找文件,-n 指 n 天以内,+n 指 n 天以前
-atime     -n +n              #按文件访问时间查找文件
```

```
-ctime      -n +n                    #按文件创建时间查找文件
-mmin       -n +n                    #按文件更改时间查找文件,-n 指 n min 以内,+n
                                     #指 n min 以前
-amin       -n +n                    #按文件访问时间查找文件
-cmin       -n +n                    #按文件创建时间查找文件
-nogroup                             #查无有效属组的文件
-nouser                              #查无有效属主的文件
-depth                               #先查找本目录,再进入子目录查找
-mount                               #查文件时不跨越文件系统 mount 点
-follow                              #如果遇到符号链接文件,就跟踪链接所指的文件
-prune                               #忽略某个目录
-maxdepth                            #查找目录级别深度
```

(1) find 工具-name 参数案例如下。

```
find   /data/    -name   "*.txt"         #查找/data/目录下以.txt 结尾的文件
find   /data/    -name   "[A-Z]*"        #查找/data/目录下以大写字母开头的文件
find   /data/    -name   "test*"         #查找/data/目录下以 test 开头的文件
```

(2) find 工具-type 参数案例如下。

```
find   /data/    -type   d                       #查找/data/目录下的文件夹
find   /data/    !  -type   d                    #查找/data/目录下的非文件夹
find   /data/    -type   l                       #查找/data/目录下的链接文件
find   /data/ -type d|xargs chmod 755 -R         #查目录类型并将权限设置为 755
find   /data/ -type f|xargs chmod 644 -R         #查文件类型并将权限设置为 644
```

(3) find 工具-size 参数案例如下。

```
find   /data/    -size   +1M             #查找文件大小大于 1 MB 的文件
find   /data/    -size   10M             #查找文件大小大小为 10 MB 的文件
find   /data/    -size   -1M             #查找文件大小小于 1 MB 的文件
```

(4) find 工具-perm 参数案例如下。

```
find   /data/    -perm   755             #查找/data/目录下权限为 755 的文件或目录
find   /data/    -perm   -007            #查找/data/目录下,权限为所有权限的文件或目录
                                         # -perm-007 与-perm 777 相同,表示所有权限
find   /data/    -perm   +644            #查找/data/目录下文件权限符号 644 以上的文件或目录
```

(5) find 工具-mtime 参数案例如下。

```
atime,access time                        #文件被读取或者执行的时间
ctime,change time                        #文件状态改变的时间
mtime,modify time                        #文件内容被修改的时间
find /data/ -mtime +30 -name   "*.log"   #查找 30 天以前的.log 文件
find /data/ -mtime -30 -name   "*.txt"   #查找 30 天以内的.txt 文件
```

```
find /data/ -mtime 30   -name   "*.txt"    #查找第 30 天的.txt 文件
find /data/ -mmin +30   -name   "*.log"    #查找 30 min 以前修改的.log 文件
find /data/ -amin -30   -name   "*.txt"    #查找 30 min 以内被访问的.txt 文件
find /data/ -cmin 30    -name   "*.txt"    #查找第 30 min 改变的.txt 文件
```

(6) find 工具参数综合案例如下。

```
#查找/data 目录下以.log 结尾、文件大小大于 10 KB 的文件,同时复制到/tmp 目录
find /data/ -name "*.log"  -type f -size +10k -exec cp {} /tmp/ \;
#查找/data 目录下以.log 结尾、文件大小大于 10KB、权限为 644 的文件并删除该文件
find /data/ -name "*.log"  -type f -size +10k -m perm 644 -exec rm -rf {} \;
#查找/data 目录下以.log 结尾、30 天以前的、文件大小大于 10MB 的文件并移动到/tmp 目录
find /data/ -name "*.log"  -type f -mtime +30 -size +10M -exec mv {} /tmp/ \;
```

1.13.2 sed

sed 是一个非交互式文本编辑器,它可以对文本文件和标准输入进行编辑,标准输入可以来自键盘输入、文本重定向、字符串、变量,甚至可以来自管道的文本。与 Vim 编辑器类似,它一次处理一行内容,可以编辑一个或多个文件,简化对文件的反复操作、编写转换程序等。

在处理文本时,把当前处理的行存储在临时缓冲区中,这个临时缓冲区称为"模式空间"(pattern space),紧接着用 sed 命令处理缓冲区中的内容,处理完成后再把缓冲区中的内容输出到屏幕上或写入文件中,这样逐行处理直到文件末尾。然而如果是打印到屏幕上,实质文件内容并没有改变,除非使用重定向存储输出或写入文件。其语法参数格式如下。

```
sed   [-Options]    ['Commands']    filename;
#sed 工具默认处理文本,文本内容输出到屏幕已经修改,但是文件内容其实并没有修改,需要加-i
#参数对文件进行彻底修改
x                           #x 为指定行号
x,y                         #指定从 x 行号到 y 行号的行号范围
/pattern/                   #查询包含模式的行
/pattern/pattern/           #查询包含两个模式的行
/pattern/,x                 #从与 pattern 的匹配行到 x 号行之间的行
x,/pattern/                 #从 x 号行到与 pattern 的匹配行之间的行
x,y!                        #查询不包括 x 行号和 y 行号的行
r                           #从另一个文件中读取文件
w                           #将文本写入到一个文件
y                           #变换字符
q                           #第一个模式匹配完成后退出
l                           #显示与八进制 ASCII 码等价的控制字符
{}                          #在定位行执行的命令组
p                           #打印匹配行
=                           #打印文件行号
```

```
a\                              #在定位行号之后追加文本信息
i\                              #在定位行号之前插入文本信息
d                               #删除定位行
c\                              #用新文本替换定位文本
s                               #使用替换模式替换相应模式
n                               #读取下一个输入行,用下一个命令处理新的行
N                               #将当前读入行的下一行读取到当前模式空间
h                               #将模式缓冲区的文本复制到保持缓冲区
H                               #将模式缓冲区的文本追加到保持缓冲区
x                               #互换模式缓冲区和保持缓冲区的内容
g                               #将保持缓冲区的内容复制到模式缓冲区
G                               #将保持缓冲区的内容追加到模式缓冲区
```

常用 sed 工具的企业演练案例如下。

(1) 替换 jfedu.txt 文件中的 old 为 new。

```
sed     's/old/new/g'           jfedu.txt
```

(2) 打印 jfedu.txt 文件中的第 1 行~第 3 行。

```
sed     -n '1,3p'               jfedu.txt
```

(3) 打印 jfedu.txt 文件中的第 1 行与最后一行。

```
sed     -n '1p;$p'              jfedu.txt
```

(4) 删除 jfedu.txt 文件中的第 1 行~第 3 行,删除匹配 jfedu 模式的行至最后一行。

```
sed     '1,3d'                  jfedu.txt
sed     '/jfedu/,$d'            jfedu.txt
```

(5) 删除 jfedu.txt 文件中的最后 6 行后,再删除剩余文件中的最后一行。

```
for  i  in 'seq 1 6';do  sed  -i  '$d'  jfedu.txt ;done
sed     '$d'                    jfedu.txt
```

(6) 删除 jfedu.txt 文件中的最后一行。

```
sed     '$d'                    jfedu.txt
```

(7) 在 jfedu.txt 文件中查找 jfedu 字符串的所在行,并在其下一行添加 word 字符串,a 表示在其下一行添加字符串。

```
sed     '/jfedu/aword'          jfedu.txt
```

(8) 在 jfedu.txt 文件中查找 jfedu 字符串的所在行,并在其上一行添加 word 字符串,i 表示在其上一行添加字符串。

```
sed     '/jfedu/iword'          jfedu.txt
```

（9）在 jfedu.txt 文件中查找以 test 结尾的行，并在其行尾添加 word 字符串，$表示结尾标识，&表示添加。

```
sed    's/test$/&word/g'    jfedu.txt
```

（10）在 jfedu.txt 文件中查找 www 的所在行，并在其行首添加 word 字符串，^表示起始标识，&表示添加。

```
sed    '/www/s/^/&word/'    jfedu.txt
```

（11）多个 sed 命令组合，使用-e 参数。

```
sed  -e  '/www.jd.com/s/^/&1./'  -e  's/www.jd.com$/&./g'  jfedu.txt
```

（12）多个 sed 命令组合，使用分号";"分隔。

```
sed  -e  '/www.jd.com/s/^/&1.;s/www.jd.com$/&./g'  jfedu.txt
```

（13）sed 读取系统变量，并替换变量。

```
WEBSITE=WWW.JFEDU.NET
sed  "s/www.jd.com/$WEBSITE/g"  jfedu.txt
```

（14）修改 SELinux 策略 enforcing 为 disabled。第一行命令表示查找/SELINUX/行，并将其行 enforcing 值改成 disabled。再执行第二行命令，其中!s 表示不包括/SELINUX/行。

```
sed  -i    '/SELINUX/s/enforcing/disabled/g'  /etc/selinux/config
sed  -i    '/SELINUX/!s/enforcing/disabled/g'  /etc/selinux/config
```

通常，sed 将待处理的行读入模式空间，脚本中的命令逐行进行处理，直到脚本执行完毕，然后该行被输出，模式空间清空；重复上述动作，文件中新的一行被读入，直到文件处理完毕。

如果希望在某个条件下脚本中的某个命令被执行，或者希望模式空间得到保留以便下一次处理，都可以使 sed 在处理文件时不按照正常的流程来进行，这时可以使用 sed 高级命令。sed 高级命令分为 3 种功能。

（1）N、D、P：处理多行模式空间的问题。

（2）H、h、G、g、x：将模式空间的内容放入存储空间以便接下来编辑。

（3）:、b、t：在脚本中实现分支与条件结构。

① 在 jfedu.txt 文件中的每行后插入一行空行。

```
sed    '/^$/d;G'            jfedu.txt
```

② 将 jfedu.txt 文件中的偶数行删除及隔两行删除一行。

```
sed    'n;d'                jfedu.txt
sed    'n;n;d'              jfedu.txt
```

③ 在 jfedu.txt 文件中的匹配行的前一行、后一行插入空行,同时在匹配前后插入空行。

```
sed '/jfedu/{x;p;x;}'       jfedu.txt
sed '/jfedu/G'              jfedu.txt
sed '/jfedu/{x;p;x;G;}'     jfedu.txt
```

④ 在 jfedu.txt 文件中的每行后加入空行,即每行占据两行空间。

```
sed '/^$/d;G'               jfedu.txt
```

⑤ 在 jfedu.txt 文件中的每行前加入顺序数字序号,加上制表符\t 及符号\.。

```
sed = jfedu.txt| sed 'N;s/\n/ /'
sed = jfedu.txt| sed 'N;s/\n/\t/'
sed = jfedu.txt| sed 'N;s/\n/\./'
```

⑥ 删除 jfedu.txt 文件中的行前和行尾的任意空格。

```
sed 's/^[ \t]*//;s/[ \t]*$//'  jfedu.txt
```

⑦ 打印 jfedu.txt 文件中的关键词 old 与 new 之间的内容。

```
sed -n '/old/,/new/'p       jfedu.txt
```

⑧ 打印及删除 jfedu.txt 文件中的最后两行。

```
sed '$!N;$!D'               jfedu.txt
sed 'N;$!P;$!D;$d'          jfedu.txt
```

⑨ 合并上下两行,即两行合并。

```
sed '$!N;s/\n/ /'           jfedu.txt
sed 'N;s/\n/ /'             jfedu.txt
```

1.13.3 awk

awk 是一个优良的文本处理工具,是 Linux 操作系统及 UNIX 操作系统环境中现有的功能最强大的数据处理引擎之一,以 Aho、Weinberger、Kernighan 三位发明者名字首字母命名。awk 是一个行级文本高效处理工具。awk 经过改进后生成的新版本有 Nawk、Gawk,一般 Linux 操作系统默认为 Gawk,Gawk 是 awk 的 GNU 开源免费版本。

awk 的基本原理是逐行处理文件中的数据,查找与命令行中所给定内容相匹配的模式,如果发现匹配内容,则进行下一个编程步骤;如果找不到匹配内容,则继续处理下一行。

awk 常用参数、变量、函数等详解如下。

```
awk    'pattern + {action}'    file
```

(1) awk 基本语法参数详解。

① 单引号' '是为了和 Shell 命令区分开。

② 大括号{ }表示一个命令分组。

③ pattern 是一个过滤器，表示匹配 pattern 条件的行才会进行 action 处理。

④ action 是处理动作，常见动作为 print。

⑤ 使用#作为注释，pattern 和 action 可以只有其一，但不能二者都没有。

（2）awk 内置变量详解。

```
FS                          #分隔符,默认是空格
OFS                         #输出分隔符
NR                          #当前行数,从 1 开始
NF                          #当前记录字段个数
$0                          #当前记录
$1~$n                       #当前记录第 n 个字段(列)
```

（3）awk 内置函数详解。

```
gsub(r,s)                   #在$0 中用 s 代替 r
index(s,t)                  #返回 s 中 t 的第一个位置
length(s)                   #s 的长度
match(s,r)                  #s 是否匹配 r
split(s,a,fs)               #在 fs 上将 s 分成序列 a
substr(s,p)                 #返回 s 从 p 开始的子串
```

（4）awk 常用操作符、运算符及判断符详解。

```
++ --                       #增加与减少(前置或后置)
^ **                        #指数(右结合性)
! + -                       #非、一元(unary) 加号、一元减号
+ - * / %                   #加、减、乘、除、余数
< <= == != > >=             #数字比较
&&                          #逻辑 and
||                          #逻辑 or
= += -= *= /= %= ^= **=     #赋值
```

（5）awk 与流程控制语句。

```
if(condition) { } else { }
while { }
do{ }while(condition)
for(init;condition;step){ }
break/continue
```

常用 awk 工具的企业演练案例如下。

（1）awk 打印硬盘设备名称，默认以空格分隔。

```
df   -h|awk '{print $1}'
```

（2）awk 以空格、冒号、\t、分号分隔。

```
awk -F '[ :\t;]' '{print $1}'              jfedu.txt
```

（3）awk 以冒号分隔，打印第 1 列，同时将内容追加到/tmp/awk.log 文件的后面。

```
awk -F: '{print $1 >>"/tmp/awk.log"}'        jfedu.txt
```

（4）打印 jfedu.txt 文件中的第 3 行~第 5 行，NR 表示打印行，$0 表示文本所有域。

```
awk 'NR==3,NR==5 {print}'             jfedu.txt
awk 'NR==3,NR==5 {print $0}'          jfedu.txt
```

（5）打印 jfedu.txt 文件中的第 3 行~第 5 行的第 1 列与最后一列。

```
awk 'NR==3,NR==5 {print $1,$NF}'      jfedu.txt
```

（6）打印 jfedu.txt 文件中长度大于 80 的行号。

```
awk 'length($0)>80 {print NR}'        jfedu.txt
```

（7）awk 引用 Shell 变量，使用-v 或双引号+单引号。

```
awk -v STR=hello '{print STR,$NF}'    jfedu.txt
STR="hello";echo| awk '{print "'${STR}'";}'
```

（8）awk 以冒号分隔，打印第 1 列并且只显示前 5 行。

```
cat /etc/passwd|head -5|awk -F: '{print $1}'
awk -F: 'NR>=1&&NR<=5 {print $1}' /etc/passwd
```

（9）awk 指定文件 jfedu.txt 中第 1 列的总和。

```
cat jfedu.txt |awk '{sum+=$1}END{print sum}'
```

（10）awk NR 行号除以 2 后若余数为 0 则跳过该行，继续执行下一行，并打印到屏幕上。

```
awk -F: 'NR%2==0 {next} {print NR,$1}' /etc/passwd
```

（11）awk 添加自定义字符。

```
ifconfig eth0|grep "Bcast"|awk '{print "ip_"$2}'
```

（12）awk 格式化输出 passwd 内容，printf 表示打印字符串，%表示格式化输出分隔符，s 表示字符串类型，-12 表示 12 个字符，-6 表示 6 个字符。

```
awk -F: '{printf "%-12s %-6s %-8s\n",$1,$2,$NF}' /etc/passwd
```

（13）awk OFS 输出格式化\t。

```
netstat -an|awk '$6 ~ /LISTEN/&&NR>=1&&NR<=10 {print NR,$4,$5,$6}' OFS="\t"
```

（14）awk 与 if 组合实战，判断数字比较。

```
echo 3 2 1 | awk '{ if(($1>$2)||($1>$3)) { print $2} else {print $1} }'
```

（15）awk 与数组组合实战，统计 passwd 文件用户数。

```
awk -F ':' 'BEGIN {count=0;} {name[count] = $1;count++;}; END{for (i = 0; i < NR; i++) print i, name[i]}' /etc/passwd
```

（16）awk 分析 Nginx 访问日志中状态码为 404、502 等的错误信息页面，并统计次数大于 20 的 IP 地址。

```
awk '{if ($9~/502|499|500|503|404/) print $1,$9}' access.log|sort|uniq -c|sort -nr | awk '{if($1>20) print $2}'
```

（17）用/etc/shadow 文件中的密文部分替换/etc/passwd 文件中的"x"位置，并生成新的 /tmp/passwd 文件。

```
awk 'BEGIN{OFS=FS=":"} NR==FNR{a[$1]=$2}NR>FNR{$2=a[$1];print >>"/tmp/passwd"}' /etc/shadow /etc/passwd
```

（18）awk 统计服务器状态连接数。

```
netstat -an | awk '/tcp/ {s[$NF]++} END {for(a in s) {print a,s[a]}}'
netstat -an | awk '/tcp/ {print $NF}' | sort | uniq -c
```

1.13.4　grep

全面搜索正则表达式（grep）是一种强大的文本搜索工具，它能使用正则表达式搜索文本，并把匹配的行打印出来。

UNIX/Linux 操作系统的 grep 家族包括 grep、egrep 和 fgrep，其中，egrep 和 fgrep 的命令与 grep 的命令有细微的区别：egrep 是 grep 的扩展，支持更多的 re 元字符；fgrep 是 fixed grep 或 fast grep 的简写，它们把所有的字母都看作单词，正则表达式中的元字符表示其自身的字面意义，不再有其他特殊的含义，一般使用比较少。

目前 Linux 操作系统默认使用 GNU 版本的 grep。它的功能更强，可以通过-G、-E、-F 命令行选项使用 egrep 和 fgrep 的功能。其语法格式及常用参数详解如下。

```
grep     -[acinv]    'word'    Filename
```

grep 常用参数详解如下。

```
-a              #以文本文件方式搜索
-c              #计算找到符合行的次数
-i              #忽略大小写
-n              #顺便输出行号
-v              #反向选择，即显示不包含匹配文本的所有行
-h              #查询多文件时不显示文件名
-l              #查询多文件时只输出包含匹配字符的文件名
-s              #不显示不存在或无匹配文本的错误信息
-E              #允许使用 egrep 扩展模式匹配
```

学习 grep 时，需要了解通配符、正则表达式两个概念。通配符主要用在 Linux 操作系统的 Shell 命令中，常用于文件或文件名称的操作，而正则表达式则用于文本内容中的字符串搜索和

替换，常用在 awk、grep、sed、Vim 工具中对文本的操作。

通配符类型详解如下。

```
*                        #0 个或者多个字符、数字
?                        #匹配任意一个字符
#                        #表示注解
|                        #管道符号
;                        #多个命令连续执行
&                        #后台运行指令
!                        #逻辑运算非
[ ]                      #内容范围,匹配括号中的内容
{ }                      #命令块,多个命令匹配
```

正则表达式详解如下。

```
*                        #前一个字符匹配 0 次或多次
.                        #匹配除了换行符以外任意一个字符
.*                       #代表任意字符
^                        #匹配行首,即以某个字符开头
$                        #匹配行尾,即以某个字符结尾
\(..\)                   #标记匹配字符
[]                       #匹配中括号里的任意指定字符,但只匹配一个字符
[^]                      #匹配除中括号以外的任意一个字符
\                        #转义符,取消特殊含义
\<                       #锚定单词的开始
\>                       #锚定单词的结束
{n}                      #匹配字符出现 n 次
{n,}                     #匹配字符出现大于等于 n 次
{n,m}                    #匹配字符至少出现 n 次,最多出现 m 次
\w                       #匹配文字和数字字符
\W                       #\w 的反置形式,匹配一个或多个非单词字符
\b                       #单词锁定符
\s                       #匹配任何空白字符
\d                       #匹配一个数字字符,等价于[0-9]
```

常用 grep 工具的企业演练案例如下。

```
grep -c "test"      jfedu.txt    #统计包含 test 字符串的总行数
grep -i "TEST"      jfedu.txt    #不区分大小写查找包含 test 字符串所有的行
grep -n "test"      jfedu.txt    #打印包含 test 字符串的行及行号
grep -v "test"      jfedu.txt    #不打印包含 test 字符串的行
grep "test[53]"     jfedu.txt    #以 test 字符开头,接 5 或 3 的行
grep "^[^test]"     jfedu.txt    #显示输出行首不是 test 字符串的行
grep "[Mm]ay"       jfedu.txt    #匹配以 M 字符或 m 字符开头的行
```

```
grep    "K…D"              jfedu.txt        #匹配以 K 字符开头,后接 3 个任意字符,再紧接 D
                                            #字符的行
grep    "[A-Z][9]D"         jfedu.txt        #匹配以大写字母开头,紧跟 9D 字符串的行
grep    "T\{2,\}"           jfedu.txt        #打印 T 字符连续出现 2 次以上的行
grep    "T\{4,6\}"          jfedu.txt        #打印 T 字符连续出现 4 次及 6 次的行
grep    -n "^$"             jfedu.txt        #打印空行所在的行号
grep    -vE "#|^$"          jfedu.txt        #不匹配文件中的#和空行
grep    --color -ra -E "db|config|sql"  *    #匹配包含 db 或 config 或 sql 的文件
grep    --color -E "\<([0-9]{1,3}\.){3}([0-9]{1,3})\>"   #jfedu.txt 匹配
                                                         #IPv4 地址
```

1.14 Shell 数组编程

数组是相同数据类型的元素按一定顺序排列的集合,把有限个类型相同的变量用一个名字命名,然后用编号区分各变量,这个名字称为数组名,编号称为下标。Shell 编程中常用一维数组。

数组的设计其实是为了处理方便,把具有相同类型的若干变量按有序的形式组织起来,可以减少重复频繁的单独定义。如图 1-3 所示为三维数组。

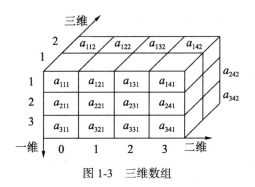

图 1-3 三维数组

数组一般以小括号的方式定义,数组的值可以随机指定。以下为一维数组的定义、统计、引用和删除操作。

(1)一维数组的定义及创建。

```
JFTEST=(
        test1
        test2
        test3
)
LAMP=(httpd  php  php-devel php-mysql mysql mysql-server)
```

（2）数组下标一般从 0 开始，以下为引用数组的方法。

```
echo        ${JFTEST[0]}                    #引用第 1 个数组变量,结果打印 test1
echo        ${JFTEST[1]}                    #引用第 2 个数组变量,结果打印 test2
echo        ${JFTEST[@]}                    #显示该数组的所有参数
echo        ${#JFTEST[@]}                   #显示该数组的参数个数
echo        ${#JFTEST[0]}                   #显示 test1 字符串的长度
echo        ${JFTEST[@]:0}                  #打印该数组所有的值
echo        ${JFTEST[@]:1}                  #打印该数组的第 2 个值开始的所有值
echo        ${JFTEST[@]:0:2}                #打印该数组的第 1 个值与第 2 个值
echo        ${JFTEST[@]:1:2}                #打印该数组的第 2 个值与第 3 个值
```

（3）数组替换操作。

```
JFTEST=( [0]=www1 [1]=www2 [2]=www3 )            #数组赋值
echo ${JFTEST[@]/test/jfedu}                     #将数组中值为 test 的替换为 jfedu
NEWJFTEST='echo ${JFTEST[@]/test/jfedu}'         #将结果赋值给新数组
```

（4）数组删除操作。

```
unset array[0]                                   #删除数组第 1 个值
unset array[1]                                   #删除数组第 2 个值
unset array                                      #删除整个数组
```

数组 Shell 脚本的企业案例 1：网卡绑定脚本。

```
#!/bin/bash
#Auto Make KVM Virtualization
#Auto config bond scripts
#By author jfedu.net 2021
eth_bond()
{
NETWORK=(
  HWADDR=`ifconfig eth0 |egrep "HWaddr|Bcast" |tr "\n" " "|awk '{print $5,$7,$NF}'|sed -e 's/addr://g' -e 's/Mask://g'|awk '{print $1}'`
  IPADDR=`ifconfig eth0 |egrep "HWaddr|Bcast" |tr "\n" " "|awk '{print $5,$7,$NF}'|sed -e 's/addr://g' -e 's/Mask://g'|awk '{print $2}'`
  NETMASK=`ifconfig eth0 |egrep "HWaddr|Bcast" |tr "\n" " "|awk '{print $5,$7,$NF}'|sed -e 's/addr://g' -e 's/Mask://g'|awk '{print $3}'`
  GATEWAY=`route -n|grep "UG"|awk '{print $2}'`
)
cat >ifcfg-bond0<<EOF
DEVICE=bond0
BOOTPROTO=static
${NETWORK[1]}
${NETWORK[2]}
${NETWORK[3]}
```

```
ONBOOT=yes
TYPE=Ethernet
NM_CONTROLLED=no
EOF
```

数组 Shell 脚本的企业案例 2：定义 IPv4 值。

```
#!/bin/bash
#auto Change ip netmask gateway scripts
#By author jfedu.net 2021
ETHCONF=/etc/sysconfig/network-scripts/ifcfg-eth0
HOSTS=/etc/hosts
NETWORK=/etc/sysconfig/network
DIR=/data/backup/`date +%Y%m%d`
NETMASK=255.255.255.0
echo "--------------------------"
count_ip(){
        count=(`echo $IPADDR|awk -F. '{print $1,$2,$3,$4}'`)
        IP1=${count[0]}
        IP2=${count[1]}
        IP3=${count[2]}
        IP4=${count[3]}
}
```

第 2 章　Shell 编程高级企业实战

企业生产环境中，服务器规模成百上千，如果依靠人工管理和维护，将非常吃力。Shell 编程脚本使管理和维护服务器变得简单、从容，对企业自动化运维之路的建设起到极大的推动作用。

本章将介绍企业生产环境 Shell 编程案例、自动化备份 MySQL 数据、服务器信息收集、防止恶意 IP 地址访问、LAMP+MySQL 主从实战、千台服务器 IP 地址修改、Nginx+Tomcat 高级自动化部署脚本、Nginx 虚拟主机配置、Docker 管理平台等内容。

2.1　Shell 编程 Linux 操作系统备份脚本

在日常企业运维中，需要备份 Linux 操作系统中重要的文件，例如，/etc 分区、/boot 分区、重要网站数据等。在备份时，由于数据量非常大，需要指定高效的备份方案，以下为常用的备份数据方案。

（1）每星期日进行完整备份，星期一至星期六使用增量备份。
（2）每星期六进行完整备份，星期日至星期五使用增量备份。

企业备份数据的工具主要有 tar、cp、rsync、scp、sersync、dd 等。以下为基于开源 tar 工具实现系统数据备份方案。

tar 工具手动完整备份网站，-g 参数指定新的快照文件。

```
tar -g  /tmp/snapshot  -czvf  /tmp/2021_full_system_data.tar.gz  /data/sh/
```

tar 工具手动增量备份网站，-g 参数指定完整备份已生成的快照文件，后续增量备份基于上一个增量备份快照文件。

```
tar -g  /tmp/snapshot  -czvf  /tmp/2021_add01_system_data.tar.gz  /data/sh/
```

tar 工具完整备份、增量备份网站，Shell 编程实现自动打包备份脚本，编写思路如下。

(1) 系统备份数据按天存放。

(2) 创建完整备份函数块。

(3) 创建增量备份函数块。

(4) 根据星期数判断完整备份或增量备份。

(5) 将脚本加入 Crontab 实现自动备份。

tar 工具完整备份、增量备份网站，Shell 编程实现自动打包备份脚本，相关代码如下。

```bash
#!/bin/bash
#Auto Backup Linux System Files
#By author jfedu.net 2021
#Define Path variables
SOURCE_DIR=(
    $*
)
TARGET_DIR=/data/backup/
YEAR=`date +%Y`
MONTH=`date +%m`
DAY=`date +%d`
WEEK=`date +%u`
A_NAME=`date +%H%M`
FILES=system_backup.tgz
CODE=$?
if
   [ -z "$*" ];then
   echo -e "\033[32mUsage:\nPlease Enter Your Backup Files or Directories\
n---------------------------------------------\n\nUsage: { $0 /boot /etc}\
033[0m"
   exit
fi
#Determine Whether the Target Directory Exists
if
   [ ! -d $TARGET_DIR/$YEAR/$MONTH/$DAY ];then
   mkdir -p $TARGET_DIR/$YEAR/$MONTH/$DAY
   echo -e "\033[32mThe $TARGET_DIR Created Successfully !\033[0m"
fi
#EXEC Full_Backup Function Command
Full_Backup()
{
if
   [ "$WEEK" -eq "7" ];then
   rm -rf $TARGET_DIR/snapshot
   cd $TARGET_DIR/$YEAR/$MONTH/$DAY ;tar -g $TARGET_DIR/snapshot -czvf
$FILES ${SOURCE_DIR[@]}
   [ "$CODE" == "0" ]&&echo -e "----------------------------------------
```

```
-----\n\033[32mThese Full_Backup System Files Backup Successfully !\033[0m"
        fi
}
#Perform incremental BACKUP Function Command
Add_Backup()
{
    if
        [ $WEEK -ne "7" ];then
        cd $TARGET_DIR/$YEAR/$MONTH/$DAY ;tar -g $TARGET_DIR/snapshot -czvf 
$A_NAME$FILES ${SOURCE_DIR[@]}
        [ "$CODE" == "0" ]&&echo -e "------------------------------------
-----\n\033[32mThese Add_Backup System Files $TARGET_DIR/$YEAR/$MONTH/
$DAY/${YEAR}_$A_NAME$FILES Backup Successfully !\033[0m"
        fi
}
sleep 3
Full_Backup;Add_Backup
```

在 Crontab 任务计划中添加如下语句, 每天凌晨 1 点整执行备份脚本。

```
0 1 * * * /bin/sh /data/sh/auto_backup.sh /boot /etc/ >> /tmp/back.log 2>&1
```

2.2 Shell 编程收集服务器信息脚本

在企业生产环境中, 经常需要对服务器资产进行统计存档, 单台服务器可以手动统计服务器的 CPU 型号、内存大小、硬盘容量、网卡流量等, 但如果服务器数量超过百台、千台, 使用手动方式将变得非常吃力。

Shell 编程实现自动化收集服务器的硬件信息, 并将收集的内容存放在数据库中, 能更快、更高效地实现对服务器资产信息的管理。Shell 编程实现服务器信息自动收集脚本, 编写思路如下。

(1) 创建数据库和表用来存储服务器信息。
(2) 使用 Shell "四剑客"(awk、find、sed、grep) 获取服务器信息。
(3) 将获取的信息写成 SQL 语句。
(4) 定期对 SQL 数据进行备份。
(5) 将脚本加入 Crontab 实现自动备份。

创建数据库表的 SQL 语句如下。

```
CREATE TABLE 'audit_system' (
  'id' int(11) NOT NULL AUTO_INCREMENT,
  'ip_info' varchar(50) NOT NULL,
  'serv_info' varchar(50) NOT NULL,
  'cpu_info' varchar(50) NOT NULL,
```

```
  'disk_info' varchar(50) NOT NULL,
  'mem_info' varchar(50) NOT NULL,
  'load_info' varchar(50) NOT NULL,
  'mark_info' varchar(50) NOT NULL,
  PRIMARY KEY ('id'),
  UNIQUE KEY 'ip_info' ('ip_info'),
  UNIQUE KEY 'ip_info_2' ('ip_info')
);
```

Shell 编程实现服务器信息自动收集脚本，相关代码如下。

```
#!/bin/bash
#Auto get system info
#By author jfedu.net 2021
#Define Path variables
echo -e "\033[34m \033[1m"
cat <<EOF
++++++++++++++++++++++++++++++++++++++++++++
++++++++Welcome to use system Collect+++++++++
++++++++++++++++++++++++++++++++++++++++++++
EOF
ip_info=`ifconfig |grep "Bcast"|tail -1 |awk '{print $2}'|cut -d: -f 2`
cpu_info1=`cat /proc/cpuinfo |grep 'model name'|tail -1 |awk -F: '{print $2}'|sed 's/^ //g'|awk '{print $1,$3,$4,$NF}'`
cpu_info2=`cat /proc/cpuinfo |grep "physical id"|sort |uniq -c|wc -l`
serv_info=`hostname |tail -1`
disk_info=`fdisk -l|grep "Disk"|grep -v "identifier"|awk '{print $2,$3,$4}'|sed 's/,//g'`
mem_info=`free -m |grep "Mem"|awk '{print "Total",$1,$2"M"}'`
load_info=`uptime |awk '{print "Current Load: "$(NF-2)}'|sed 's/\,//g'`
mark_info='BeiJing_IDC'
echo -e "\033[32m--------------------------------------------\033[1m"
echo IPADDR:${ip_info}
echo HOSTNAME:$serv_info
echo CPU_INFO:${cpu_info1} X${cpu_info2}
echo DISK_INFO:$disk_info
echo MEM_INFO:$mem_info
echo LOAD_INFO:$load_info
echo -e "\033[32m--------------------------------------------\033[0m"
echo -e -n "\033[36mYou want to write the data to the databases? \033[1m" ;
read ensure
if [ "$ensure" == "yes" -o "$ensure" == "y" -o "$ensure" == "Y" ];then
    echo "--------------------------------------------"
    echo -e '\033[31mmysql -uaudit -p123456 -D audit -e ''' "insert into audit_system values('','${ip_info}','$serv_info','${cpu_info1} X${cpu_info2}','$disk_info','$mem_info','$load_info','$mark_info')" ''' \033[0m '
    mysql -uroot -p123456 -D test -e "insert into audit_system values
```

```
('','${ip_info}','$serv_info','${cpu_info1} X${cpu_info2}','${disk_info},
'$mem_info','$load_info','$mark_info')"
else
    echo "Please wait,exit......"
    exit
fi
```

手动读取数据库服务器信息的命令如下。

```
mysql -uroot -p123 -e 'use wugk1 ;select * from audit_audit_system;'|sed
's/-//g'|grep -v "id"
```

2.3 Shell 编程拒绝恶意 IP 地址登录脚本

企业服务器暴露在外网，每天会有大量的人使用各种用户名和密码尝试登录服务器，如果用户一直尝试，难免会猜出密码。通过开发 Shell 脚本，可以自动将尝试登录服务器错误密码达到一定次数的 IP 地址列表加入防火墙配置。

Shell 编程实现服务器拒绝恶意 IP 地址登录脚本，编写思路如下。

（1）登录服务器日志/var/log/secure。

（2）检查日志中认证失败达到一定次数的行并打印其 IP 地址。

（3）将 IP 地址写入防火墙。

（4）禁止该 IP 地址访问服务器 SSH 22 端口。

（5）将脚本加入 Crontab 实现自动禁止恶意 IP 地址登录。

Shell 编程实现服务器拒绝恶意 IP 地址登录脚本，相关代码如下。

```
#!/bin/bash
#Auto drop ssh failed IP address
#By author jfedu.net 2021
#Define Path variables
SEC_FILE=/var/log/secure
IP_ADDR=`awk '{print $0}'  /var/log/secure|grep -i  "fail"| egrep -o "
([0-9]{1,3}\.){3}[0-9]{1,3}" | sort -nr | uniq -c |awk '$1>=15 {print $2}'`
IPTABLE_CONF=/etc/sysconfig/iptables
echo
cat <<EOF
++++++++++++welcome to use ssh login drop failed ip++++++++++++++++
++++++++++++++++++++++++++++++++++++++++++++++++++++++++++++++++++
++++++++++++++-----------------------------------++++++++++++++++
EOF
echo
for ((j=0;j<=6;j++)) ;do echo -n "-";sleep 1 ;done
echo
```

```
for i in 'echo $IP_ADDR'
do
    cat $IPTABLE_CONF |grep $i >/dev/null
if
    [ $? -ne 0 ];then
    sed -i "/lo/a -A INPUT -s $i -m state --state NEW -m tcp -p tcp --dport 22
-j DROP" $IPTABLE_CONF
fi
done
NUM='find /etc/sysconfig/ -name iptables -a -mmin -1|wc -l'
        if [ $NUM -eq 1 ];then
            /etc/init.d/iptables restart
        fi
```

2.4 Shell 编程 LAMP 部署脚本

LAMP 是目前互联网主流 Web 网站架构,通过源码安装、维护和管理单台服务器时很轻松,但如果服务器数量较多,手动管理就非常困难,Shell 脚本可以更快速地维护 LAMP 架构。

Shell 编程实现服务器 LAMP 一键源码安装配置脚本,编写思路如下。

(1) 利用脚本安装 LAMP 环境。

(2) Apache 安装配置,安装 MySQL、PHP。

(3) 源码 LAMP 整合配置。

(4) 启动数据库,创建数据库并授权。

(5) 重启 LAMP 所有服务,验证访问。

Shell 编程实现服务器 LAMP 一键源码安装配置脚本,相关代码如下。

```
#!/bin/bash
#Auto install LAMP
#By author jfedu.net 2021
#Define Path variables
#Httpd define path variable
H_FILES=httpd-2.2.32.tar.bz2
H_FILES_DIR=httpd-2.2.32
H_URL=http://mirrors.cnnic.cn/apache/httpd/
H_PREFIX=/usr/local/apache2/
#MySQL define path variable
M_FILES=mysql-5.5.20.tar.gz
M_FILES_DIR=mysql-5.5.20
M_URL=http://down1.chinaunix.net/distfiles/
M_PREFIX=/usr/local/mysql/
#PHP define path variable
P_FILES=php-5.3.28.tar.bz2
```

```bash
P_FILES_DIR=php-5.3.28
P_URL=http://mirrors.sohu.com/php/
P_PREFIX=/usr/local/php5/
function httpd_install(){
if [[ "$1" -eq "1" ]];then
    wget -c $H_URL/$H_FILES && tar -jxvf $H_FILES && cd $H_FILES_DIR &&./configure --prefix=$H_PREFIX
    if [ $? -eq 0 ];then
        make && make install
    fi
fi
}
function mysql_install(){
if [[ "$1" -eq "2" ]];then
wget -c $M_URL/$M_FILES && tar -xzvf $M_FILES && cd $M_FILES_DIR &&yum install cmake ncurses-devel -y ;cmake . -DCMAKE_INSTALL_PREFIX=$M_PREFIX \
-DMYSQL_UNIX_ADDR=/tmp/mysql.sock \
-DMYSQL_DATADIR=/data/mysql \
-DSYSCONFDIR=/etc \
-DMYSQL_USER=mysql \
-DMYSQL_TCP_PORT=3306 \
-DWITH_XTRADB_STORAGE_ENGINE=1 \
-DWITH_INNOBASE_STORAGE_ENGINE=1 \
-DWITH_PARTITION_STORAGE_ENGINE=1 \
-DWITH_BLACKHOLE_STORAGE_ENGINE=1 \
-DWITH_MYISAM_STORAGE_ENGINE=1 \
-DWITH_READLINE=1 \
-DENABLED_LOCAL_INFILE=1 \
-DWITH_EXTRA_CHARSETS=1 \
-DDEFAULT_CHARSET=utf8 \
-DDEFAULT_COLLATION=utf8_general_ci \
-DEXTRA_CHARSETS=all \
-DWITH_BIG_TABLES=1 \
-DWITH_DEBUG=0
if [ $? -eq 0 ];then
    make && make install
    echo -e "\n\033[32m--------------------------------------------------\033[0m"
            echo -e "\033[32mThe $M_FILES_DIR Server Install Success !\033[0m"
    else
            echo -e "\033[32mThe $M_FILES_DIR Make or Make install ERROR,Please Check......"
            exit 0
fi
/bin/cp support-files/my-small.cnf  /etc/my.cnf
```

```bash
/bin/cp support-files/mysql.server /etc/init.d/mysqld
chmod +x /etc/init.d/mysqld
chkconfig --add mysqld
chkconfig mysqld on
fi
}
function php_install(){
if [[ "$1" -eq "3" ]];then
        yum install libxml2-devel perl-devel perl libtool* -y
        wget -c $P_URL/$P_FILES && tar -jxvf $P_FILES && cd $P_FILES_DIR
&&./configure --prefix=$P_PREFIX --with-config-file-path=$P_PREFIX/etc
--with-mysql=$M_PREFIX --with-apxs2=$H_PREFIX/bin/apxs
        if [ $? -eq 0 ];then
                make ZEND_EXTRA_LIBS='-liconv' && make install
                echo -e "\n\033[32m----------------------------------------
--------\033[0m"
                echo -e "\033[32mThe $P_FILES_DIR Server Install Success !\
033[0m"
        else
                echo -e "\033[32mThe $P_FILES_DIR Make or Make install
ERROR,Please Check......"
                exit 0
        fi
fi
}
function lamp_config(){
if [[ "$1" -eq "4" ]];then
    sed -i '/DirectoryIndex/s/index.html/index.php index.html/g' $H_PREFIX/
conf/httpd.conf
    $H_PREFIX/bin/apachectl restart
    echo "AddType     application/x-httpd-php .php" >>$H_PREFIX/conf/httpd.
conf
    IP=`ifconfig eth0|grep "Bcast"|awk '{print $2}'|cut -d: -f2`
    echo "You can to access http://$IP/"

cat >$H_PREFIX/htdocs/index.php<<EOF
<?php
phpinfo();
?>
EOF
fi
}
PS3="Please enter you select install menu:"
select i in http mysql php config quit
do
```

```
case $i in
   http)
     httpd_install 1
     ;;
   mysql)
     mysql_install 2
     ;;
   php)
     php_install 3
     ;;
   config)
     lamp_config 4
     ;;
   quit)
     exit
esac
done
```

2.5 Shell 编程 LNMP 部署脚本

Shell 编程实现服务器 LNMP 部署安装脚本，编写思路如下。

（1）利用脚本安装 LNMP 环境。

（2）Nginx 安装配置，安装 MySQL、PHP。

（3）源码 LNMP 整合配置。

（4）启动数据库，创建数据库并授权。

（5）重启 LNMP 所有服务，验证访问。

Shell 编程实现服务器 LNMP 部署安装脚本，相关代码如下。

```
#!/bin/bash
#auto install lnmp web.
#by author www.jfedu.net
#######################
if [ $1 -eq 1 ];then
    #Install Nginx WEB.
    yum install -y wget gzip tar make gcc
    yum install -y pcre pcre-devel zlib-devel
    wget -c http://nginx.org/download/nginx-1.16.0.tar.gz
    tar zxf nginx-1.16.0.tar.gz
    cd nginx-1.16.0
    useradd -s /sbin/nologin www -M
    ./configure --user=www --group=www --prefix=/usr/local/nginx
    make && make install
    /usr/local/nginx/sbin/nginx
```

```bash
    setenforce 0
    systemctl stop firewalld.service
fi

if [ $1 -eq 2 ];then
    #Install MySQL Database.
    cd ../
    yum install -y gcc-c++ ncurses-devel cmake make perl gcc autoconf
    yum install -y automake zlib libxml2 libxml2-devel libgcrypt libtool bison
    wget -c http://mirrors.163.com/mysql/Downloads/MySQL-5.6/mysql-5.6.45.tar.gz
    tar -xzf mysql-5.6.45.tar.gz
    cd mysql-5.6.45
    cmake . -DCMAKE_INSTALL_PREFIX=/usr/local/mysql56/ \
    -DMYSQL_UNIX_ADDR=/tmp/mysql.sock \
    -DMYSQL_DATADIR=/data/mysql \
    -DSYSCONFDIR=/etc \
    -DMYSQL_USER=mysql \
    -DMYSQL_TCP_PORT=3306 \
    -DWITH_XTRADB_STORAGE_ENGINE=1 \
    -DWITH_INNOBASE_STORAGE_ENGINE=1 \
    -DWITH_PARTITION_STORAGE_ENGINE=1 \
    -DWITH_BLACKHOLE_STORAGE_ENGINE=1 \
    -DWITH_MYISAM_STORAGE_ENGINE=1 \
    -DWITH_READLINE=1 \
    -DENABLED_LOCAL_INFILE=1 \
    -DWITH_EXTRA_CHARSETS=1 \
    -DDEFAULT_CHARSET=utf8 \
    -DDEFAULT_COLLATION=utf8_general_ci \
    -DEXTRA_CHARSETS=all \
    -DWITH_BIG_TABLES=1 \
    -DWITH_DEBUG=0
    make
    make install
    #Config MySQL Set System Service
    cd /usr/local/mysql56/
    \cp support-files/my-large.cnf /etc/my.cnf
    \cp support-files/mysql.server /etc/init.d/mysqld
    chkconfig --add mysqld
    chkconfig --level 35 mysqld on
    mkdir -p /data/mysql
    useradd mysql
    /usr/local/mysql56/scripts/mysql_install_db --user=mysql --datadir=/data/mysql/ --basedir=/usr/local/mysql56/
    ln -s /usr/local/mysql56/bin/* /usr/bin/
    service mysqld restart
```

```
fi

if [ $1 -eq 3 ];then
    #Install PHP WEB 2018
    cd ../../
    yum install libxml2 libxml2-devel -y
    wget http://mirrors.sohu.com/php/php-5.6.28.tar.bz2
    tar jxf php-5.6.28.tar.bz2
    cd php-5.6.28
    ./configure --prefix=/usr/local/php5 --with-config-file-path=/usr/local/php5/etc --with-mysql=/usr/local/mysql56/ --enable-fpm
    make
    make install
fi

if [ $1 -eq 4 ];then
#Config LNMP WEB and Start Server.
cp php.ini-development   /usr/local/php5/etc/php.ini
cp  /usr/local/php5/etc/php-fpm.conf.default  /usr/local/php5/etc/php-fpm.conf
cp sapi/fpm/init.d.php-fpm /etc/init.d/php-fpm
chmod o+x /etc/init.d/php-fpm
/etc/init.d/php-fpm start
echo "
worker_processes  1;
events {
    worker_connections  1024;
}
http {
    include       mime.types;
    default_type  application/octet-stream;
    sendfile        on;
    keepalive_timeout  65;
    server {
        listen       80;
        server_name  localhost;
        location / {
            root   html;
            fastcgi_pass   127.0.0.1:9000;
            fastcgi_index  index.php;
            fastcgi_param  SCRIPT_FILENAME  \$document_root\$fastcgi_script_name;
            include        fastcgi_params;
        }
    }
}" >/usr/local/nginx/conf/nginx.conf
```

```
echo "
<?php
phpinfo();
?>">/usr/local/nginx/html/index.php
/usr/local/nginx/sbin/nginx -s reload
fi
```

2.6 Shell 编程 MySQL 主从复制脚本

MySQL 数据库服务器主要应用于动态网站,存放网站必要的数据,如订单、交易、员工表、薪资等记录。为了实现数据备份,需引入 MySQL 主从架构,MySQL 主从架构脚本可以实现自动化安装、配置和管理。

Shell 编程实现服务器 MySQL 一键 YUM 安装配置脚本,编写思路如下。

(1) MySQL 主库的操作。

① 在主库上安装 MySQL,并设置参数 server-id、bin-log。

② 授权复制同步的用户,对客户端授权。

③ 确认 bin-log 文件名及 position 位置点。

(2) MySQL 从库的操作。

① 在从库上安装 MySQL,设置参数 server-id。

② change master:指定主库、bin-log 文件名和 position 位置点。

③ start slave:启动从库 I/O 线程。

④ show slave status\G:查看主从的状态。

Shell 编程实现服务器 MySQL 一键 YUM 安装配置脚本,需要提前手动授权主库免密码登录从库服务器,相关代码如下。

```
#!/bin/bash
#Auto install Mysql AB Replication
#By author jfedu.net 2021
#Define Path variables
MYSQL_SOFT="mysql mysql-server mysql-devel php-mysql mysql-libs"
NUM=`rpm -qa |grep -i mysql |wc -l`
INIT="/etc/init.d/mysqld"
CODE=$?
#Mysql To Install 2021
if [ $NUM -ne 0 -a -f $INIT ];then
    echo -e "\033[32mThis Server Mysql already Install.\033[0m"
    read -p "Please ensure yum remove Mysql Server,YES or NO": INPUT
    if [ $INPUT == "y" -o $INPUT == "yes" ];then
```

```
            yum remove $MYSQL_SOFT -y ;rm -rf /var/lib/mysql /etc/my.cnf
            yum install $MYSQL_SOFT -y
        else
            echo
        fi
else
    yum remove $MYSQL_SOFT -y ;rm -rf /var/lib/mysql /etc/my.cnf
    yum install $MYSQL_SOFT -y
    if [ $CODE -eq 0 ];then
        echo -e "\033[32mThe Mysql Install Successfully.\033[0m"
    else
        echo -e "\033[32mThe Mysql Install Failed.\033[0m"
        exit 1
    fi
fi
my_config(){
cat >/etc/my.cnf<<EOF
[mysqld]
datadir=/var/lib/mysql
socket=/var/lib/mysql/mysql.sock
user=mysql
symbolic-links=0
log-bin=mysql-bin
server-id = 1
auto_increment_offset=1
auto_increment_increment=2
[mysqld_safe]
log-error=/var/log/mysqld.log
pid-file=/var/run/mysqld/mysqld.pid
EOF
}
my_config
/etc/init.d/mysqld restart
ps -ef |grep mysql
MYSQL_CONFIG(){
#Master Config Mysql
mysql -e "grant replication slave on *.* to 'tongbu'@'%' identified by '123456';"
MASTER_FILE=`mysql -e "show master status;"|tail -1|awk '{print $1}'`
MASTER_POS=`mysql -e "show master status;"|tail -1|awk '{print $2}'`
MASTER_IPADDR=`ifconfig eth0|grep "Bcast"|awk '{print $2}'|cut -d: -f2`
read -p "Please Input Slave IPaddr: " SLAVE_IPADDR
#Slave Config Mysql
ssh -l root $SLAVE_IPADDR "yum remove $MYSQL_SOFT -y ;rm -rf /var/lib/mysql /etc/my.cnf ;yum install $MYSQL_SOFT -y"
ssh -l root $SLAVE_IPADDR "$my_config"
```

```
#scp -r /etc/my.cnf root@192.168.111.129:/etc/
ssh -l root $SLAVE_IPADDR "sed -i 's#server-id = 1#server-id = 2#g'
/etc/my.cnf"
ssh -l root $SLAVE_IPADDR "sed -i '/log-bin=mysql-bin/d' /etc/my.cnf"
ssh -l root $SLAVE_IPADDR "/etc/init.d/mysqld restart"
ssh -l root $SLAVE_IPADDR "mysql -e \"change master to master_host=
'$MASTER_IPADDR',master_user='tongbu',master_password='123456',master_
log_file='$MASTER_FILE',master_log_pos=$MASTER_POS;\""
ssh -l root $SLAVE_IPADDR "mysql -e \"slave start;\""
ssh -l root $SLAVE_IPADDR "mysql -e \"show slave status\G;\""
}

read -p "Please ensure your Server is Master and you will config mysql
Replication?yes or no": INPUT
if [ $INPUT == "y" -o $INPUT == "yes" ];then
    MYSQL_CONFIG
else
    exit 0
fi
```

2.7 Shell 编程修改 IP 地址及主机名脚本

在企业中,服务器 IP 地址系统通过自动化工具安装,IP 地址均是自动获取的,而服务器要求固定的静态 IP 地址,手动配置上百台服务器的静态 IP 地址是不可取的,可以利用 Shell 脚本自动修改 IP 地址、主机名等信息。

Shell 编程实现服务器 IP、主机名自动修改及配置脚本,编写思路如下。

(1)静态 IP 地址修改。

(2)动态 IP 地址修改。

(3)根据 IP 地址生成主机名并配置。

(4)修改 DNS 域名解析。

Shell 编程实现服务器 IP、主机名自动修改及配置脚本,相关代码如下。

```
#!/bin/bash
#Auto Change ip netmask gateway scripts
#By author jfedu.net 2021
#Define Path variables
ETHCONF=/etc/sysconfig/network-scripts/ifcfg-eth0
HOSTS=/etc/hosts
NETWORK=/etc/sysconfig/network
DIR=/data/backup/`date +%Y%m%d`
NETMASK=255.255.255.0
echo "-----------------------------"
```

```bash
judge_ip(){
    read -p "Please enter ip Address,example 192.168.0.11 ip": IPADDR
    echo $IPADDR|grep -v "[Aa-Zz]"|grep --color -E "([0-9]{1,3}\.){3}[0-9]{1,3}"
}
count_ip(){
    count=('echo $IPADDR|awk -F. '{print $1,$2,$3,$4}'')
    IP1=${count[0]}
    IP2=${count[1]}
    IP3=${count[2]}
    IP4=${count[3]}
}
ip_check()
{
judge_ip
while [ $? -ne 0 ]
do
    judge_ip
done
count_ip
while [ "$IP1" -lt 0 -o "$IP1" -ge 255 -o "$IP2" -ge 255 -o "$IP3" -ge 255 -o "$IP4" -ge 255 ]
do
    judge_ip
    while [ $? -ne 0 ]
    do
        judge_ip
    done
    count_ip
done
}
change_ip()
{
if [ ! -d $DIR ];then
    mkdir -p $DIR
fi
echo "The Change ip address to Backup Interface eth0"
cp $ETHCONF  $DIR
grep "dhcp" $ETHCONF
if [ $? -eq 0 ];then
    read -p "Please enter ip Address:" IPADDR
    sed -i 's/dhcp/static/g' $ETHCONF
    echo -e "IPADDR=$IPADDR\nNETMASK=$NETMASK\nGATEWAY='echo $IPADDR|awk -F. '{print $1"."$2"."$3}''.2" >>$ETHCONF
    echo "The IP configuration success. !"
else
```

```bash
        echo -n "Static IP has been configured,please confirm whether to
modify,yes or No":
        read i
    fi
    if [ "$i" == "y" -o "$i" == "yes" ];then
        ip_check
        sed -i -e '/IPADDR/d' -e '/NETMASK/d' -e '/GATEWAY/d' $ETHCONF
        echo -e "IPADDR=$IPADDR\nNETMASK=$NETMASK\nGATEWAY='echo $IPADDR|awk
-F. '{print $1"."$2"."$3}''.2" >>$ETHCONF
        echo "The IP configuration success. !"
        echo
    else
        echo "Static IP already exists,please exit."
        exit $?
    fi
}
change_hosts()
{

if [ ! -d $DIR ];then
    mkdir -p $DIR
fi
cp $HOSTS $DIR
ip_check
host=' echo $IPADDR|sed 's/\./-/g'|awk '{print "BJ-IDC-"$0"-jfedu.net"}''
cat $HOSTS |grep "$host"
if [ $? -ne 0 ];then
    echo "$IPADDR    $host" >> $HOSTS
    echo "The hosts modify success "
fi
grep "$host" $NETWORK
if [ $? -ne 0 ];then
    sed -i "s/^HOSTNAME/#HOSTNAME/g" $NETWORK
    echo "NETWORK=$host" >>$NETWORK
    hostname $host;su
fi
}
PS3="Please Select configuration ip or configuration host:"
select i in "modify_ip" "modify_hosts" "exit"
do
    case $i in
            modify_ip)
            change_ip
        ;;
            modify_hosts)
            change_hosts
```

```
            ;;
        exit)
            exit
        ;;
        *)
            echo -e "1) modify_ip\n2) modify_ip\n3)exit"
    esac

done
```

2.8 Shell 编程 Zabbix 安装配置脚本

Zabbix 是一款分布式监控系统,基于客户端/服务器(Client/Server,C/S)模式,需在服务器安装 Zabbix_server,在客户端安装 Zabbix_agent。通过 Shell 脚本可以更快速地实现该需求。

Shell 编程实现 Zabbix 服务器和客户端自动安装脚本,编写思路如下。

(1)确定 Zabbix 软件的版本源码安装路径,启用服务器和代理服务器。

(2)cp zabbix_agentd 启动进程,-/etc/init.d/zabbix_agentd 执行权限。

(3)配置 zabbix_agentd.conf 文件,指定 server IP 变量。

(4)指定客户端的主机名称,可以等于客户端 IP 地址。

(5)启动 zabbix_agentd 服务,创建 zabbix user。

Shell 编程实现 Zabbix 服务器和客户端自动安装脚本,相关代码如下。

```
#!/bin/bash
#Auto install zabbix server and client
#By author jfedu.net 2021
#Define Path variables
ZABBIX_SOFT="zabbix-4.0.26.tar.gz"
INSTALL_DIR="/usr/local/zabbix/"
SERVER_IP="192.168.111.128"
IP=`ifconfig|grep Bcast|awk '{print $2}'|sed 's/addr://g'`
SERVER_INSTALL(){
yum -y install curl curl-devel net-snmp net-snmp-devel perl-DBI
groupadd zabbix ;useradd -g zabbix zabbix;usermod -s /sbin/nologin zabbix
tar -xzf $ZABBIX_SOFT;cd `echo $ZABBIX_SOFT|sed 's/.tar.*//g'`
./configure  --prefix=/usr/local/zabbix --enable-server --enable-agent
--with-mysql --enable-ipv6 --with-net-snmp --with-libcurl &&make install
if [ $? -eq 0 ];then
    ln -s /usr/local/zabbix/sbin/zabbix_* /usr/local/sbin/
fi
cd - ;cd zabbix-4.0.26
cp  misc/init.d/tru64/{zabbix_agentd,zabbix_server}  /etc/init.d/ ;chmod
o+x /etc/init.d/zabbix_*
```

```
mkdir -p /var/www/html/zabbix/;cp -a  frontends/php/*  /var/www/html/zabbix/
#config zabbix server
cat >$INSTALL_DIR/etc/zabbix_server.conf<<EOF
LogFile=/tmp/zabbix_server.log
DBHost=localhost
DBName=zabbix
DBUser=zabbix
DBPassword=123456
EOF
#config zabbix agentd
cat >$INSTALL_DIR/etc/zabbix_agentd.conf<<EOF
LogFile=/tmp/zabbix_agentd.log
Server=$SERVER_IP
ServerActive=$SERVER_IP
Hostname = $IP
EOF
#start zabbix agentd
/etc/init.d/zabbix_server restart
/etc/init.d/zabbix_agentd restart
/etc/init.d/iptables stop
setenforce 0
}
AGENT_INSTALL(){
yum -y install curl curl-devel net-snmp net-snmp-devel perl-DBI
groupadd zabbix ;useradd -g zabbix zabbix;usermod -s /sbin/nologin zabbix

tar -xzf $ZABBIX_SOFT;cd 'echo $ZABBIX_SOFT|sed 's/.tar.*//g''
./configure  --prefix=/usr/local/zabbix  --enable-agent&&make install
if [ $? -eq 0 ];then
    ln -s /usr/local/zabbix/sbin/zabbix_* /usr/local/sbin/
fi
cd - ;cd zabbix-4.0.26
cp misc/init.d/tru64/zabbix_agentd  /etc/init.d/zabbix_agentd ;chmod o+x /etc/init.d/zabbix_agentd
#config zabbix agentd
cat >$INSTALL_DIR/etc/zabbix_agentd.conf<<EOF
LogFile=/tmp/zabbix_agentd.log
Server=$SERVER_IP
ServerActive=$SERVER_IP
Hostname = $IP
EOF
#start zabbix agentd
/etc/init.d/zabbix_agentd restart
/etc/init.d/iptables stop
setenforce 0
}
```

```
read -p "Please confirm whether to install Zabbix Server,yes or no? " INPUT
if [ $INPUT == "yes" -o $INPUT == "y" ];then
    SERVER_INSTALL
else
    AGENT_INSTALL
fi
```

2.9　Shell 编程 Nginx 虚拟主机脚本

Nginx Web 服务器的最大特点在于 Nginx 常被用于负载均衡和反向代理，单台 Nginx 服务器配置多个虚拟主机时，使用 Shell 脚本更加高效。

Shell 编程实现 Nginx 自动安装及虚拟主机的维护脚本，编写思路如下。

（1）脚本指定参数 v1.jfedu.net。

（2）创建 v1.jfedu.net 的同时创建目录/var/www/v1。

（3）将 Nginx 虚拟主机配置定向到新的目录。

（4）重复虚拟主机不再添加。

Shell 编程实现 Nginx 自动安装及虚拟主机的维护脚本，相关代码如下。

```
#!/bin/bash
#Auto config Nginx virtual Hosts
#By author jfedu.net 2021
#Define Path variables
NGINX_CONF="/usr/local/nginx/conf/"
NGINX_MAKE="--user=www --group=www --prefix=/usr/local/nginx --with-http_stub_status_module --with-http_ssl_module"
NGINX_SBIN="/usr/local/nginx/sbin/nginx"
NGINX_INSTALL(){
#Install Nginx server
NGINX_FILE=nginx-1.16.0.tar.gz
NGINX_DIR=`echo $NGINX_FILE|sed 's/.tar*.*//g'`
if [ ! -e /usr/local/nginx/ -a ! -e /etc/nginx/ ];then
    pkill nginx
    wget -c http://nginx.org/download/$NGINX_FILE
    yum install pcre-devel pcre -y
    rm -rf $NGINX_DIR ;tar xf $NGINX_FILE
    cd $NGINX_DIR;useradd www;./configure $NGINX_MAKE
    make &&make install
    grep -vE "#|^$" $NGINX_CONF/nginx.conf >$NGINX_CONF/nginx.conf.swp
    \mv  $NGINX_CONF/nginx.conf.swp $NGINX_CONF/nginx.conf
    for i in `seq 1 6`;do sed -i '$d' $NGINX_CONF/nginx.conf;done
    echo "}" >>$NGINX_CONF/nginx.conf
```

```
        cd ../
    fi
}
NGINX_CONFIG(){
#config tomcat nginx vhosts
grep "include domains" $NGINX_CONF/nginx.conf >>/dev/null
if [ $? -ne 0 ];then
    #sed -i '$d' $NGINX_CONF/nginx.conf
    echo -e "\ninclude domains/*;\n}" >>$NGINX_CONF/nginx.conf
    mkdir -p $NGINX_CONF/domains/
fi
VHOSTS=$1
ls $NGINX_CONF/domains/$VHOSTS>>/dev/null 2>&1
if [ $? -ne 0 ];then
    #cp -r xxx.jfedu.net $NGINX_CONF/domains/$VHOSTS
    #sed -i "s/xxx/$VHOSTS/g" $NGINX_CONF/domains/$VHOSTS
    cat>$NGINX_CONF/domains/$VHOSTS<<EOF
    #vhost server $VHOSTS
    server {
        listen       80;
        server_name  $VHOSTS;
        location / {
            root   /data/www/$VHOSTS/;
            index  index.html index.htm;
        }
    }
EOF
    mkdir -p /data/www/$VHOSTS/
    cat>/data/www/$VHOSTS/index.html<<EOF
    <html>
    <h1><center>The First Test Nginx page.</center></h1>
    <hr color="red">
    <h2><center>$VHOSTS</center></h2>
    </html>
EOF
    echo -e "\033[32mThe $VHOSTS Config success,You can to access http://$VHOSTS/\033[0m"
    NUM=`ps -ef |grep nginx|grep -v grep|grep -v auto|wc -l`
    $NGINX_SBIN -t >>/dev/null 2>&1
    if [ $? -eq 0 -a $NUM -eq 0 ];then
        $NGINX_SBIN
    else
        $NGINX_SBIN -t >>/dev/null 2>&1
        if [ $? -eq 0 ];then
            $NGINX_SBIN -s reload
        fi
```

```
        fi
else
    echo -e "\033[32mThe $VHOSTS has been config,Please exit.\033[0m"
fi
}
if [ -z $1 ];then
    echo -e "\033[32m---------------------\033[0m"
    echo -e "\033[32mPlease enter sh $0 xx.jf.com.\033[0m"
    exit 0
fi
NGINX_INSTALL
NGINX_CONFIG $1
```

2.10 Shell 编程 Nginx、Tomcat 脚本

Tomcat 用于发布 JSP Web 页面，根据企业实际需求，会在单台服务器上配置多个 Tomcat 实例，同时手动将 Tomcat 创建后的实例加入 Nginx 虚拟主机，并重启 Nginx。Nginx、Tomcat 自动创建 Tomcat 实例及 Nginx 虚拟主机管理脚本能大大减轻人工的干预，实现快速交付。

Shell 编程实现 Nginx 自动安装、虚拟主机及自动将 Tomcat 加入虚拟主机脚本，编写思路如下。

（1）手动将 Tomcat 复制到和脚本在同一目录下（可自动修改）。

（2）手动修改 Tomcat 端口为 6001、7001、8001（可自动修改）。

（3）脚本指定参数 v1.jfedu.net。

（4）创建 v1.jfedu.net Tomcat 实例。

（5）修改 Tomcat 实例端口，保证 Port 唯一。

（6）将 Tomcat 实例加入 Nginx 虚拟主机。

（7）重复创建 Tomcat 实例，端口自动增加，并加入原 Nginx 虚拟主机，实现负载均衡。

Shell 编程实现 Nginx 自动安装、虚拟主机及自动将 Tomcat 加入虚拟主机脚本，相关代码如下。

```
#!/bin/bash
#Auto config Nginx and tomcat cluster
#By author jfedu.net 2021
#Define Path variables
NGINX_CONF="/usr/local/nginx/conf/"
install_nginx(){
    NGINX_FILE=nginx-1.10.2.tar.gz
    NGINX_DIR=`echo $NGINX_FILE|sed 's/.tar*.*//g'`
    wget -c http://nginx.org/download/$NGINX_FILE
    yum install pcre-devel pcre -y
```

```
        rm -rf $NGINX_DIR ;tar xf $NGINX_FILE
        cd $NGINX_DIR;useradd www;./configure --user=www --group=www --prefix=
/usr/local/nginx2 --with-http_stub_status_module --with-http_ssl_module
        make &&make install
        cd ../
}
install_tomcat(){
    JDK_FILE="jdk1.8.0_131.tar.gz"
    JDK_DIR=`echo $JDK_FILE|sed 's/.tar.*//g'`
    tar -xzf $JDK_FILE ;mkdir -p /usr/java/ ;mv $JDK_DIR /usr/java/
    sed -i '/JAVA_HOME/d;/JAVA_BIN/d;/JAVA_OPTS/d' /etc/profile
    cat >> /etc/profile <<EOF
    export JAVA_HOME=/export/servers/$JAVA_DIR
    export JAVA_BIN=/export/servers/$JAVA_DIR/bin
    export PATH=\$JAVA_HOME/bin:\$PATH
    export CLASSPATH=.:\$JAVA_HOME/lib/dt.jar:\$JAVA_HOME/lib/tools.jar
    export JAVA_HOME JAVA_BIN PATH CLASSPATH
EOF
    source /etc/profile;java -version
    #install tomcat start
    ls tomcat
}
config_tomcat_nginx(){
    #config tomcat nginx vhosts
    grep "include domains" $NGINX_CONF/nginx.conf >>/dev/null
    if [ $? -ne 0 ];then
        sed -i '$d' $NGINX_CONF/nginx.conf
        echo -e "\ninclude domains/*;\n}" >>$NGINX_CONF/nginx.conf
        mkdir -p $NGINX_CONF/domains/
    fi
    VHOSTS=$1
    NUM=`ls /usr/local/|grep -c tomcat`
    if [ $NUM -eq 0 ];then
        cp -r tomcat /usr/local/tomcat_$VHOSTS
        cp -r xxx.jfedu.net $NGINX_CONF/domains/$VHOSTS
        #sed -i "s/VHOSTS/$VHOSTS/g" $NGINX_CONF/domains/$VHOSTS
        sed -i "s/xxx/$VHOSTS/g" $NGINX_CONF/domains/$VHOSTS
        exit 0
    fi
    #-------------------------------
    #VHOSTS=$1
    VHOSTS_NUM=`ls $NGINX_CONF/domains/|grep -c $VHOSTS`
    SERVER_NUM=`grep -c "127" $NGINX_CONF/domains/$VHOSTS`
    SERVER_NUM_1=`expr $SERVER_NUM + 1`
    rm -rf /tmp/.port.txt
    for i in `find /usr/local/ -maxdepth 1 -name "tomcat*"`;do
```

```
            grep "port" $i/conf/server.xml |egrep -v "\--|8080|SSLEnabled"|awk
'{print $2}'|sed 's/port=//g;s/\"//g'|sort -nr >>/tmp/.port.txt
    done
    MAX_PORT=`cat /tmp/.port.txt|grep -v 8443|sort -nr|head -1`
    PORT_1=`expr $MAX_PORT - 2000 + 1`
    PORT_2=`expr $MAX_PORT - 1000 + 1`
    PORT_3=`expr $MAX_PORT + 1`
    if [ $VHOSTS_NUM -eq 1 ];then
        read -p "The $VHOSTS is exists,You sure create mulit Tomcat for the
$VHOSTS? yes or no " INPUT
        if [ $INPUT == "YES" -o $INPUT == "Y" -o $INPUT == "yes" ];then
            cp -r tomcat /usr/local/tomcat_${VHOSTS}_${SERVER_NUM_1}
            sed -i "s/6001/$PORT_1/g" /usr/local/tomcat_${VHOSTS}_${SERVER_NUM_1}/conf/server.xml
            sed -i "s/7001/$PORT_2/g" /usr/local/tomcat_${VHOSTS}_${SERVER_NUM_1}/conf/server.xml
            sed -i "s/8001/$PORT_3/g" /usr/local/tomcat_${VHOSTS}_${SERVER_NUM_1}/conf/server.xml
            sed -i "/^upstream/a    server 127.0.0.1:${PORT_2} weight=1 max_fails=2 fail_timeout=30s;" $NGINX_CONF/domains/$VHOSTS
            exit 0
        fi
        exit
    fi
    cp -r tomcat /usr/local/tomcat_$VHOSTS
        cp -r xxx.jfedu.net $NGINX_CONF/domains/$VHOSTS
        sed -i "s/VHOSTS/$VHOSTS/g" $NGINX_CONF/domains/$VHOSTS
        sed -i "s/xxx/$VHOSTS/g" $NGINX_CONF/domains/$VHOSTS
    sed -i "s/7001/${PORT_2}/g" $NGINX_CONF/domains/$VHOSTS
    #######config tomcat
        sed -i "s/6001/$PORT_1/g" /usr/local/tomcat_${VHOSTS}/conf/server.xml
        sed -i "s/7001/$PORT_2/g" /usr/local/tomcat_${VHOSTS}/conf/server.xml
        sed -i "s/8001/$PORT_3/g" /usr/local/tomcat_${VHOSTS}/conf/server.xml
}
if [ ! -d $NGINX_CONF -o ! -d /usr/java/$JDK_DIR ];then
    install_nginx
    install_tomcat
fi
config_tomcat_nginx $1
```

2.11 Shell 编程管理 Linux 操作系统的系统用户和系统组脚本

Shell 编程实现管理 Linux 操作系统的系统用户和系统组脚本，编写思路如下。

（1）支持创建普通用户。

（2）支持创建多个用户或者列表用户添加。

（3）支持 Linux 操作系统的系统用户删除。

（4）支持 Linux 操作系统的系统组删除。

（5）支持对某个用户修改密码。

Shell 编程实现管理 Linux 操作系统的系统用户和系统组脚本，相关代码如下。

```
#!/bin/bash
#auto manager linux user
#by author www.jfedu.net
##########################
USR="$*"
if [ $UID -ne 0 ];then
    echo -e "\033[32m-----------------\033[0m"
    echo -e "\033[32mThe script must be executed using the root user.\033[0m"
    exit 1
fi
add_user(){
    read -p "Please enter the user name you need to create? " USR
    for USR in $USR
    do
        id $USR
        if [ $? -ne 0 ];then
            useradd -s /bin/bash $USR -d /home/$USR
            echo ${USR}_123456|passwd --stdin $USR
            if [ $? -eq 0 ];then
                echo -e "\033[32m-----------------\033[0m"
                echo -e "\033[32mThe $USR user created successfully\033[0m"
                echo -e "User,Password"
                echo -e "$USR,${USR}_123"
                echo
                tail -n 5 /etc/passwd
            fi
        else
            echo -e "\033[32m-----------------\033[0m"
            echo -e "\033[32mThis $USR user already exists, please exit\033[0m"
            exit 1
        fi
    done
```

```bash
}
add_user_list(){
    G
            useradd -s /bin/bash $USR -d /home/$USR
            echo ${USR}_123456|passwd --stdin $USR
            if [ $? -eq 0 ];then
                echo -e "\033[32m------------------\033[0m"
                echo -e "\033[32mThe $USR user created successfully\033[0m"
                echo -e "User,Password"
                echo -e "$USR,${USR}_123"
                echo
                tail -n 5 /etc/passwd
            fi
        done

    else
        echo -e "\033[32m------------------\033[0m"
        echo -e "\033[32mThe user list file must be entered. The reference
content format is as follows:\033[0m"
        echo "jfedu1"
        echo "jfedu2"
        echo "jfedu3"
        echo "jfedu4"
        echo "......"
    fi
}

remove_user(){
        for USR in $USR
        do
        userdel -r $USR
        groupdel $USR
            if [ $? -eq 0 ];then
                echo -e "\033[32m------------------\033[0m"
                echo -e "\033[32mThe $USR user delete successfully\
033[0m"
                echo
                tail -n 5 /etc/passwd
            fi
        done
}

remove_group(){
        for USR in $USR
        do
        groupdel $USR
```

```
                    if [ $? -eq 0 ];then
                            echo -e "\033[32m------------------\033[0m"
                            echo -e "\033[32mThe $USR group delete successfully\033[0m"
                            echo
                            tail -n 5 /etc/passwd
            else
                grep "$USR" /etc/group
                if [ $? -eq 0 ];then
                    echo -e "\033[32m------------------\033[0m"
                    echo -e "\033[32mThe $USR group delete falied,cannot remove the primary group of user $USR\033[0m"
                    read -p "Are you sure you want to delete the $USR user? yes or no " INPUT
                    if [ $INPUT == "y" -o $INPUT == "Y" -o $INPUT == "yes" -o $INPUT == "YES" ];then
                            userdel -r $USR
                            groupdel $USR >>/dev/null 2>&1
                            echo -e "\033[32m------------------\033[0m"
                            echo -e "\033[32mThe $USR user delete successfully\033[0m"
                            echo -e "\033[32mThe $USR group delete successfully\033[0m"
                    fi
                fi
            fi
        done
}

change_user_passwd(){
    read -p "Please enter your user name and new password: username password: " INPUT
    if ['echo $INPUT|sed 's/ /\n/g'|wc -l' -eq 2 ];then
        USR='echo $INPUT|awk '{print $1}''
        PAS='echo $INPUT|awk '{print $2}''
        for USR in $USR
        do
        echo $PAS|passwd --stdin $USR
            if [ $? -eq 0 ];then
                    echo -e "\033[32m------------------\033[0m"
                    echo -e "\033[32mThe password of $USR user was modified successfully\033[0m"
                    echo -e "User,Password"
                    echo -e "$USR,$PAS"
                    echo
            fi
        done
    fi
```

```
}
case $1 in
    1)
    add_user
    ;;
    2)
    add_user_list
    ;;
    3)
    remove_user
    ;;
    4)
    remove_group
    ;;
    5)
    change_user_passwd
    ;;
    *)
    echo "----------------------------------------------"
    echo -e "\033[34mWelcome to system user management scripts:\033[0m"
    echo -e "\033[32m1) add user\033[0m"
    echo -e "\033[32m2) add_user_list\033[0m"
    echo -e "\033[32m3) remove_user\033[0m"
    echo -e "\033[32m4) remove_group\033[0m"
    echo -e "\033[32m5) change_user_passwd\033[0m"
    echo -e "\033[32mUsage:{/bin/sh $0 1|2|3|4|5|help}\033[0m"
    echo "----------------------------------------------"
esac
```

2.12 Shell 编程 Vsftpd 虚拟用户管理脚本

Shell 编程实现 Vsftpd 虚拟用户管理脚本，编写思路如下。

（1）实现随机添加单个用户。

（2）实现随机添加多个用户。

（3）实现文件列表批量添加用户。

（4）实现单个用户或者多个用户的删除。

Shell 编程实现 Vsftpd 虚拟用户管理脚本，相关代码如下。

```
#!/bin/bash
#auto create vsftpd for virtual user
#by author www.jfedu.net
#########################
```

```
CONF_DIR="/etc/vsftpd"
VIR_USER="$*"
SYS_USER="ftpuser"
LOGIN_DB="vsftpd_login"

if [ $# -eq 0 ];then
    echo -e "\033[32m---------------------\033[0m"
    echo -e "\033[32mUsage:{/bin/sh $0 jfedu001 jfedu002|jfedu003}\033[0m"
    exit 0
fi

if [ ! -f $CONF_DIR/vsftpd.conf ];then
    yum install vsftpd* db4* -y
else
    continue
fi

#for i in 'echo $VIR_USER'
echo $VIR_USER|sed 's/ /\n/g' >list.txt
while read i
do
grep "$i" $CONF_DIR/${SYS_USER}s.txt
if [ $? -ne 0 ];then
cat>>$CONF_DIR/${SYS_USER}s.txt<<EOF
$i
pwd_$i
EOF
fi
done <list.txt

db_load -T -t hash -f $CONF_DIR/${SYS_USER}s.txt $CONF_DIR/$LOGIN_DB.db
chmod 700 $CONF_DIR/${SYS_USER}s.txt
chmod 700 $CONF_DIR/$LOGIN_DB.db

cat>/etc/pam.d/vsftpd<<EOF
auth    sufficient    /lib64/security/pam_userdb.so    db=$CONF_DIR/$LOGIN_DB
account sufficient    /lib64/security/pam_userdb.so    db=$CONF_DIR/$LOGIN_DB
EOF
useradd -s /sbin/nologin $SYS_USER

grep "guest_" $CONF_DIR/vsftpd.conf
if [ $? -ne 0 ];then
cat>>$CONF_DIR/vsftpd.conf<<EOF
guest_enable=YES
```

```
guest_username=$SYS_USER
pam_service_name=vsftpd
user_config_dir=$CONF_DIR/vsftpd_user_conf
virtual_use_local_privs=YES
EOF
fi

#for j in 'echo $VIR_USER'
while read j
do
mkdir -p $CONF_DIR/vsftpd_user_conf/
cat>$CONF_DIR/vsftpd_user_conf/$j <<EOF
local_root=/home/$SYS_USER/$j
write_enable=YES
anon_world_readable_only=YES
anon_upload_enable=YES
anon_mkdir_write_enable=YES
EOF
mkdir -p /home/$SYS_USER/$j/
done < list.txt
chown -R $SYS_USER.$SYS_USER /home/$SYS_USER
service vsftpd restart
```

2.13 Shell 编程 Apache 多版本软件安装脚本

Shell 编程实现 Apache 多版本软件安装脚本，编写思路如下。

（1）安装不同的 Apache 版本。

（2）检测系统是否已经存在，是否可以覆盖版本。

（3）启动 Apache，并且测试访问。

Shell 编程实现 Apache 多版本软件安装脚本，相关代码如下。

```
#!/bin/bash
#by author jfedu.net
#auto install apache and vhosts
##################
H_URL="http://mirror.bit.edu.cn/apache/httpd/"
APR_URL="http://mirrors.hust.edu.cn/apache/apr/"
H_SOFT="httpd-2.4.25.tar.bz2"
APR_SOFT="apr-1.6.2.tar.bz2"
APR_UTIL_SOFT="apr-util-1.6.0.tar.bz2"
APACHE_DIR="/usr/local/apache2/"
VHOST_FILES="httpd-vhosts.conf"
DOMAINS="$1"
NUM1='grep -c "^Include conf/extra/httpd-vhosts.conf" $APACHE_DIR/conf/
```

```
httpd.conf'
NUM2=$(grep -c "$DOMAINS" $APACHE_DIR/conf/extra/httpd-vhosts.conf)

if [ $# -eq 0 ];then
    echo -e "\033[32m--------------------------\033[0m"
    echo -e "\033[32mUsage:{Please Enter $0 www.jf1.com|www.jf2.com}\033[0m"
    exit 0
fi

if [ ! -d $APACHE_DIR ];then
    wget -c $H_URL/$H_SOFT
    wget -c $APR_URL/$APR_SOFT
    wget -c $APR_URL/$APR_UTIL_SOFT
    #Install apr for apache for
    tar -jxvf $APR_SOFT
    cd apr-1.6.2
    ./configure --prefix=/usr/local/apr
    make
    make install
    #Install apr-util for apache for
    cd ..
    tar -jxvf $APR_UTIL_SOFT
    cd apr-util-1.6.0
    ./configure --prefix=/usr/local/apr-util --with-apr=/usr/local/apr/
    make
    make install

    #Install apache
    cd ..
    tar -jxvf $H_SOFT
    cd httpd-2.4.25
    ./configure --prefix=$APACHE_DIR/ --with-apr=/usr/local/apr --with-apr-util=/usr/local/apr-util
    make
    make install
    pkill httpd
    pkill nginx
    $APACHE_DIR/bin/apachectl start
fi

#config vhosts for apache
if [ $NUM1 -eq 0 ];then
    echo "Include conf/extra/$VHOST_FILES" >>$APACHE_DIR/conf/httpd.conf
fi
touch $APACHE_DIR/conf/extra/$VHOST_FILES
```

```
if [ $NUM2 -eq 0 ];then
cat >>$APACHE_DIR/conf/extra/$VHOST_FILES<<EOF
<VirtualHost *:80>
    ServerAdmin support@jfedu.net
    DocumentRoot "$APACHE_DIR/htdocs/$DOMAINS"
    ServerName $DOMAINS
    ErrorLog "logs/${DOMAINS}_error_log"
    CustomLog "logs/${DOMAINS}_access_log" common
</VirtualHost>
EOF

mkdir -p $APACHE_DIR/htdocs/$DOMAINS
touch $APACHE_DIR/htdocs/$DOMAINS/index.html
cat >$APACHE_DIR/htdocs/$DOMAINS/index.html<<EOF
<html><body>
<h1>$DOMAINS It works!</h1>
<h1><font color=\"red\">$DOMAINS</font></h1>
</body></html>
EOF
$APACHE_DIR/bin/apachectl restart
fi
```

2.14 Shell 编程局域网 IP 地址探活脚本

Shell 编程实现局域网 IP 地址探活脚本，编写思路如下。

（1）支持指定特定的网段。

（2）对特定的网段进行探活。

（3）将存活的 IP 地址写入存活的列表。

（4）将不存活的 IP 地址写入不存活的列表。

Shell 编程实现局域网 IP 地址探活脚本，相关代码如下。

```
#!/bin/bash
#auto ping check IP
#by author www.jfedu.net
#########################
INPUT="0"
IP_LIST="$*"
RES_FILE1="/tmp/available.txt"
RES_FILE2="/tmp/unavailable.txt"
#Define check function 2021
check_lan(){
```

```bash
    read -p "Please enter the LAN segment,example 192.168.1.0 (Netmask/24): " INPUT
        if ['echo $INPUT|sed 's/ /\n/g'|wc -l' -ne 0 ];then
            for IP in $(seq 1 254)
            do
        IP_PREFIX=$(echo $INPUT|awk -F\. '{print $1"."$2"."$3"."}')
                ping -c 2 -W1 ${IP_PREFIX}$IP >/dev/null 2>1
                if [ $? -eq 0 ];then
                    echo "${IP_PREFIX}$IP is up."
                    echo "${IP_PREFIX}$IP" >> $RES_FILE1
                else
                    echo "${IP_PREFIX}$IP is down."
                    echo "${IP_PREFIX}$IP" >> $RES_FILE2
                fi
            done
            echo -e "\033[32m------------------------\033[0m"
            echo -e "\033[32mPlease check the following files:\033[0m"
            echo "Available IP addresses: $RES_FILE1"
            echo "Unavailable IP addresses: $RES_FILE2"
            echo
        fi
}

check_list()
{
    read -p "Please enter the IP list to be checked,example list.txt: " INPUT
    if [ ! -z $INPUT ];then
        for IP in $(cat $INPUT)
        do
            ping -c 2 -W1 $IP >/dev/null 2>1
            if [ $? -eq 0 ];then
                echo "$IP is up."
                    echo $IP >> $RES_FILE1
            else
                echo "$IP is down."
                    echo $IP >> $RES_FILE2
            fi
        done
        echo -e "\033[32m------------------------\033[0m"
        echo -e "\033[32mPlease check the following files:\033[0m"
        echo "Available IP addresses: $RES_FILE1"
        echo "Unavailable IP addresses: $RES_FILE2"
        echo
    fi
```

```bash
}

check_ip(){
read -p "Please enter the IP to be checked,example 1.1.1.1 | 1.1.1.2: " INPUT
for INPUT in `echo $INPUT`
do
    while true
    do
        echo $INPUT|grep -E "\<([0-9]{1,3}\.){3}[0-9]{1,3}\>"
        if [ $? -eq 0 ];then
            IP=(`echo $INPUT|sed 's/\./ /g'`)
            IP1=`echo ${IP[0]}`
            IP2=`echo ${IP[1]}`
            IP3=`echo ${IP[2]}`
            IP4=`echo ${IP[3]}`
            if [ $IP1 -gt 0 -a $IP1 -le 255 -a $IP2 -ge 0 -a $IP2 -le 255 -a $IP3 -ge 0 -a $IP3 -le 255 -a $IP4 -ge 0 -a $IP4 -lt 255 ];then
                if [`echo $INPUT|sed 's/ /\n/g'|wc -l` -ne 0 ];then
                    for IP in $(echo $INPUT)
                    do
                        ping -c 2 -W1 $IP >/dev/null 2>1
                        if [ $? -eq 0 ];then
                            echo "$IP is up."
                            echo $IP >> $RES_FILE1
                        else
                            echo "$IP is down."
                            echo $IP >> $RES_FILE2
                        fi
                    done
                    echo -e "\033[32m------------------------\033[0m"
                    echo -e "\033[32mPlease check the following files:\033[0m"
                    echo "Available IP addresses: $RES_FILE1"
                    echo "Unavailable IP addresses: $RES_FILE2"
                    echo
                fi
                break;
            else
                read -p "Please Enter server IP address:" INPUT
            fi
        else
            read -p "Please Enter server IP address:" INPUT
        fi
    done
done
```

```
    done
}

case $1 in
    1)
    check_lan
    ;;
    2)
    check_ip
    ;;
    3)
    check_list
    ;;
    *)
    echo "-------------------------------------------"
    echo -e "\033[34mWelcome to LAN live scripts:\033[0m"
    echo -e "\033[32m1) check_lan\033[0m"
    echo -e "\033[32m2) check_ip\033[0m"
    echo -e "\033[32m3) check_list\033[0m"
    echo -e "\033[32mUsage:{/bin/sh $0 1|2|3|4|5|help}\033[0m"
    echo "-------------------------------------------"
esac
```

2.15 Shell 编程 Apache 虚拟主机管理脚本

Shell 编程实现 Apache 虚拟主机管理脚本，编写思路如下。

（1）检测 Apache 是否安装，并测试访问。

（2）如果已经安装，则直接添加虚拟主机。

（3）支持添加单个虚拟主机。

（4）支持添加多个虚拟主机。

（5）支持删除单个或多个虚拟主机。

Shell 编程实现 Apache 虚拟主机管理脚本，相关代码如下。

```
#!/bin/bash
#auto config httpd vhosts
#by author jfedu.net
######################
APACHE_SOFT="httpd httpd-devel httpd-tools"
BACK_DIR=/data/backup/`date +%F`
HTTP_DIR="/etc/httpd/conf"
HTTP_FILES="httpd.conf"
VHOSTS_CONF="vhosts.conf"
```

```bash
NUM1=$(grep -c "$VHOSTS_CONF" $HTTP_DIR/$HTTP_FILES)
NUM2=$(grep -c "NameVirtualHost" $HTTP_DIR/$VHOSTS_CONF)
DOMAIN="$1"

if [ $# -eq 0 ];then
    echo -e "\033[32m------------------\033[0m"
    echo -e "\033[32mUsage:{Please Enter sh $0 www.jf1.com|www.jf2.com}\
033[0m"
    exit 0
fi

yum install $APACHE_SOFT -y
mkdir -p $BACK_DIR
cp -a $HTTP_DIR/$HTTP_FILES $BACK_DIR
touch $HTTP_DIR/$VHOSTS_CONF

if [ -z $NUM1 ];then
    NUM1=0
fi
if [ $NUM1 -eq 0 ];then
    echo "Include  conf/$VHOSTS_CONF" >>$HTTP_DIR/$HTTP_FILES
fi

if [ -z $NUM2 ];then
    NUM2=0
fi
if [ $NUM2 -eq 0 ];then
    echo "NameVirtualHost *:80" >>$HTTP_DIR/$VHOSTS_CONF
fi

NUM3='grep -c "$DOMAIN" /etc/httpd/conf/vhosts.conf'
if [ $NUM3 -eq 0 ];then
echo "
<VirtualHost *:80>
    ServerAdmin wgkgood@163.com
    DocumentRoot  \"/data/webapps/$DOMAIN\"
    ServerName  $DOMAIN
  <Directory \"/data/webapps/$DOMAIN\">
    AllowOverride All
    Options -Indexes FollowSymLinks
    Order allow,deny
    Allow from all
  </Directory>
    ErrorLog  logs/error_log
    CustomLog logs/access_log common
</VirtualHost>
```

```
" >>$HTTP_DIR/$VHOSTS_CONF
fi
```

2.16　Shell 编程实现 Apache 高可用脚本

Shell 编程实现 Apache 高可用脚本，编写思路如下。

（1）部署两台 Apache 服务器，发布 Web 测试页面。

（2）通过两个 IP 地址均可以访问 Web 网页。

（3）增加第三个 IP 地址，称为 VIP，可以绑定至某一台服务器。

（4）VIP 即可访问一台 Apache Web 服务器。

（5）当该 Apache Web 服务器宕机，VIP 自动切换至另外一台 Apache Web 服务器。

（6）时刻保证不管哪台 Apache Web 服务器宕机，VIP 均可以访问 Apache Web 服务器。

Shell 编程实现 Apache 高可用脚本，相关代码如下。

```
#!/bin/bash
#auto change service VIP
#by author www.jfedu.net
#########################
ETH_NAME="ens33:1"
APA_VIP="192.168.1.188"
APA_MASK="255.255.255.0"
ETH_DIR="/etc/sysconfig/network-scripts"
APA_NUM='ps -ef|grep httpd|grep -v grep|grep -v check|wc -l'
start(){
while sleep 4
do
if [ $APA_NUM -eq 0 ];then
    ifdown $ETH_NAME
    exit 0
else
    ping -c 2 $APA_VIP >/dev/null 2>&1
    if [ $? -ne 0 ];then
cat>$ETH_DIR/ifcfg-$ETH_NAME<<EOF
TYPE="Ethernet"
BOOTPROTO="static"
DEVICE="$ETH_NAME"
IPADDR=$APA_VIP
NETMASK=$APA_MASK
ONBOOT="yes"
EOF
    ifup $ETH_NAME
    fi
```

```
    fi
    date
done
}

stop(){
    ifdown $ETH_NAME
    rm -rf $ETH_DIR/ifcfg-$ETH_NAME
}

case $1 in
    start)
    start
    ;;
    stop)
    stop
    ;;
    *)
    echo -e "\033[32m-------------------\033[0m"
    echo -e "\033[32mUsage: /bin/sh $0 {start|stop|help}\033[0m"
        exit 1
esac
```

2.17 Shell 编程拒绝黑客攻击 Linux 脚本

Shell 编程实现拒绝黑客攻击 Linux 脚本，编写思路如下。

（1）登录服务器日志/var/log/secure。

（2）检查日志中认证失败的行并打印其 IP 地址。

（3）将 IP 地址写入 Linux 服务器黑名单文件或防火墙。

（4）禁止该 IP 地址访问服务器 SSH 22 端口。

（5）将脚本加入 Crontab，实现自动禁止恶意 IP 地址。

Shell 编程实现拒绝黑客攻击 Linux 脚本，相关代码如下。

```
#!/bin/bash
#Auto drop ssh failed IP address
#By author jfedu.net 2021
#Define Path variables
SEC_FILE=/var/log/secure
IP_ADDR=`awk '{print $0}' /var/log/secure|grep -i "fail"| egrep -o
"([0-9]{1,3}\.){3}[0-9]{1,3}" | sort -nr | uniq -c |awk '$1>=1 {print $2}'`
DENY_CONF=/etc/hosts.deny
TM1=`date +%Y%m%d%H%M`
DENY_IP="/tmp/2h_deny_ip.txt"
```

```
echo
cat <<EOF
++++++++++++welcome to use ssh login drop failed ip+++++++++++++++++
++++++++++++++++++++++++++++++++++++++++++++++++++++++++++++++++++
++++++++++++++++----------------------------------+++++++++++++++++
EOF
echo
for ((j=0;j<=2;j++)) ;do echo -n "-";sleep 1 ;done
echo
for i in 'echo $IP_ADDR'
do
    cat $DENY_CONF |grep $i >/dev/null 2>&1
    if  [ $? -ne 0 ];then
        grep "$i" $DENY_IP>>/dev/null 2>&1
        if [ $? -eq 0 ];then
           TM3='date +%Y%m%d%H%M'
            IP1='awk -F"[#:]" '/'$i'/ {print $2,$4}' $DENY_IP|awk '{if ('$TM3'>=$2+2) print $1}''
            if [ ! -z $IP1 ];then
                echo "sshd:$IP1:deny #$TM1" >>$DENY_CONF
                sed -i "/$IP1/d" $DENY_IP
            fi
        else
            echo "sshd:$i:deny #$TM1" >>$DENY_CONF
        fi
    fi
done

#Allow IP to access
TM2='date +%Y%m%d%H%M'
IP2='awk -F"[#:]" '/sshd/ {print $2,$4}' $DENY_CONF|awk '{if('$TM2'>=$2+2) print $1}''
for k in 'echo $IP2'
do
    echo $k
    sed -i "/$k/d"  $DENY_CONF
    echo "sshd:$k:deny #$TM2" >>$DENY_IP
done
```

2.18 Shell 编程 mysqldump 数据库自动备份脚本

Shell 编程实现 mysqldump 数据库自动备份脚本，编写思路如下。

（1）支持 MySQL 单个库备份。

（2）支持 MySQL 多个库备份。

（3）支持 MySQL 全数据库备份。

（4）支持 MySQL 数据库定期删除数据。

Shell 编程实现 mysqldump 数据库自动备份脚本，相关代码如下。

```bash
#!/bin/bash
#auto backup mysql database
#by author www.jfedu.net
##########################
SQL_DB="$*"
SQL_USR="backup"
SQL_PWD="bak123456"
SQL_CMD="/usr/bin/mysqldump"
BAK_DIR="/data/backup/'date +%F'"
if [ $# -eq 0 ];then
    echo -e "\033[32m----------------\033[0m"
    echo -e "\033[32mUsage:{/bin/bash $0 jfedu001|jfedu002|all|help}\033[0m"
    exit
fi

if [ $UID -ne 0 ];then
    echo "Exec backup scripts,must to be use root."
    exit
fi
if [ ! -d $BAK_DIR ];then
    mkdir -p $BAK_DIR
fi

if [ $SQL_DB == "all" ];then
    for SQL_DB in $(/usr/bin/mysql -u$SQL_USR -p$SQL_PWD -e "show databases;")
    do
        $SQL_CMD -u$SQL_USR -p$SQL_PWD $SQL_DB >$BAK_DIR/${SQL_DB}.sql
        if [ $? -eq 0 ];then
            echo -e "\033[32m----------------\033[0m"
            echo -e "\033[32mThe mysql database $SQL_DB backup successfully.\033[0m"
            echo
            echo "ls -l $BAK_DIR/"
            ls -l $BAK_DIR/
        else
            echo "The mysql database $SQL_DB backup falied."
            rm -rf $BAK_DIR/${SQL_DB}.sql

        fi
```

```
        done
        exit
fi

for SQL_DB in $SQL_DB
do
    $SQL_CMD -u$SQL_USR -p$SQL_PWD $SQL_DB >$BAK_DIR/$SQL_DB.sql
    if [ $? -eq 0 ];then
        echo -e "\033[32m-----------------\033[0m"
        echo -e "\033[32mThe mysql database $SQL_DB backup successfully.\
033[0m"
        echo
        echo "ls -l $BAK_DIR/"
        ls -l $BAK_DIR/
    else
        echo "The mysql database $SQL_DB backup falied."
        rm -rf $BAK_DIR/${SQL_DB}.sql
        while true
        do
            echo
            read -p "Please retry enter database name: " INPUT
            /usr/bin/mysql -u$SQL_USR -p$SQL_PWD -e "show databases;"|grep -ai "$INPUT"
            if [ $? -eq 0 ];then
                break
            fi
        done

    fi
done
```

2.19 Shell 编程 MySQL 主从自动配置脚本

Shell 编程实现 MySQL 主从自动配置脚本，编写思路如下。

（1）在主库上安装 MySQL，设置 server-id、bin-log。

（2）授权复制同步的用户，对客户端授权。

（3）确认 bin-log 文件名、position 位置点。

（4）在从库上安装 MySQL，设置 server-id。

（5）change master：指定主库、bin-log 文件名和 position 位置点。

（6）start slave：启动从库 I/O 线程。

（7）show slave status\G：查看主从的状态。

Shell 编程实现 MySQL 主从自动配置脚本，相关代码如下。

```bash
#!/bin/bash
#auto make install Mysql AB Repliation
#by author jfudu.net wugk
MYSQL_SOFT="mariadb mariadb-server mariadb-devel mariadb-libs"
NUM=`rpm -qa |grep -i mariadb |wc -l`
INIT="mariadb.service"
CODE=$?

#Mysql To Install 2015
if [ $NUM -ne 0 -a -f /usr/lib/systemd/system/$INIT ];then
    echo -e "\033[32mThis Server Mysql already Install.\033[0m"
    read -p "Please ensure yum remove Mysql Server,YES or NO": INPUT

    if [ $INPUT == "y" -o $INPUT == "yes" ];then
       yum remove $MYSQL_SOFT -y ;rm -rf /var/lib/mysql /etc/my.cnf
       yum install $MYSQL_SOFT -y
    else
       echo
    fi
else
    yum remove $MYSQL_SOFT -y ;rm -rf /var/lib/mysql /etc/my.cnf
    yum install $MYSQL_SOFT -y
    if [ $CODE -eq 0 ];then
       echo -e "\033[32mThe Mysql Install Successfully.\033[0m"
    else
       echo -e "\033[32mThe Mysql Install Failed.\033[0m"
       exit 1
    fi
fi


cat >/etc/my.cnf<<EOF
[mysqld]
datadir=/var/lib/mysql
socket=/var/lib/mysql/mysql.sock
user=mysql
symbolic-links=0
log-bin=mysql-bin
server-id = 1
auto_increment_offset=1
auto_increment_increment=2
[mysqld_safe]
log-error=/var/log/mysqld.log
```

```
pid-file=/var/run/mysqld/mysqld.pid
EOF

chown -R mysql.mysql /var/log/
mkdir -p /var/run/mysqld
chown -R mysql.mysql /var/run/mysqld
systemctl restart mariadb.service
ps -ef |grep mysql

MYSQL_CONFIG(){

#Master Config Mysql
mysql -e "grant replication slave on *.* to 'tongbu'@'%' identified by '123456';"
MASTER_FILE='mysql -e "show master status;"|tail -1|awk '{print $1}''
MASTER_POS='mysql -e "show master status;"|tail -1|awk '{print $2}''

#MASTER_IPADDR='ifconfig eth0|grep "Bcast"|awk '{print $2}'|cut -d: -f2'
MASTER_IPADDR=$(ifconfig|grep "broadcast"|cut -d" " -f10)

read -p "Please Input Slave IPaddr: " SLAVE_IPADDR

#Slave Config Mysql
ssh -l root $SLAVE_IPADDR "yum remove $MYSQL_SOFT -y ;rm -rf /var/lib/mysql /etc/my.cnf ;yum install $MYSQL_SOFT -y"
#ssh -l root $SLAVE_IPADDR "$my_config"
scp -r /etc/my.cnf root@$SLAVE_IPADDR:/etc/
ssh -l root $SLAVE_IPADDR "sed -i 's#server-id = 1#server-id = 2#g' /etc/my.cnf"
ssh -l root $SLAVE_IPADDR "sed -i '/log-bin=mysql-bin/d' /etc/my.cnf"

ssh -l root $SLAVE_IPADDR "chown -R mysql.mysql /var/log/"
ssh -l root $SLAVE_IPADDR "mkdir -p /var/run/mysqld"
ssh -l root $SLAVE_IPADDR "chown -R mysql.mysql /var/run/mysqld"

ssh -l root $SLAVE_IPADDR "systemctl restart mariadb.service"
ssh -l root $SLAVE_IPADDR "mysql -e \"change master to master_host='$MASTER_IPADDR',master_user='tongbu',master_password='123456',master_log_file='$MASTER_FILE',master_log_pos=$MASTER_POS;\""
ssh -l root $SLAVE_IPADDR "mysql -e \"slave start;\""
ssh -l root $SLAVE_IPADDR "mysql -e \"show slave status\G;\""
}

read -p "Please ensure your Server is Master and you will config mysql Replication?yes or no": INPUT
if [ $INPUT == "y" -o $INPUT == "yes" ];then
```

```
        MYSQL_CONFIG
else
    exit 0
fi
```

2.20 Shell 编程部署 Tomcat 多实例脚本

Shell 编程实现部署 Tomcat 多实例脚本，编写思路如下。

（1）检测服务器是否部署 JDK 和 Tomcat。

（2）部署 Tomcat 实例至/usr/local/目录。

（3）Shell 脚本支持单个 Tomcat 实例添加并启动。

（4）Shell 脚本支持多个 Tomcat 实例添加并启动。

Shell 编程实现部署 Tomcat 多实例脚本，相关代码如下。

```
function config_tomcat_nginx(){
    #Install JAVA JDK
    TOMCAT_VER="8.0.50"
    JAVA_VER="1.8.0_131"
    JAVA_DIR="/usr/java"
    TOMCAT_DIR="/usr/local"
    JAVA_SOFT="jdk${JAVA_VER}.tar.gz"
    TOMCAT_SOFT="apache-tomcat-${TOMCAT_VER}.tar.gz"
    grep -ai "^export" /etc/profile|grep -ai "JAVA_HOME" >/dev/null
    if [ $? -ne 0 ];then
    #Install JAVA JDK
    ls -l $JAVA_SOFT
    tar -xzvf $JAVA_SOFT
    mkdir -p $JAVA_DIR/
    \mv jdk$JAVA_VER $JAVA_DIR/
    ls -l $JAVA_DIR/jdk$JAVA_VER
    $JAVA_DIR/jdk$JAVA_VER/bin/java -version
    cat>>/etc/profile<<-EOF
    export JAVA_HOME=$JAVA_DIR/jdk$JAVA_VER
    export CLASSPATH=\$CLASSPATH:\$JAVA_HOME/lib:\$JAVA_HOME/jre/lib
EOF
    source /etc/profile
    fi

    shift 1
    NUM=`ls /usr/local/|grep -c tomcat`
    if [ $NUM -eq 0 ];then
        cp -r tomcat /usr/local/tomcat_$*
        exit 0
```

```
    fi
#--------------------------------
#VHOSTS=$1
VHOSTS_NUM=`ls $NGINX_CONF/domains/|grep -c $*`
SERVER_NUM=`grep -c "127" $NGINX_CONF/domains/$*`
SERVER_NUM_1=`expr $SERVER_NUM + 1`
rm -rf /tmp/.port.txt
for i in `find /usr/local/ -maxdepth 1 -name "tomcat*"`;do
    grep "port" $i/conf/server.xml |egrep -v "\--|8080|SSLEnabled"|awk '{print $2}'|sed 's/port=//g;s/\"//g'|sort -nr >>/tmp/.port.txt
done
MAX_PORT=`cat /tmp/.port.txt|grep -v 8443|sort -nr|head -1`
PORT_1=`expr $MAX_PORT - 2000 + 1`
PORT_2=`expr $MAX_PORT - 1000 + 1`
PORT_3=`expr $MAX_PORT + 1`
if [ $*_NUM -eq 1 ];then
    read -p "The $* is exists,You sure create mulit Tomcat for the $*? yes or no " INPUT
    if [ $INPUT == "YES" -o $INPUT == "Y" -o $INPUT == "yes" ];then
        cp -r tomcat /usr/local/tomcat_${VHOSTS}_${SERVER_NUM_1}
        sed -i "s/6001/$PORT_1/g" /usr/local/tomcat_${VHOSTS}_${SERVER_NUM_1}/conf/server.xml
        sed -i "s/7001/$PORT_2/g" /usr/local/tomcat_${VHOSTS}_${SERVER_NUM_1}/conf/server.xml
        sed -i "s/8001/$PORT_3/g" /usr/local/tomcat_${VHOSTS}_${SERVER_NUM_1}/conf/server.xml
        sed -i "/^upstream/a    server 127.0.0.1:${PORT_2} weight=1 max_fails=2 fail_timeout=30s;" $NGINX_CONF/domains/$*
        exit 0
    fi
    exit
fi
cp -r tomcat /usr/local/tomcat_$*
cp -r xxx.jfedu.net $NGINX_CONF/domains/$*
sed -i "s/VHOSTS/$*/g" $NGINX_CONF/domains/$*
sed -i "s/xxx/$*/g" $NGINX_CONF/domains/$*
sed -i "s/7001/${PORT_2}/g" $NGINX_CONF/domains/$*
#######config tomcat
sed -i "s/6001/$PORT_1/g" /usr/local/tomcat_${VHOSTS}/conf/server.xml
sed -i "s/7001/$PORT_2/g" /usr/local/tomcat_${VHOSTS}/conf/server.xml
sed -i "s/8001/$PORT_3/g" /usr/local/tomcat_${VHOSTS}/conf/server.xml

}
```

```
if [ ! -d $NGINX_CONF -o ! -d  /usr/java/$JDK_DIR ];then
    install_nginx
    install_tomcat
fi
config_tomcat_nginx $1
```

2.21 Shell 编程 Nginx 日志切割脚本

Shell 编程实现 Nginx 日志切割脚本，编写思路如下。

（1）支持指定日志文件或多个文件切割。

（2）日志文件支持按天切割。

（3）日志文件支持按小时切割。

（4）日志文件切割之后打包并上传至 FTP 服务器。

Shell 编程实现 Nginx 日志切割脚本，相关代码如下。

```
#!/bin/bash
#auto mv nginx log shell
#by author jfedu.net
NUM=$(date +%H%M%S)
echo 'date'
if [ $NUM == "000000" ];then
        LOG_DIR="/data/logs/linux_web/"
        TIME='date -d "-1 day" +%Y%m%d'
        echo -e "\033[32mPlease wait start cut shell scripts...\033[1m"
        sleep 2
        cd $LOG_DIR
        mv access.log access_${TIME}.log
        kill -USR1 'cat /usr/local/nginx/nginx.pid'
        echo "-----------------------------------------"
        echo "The Nginx log Cutting Successfully!"
fi
```

2.22 Shell 编程 Tomcat 实例和 Nginx 均衡脚本

Shell 编程实现 Tomcat 实例和 Nginx 均衡脚本，编写思路如下。

（1）检测服务器是否部署 Nginx、JDK 和 Tomcat。

（2）部署 Tomcat 实例至/usr/local/目录。

（3）Shell 脚本支持单个 Tomcat 实例添加并启动。

（4）Shell 脚本支持多个 Tomcat 实例添加并启动。

（5）Shell 脚本除了实现单个和多个 Tomcat 之外，还将其 Tomcat 实例的端口加入 Nginx 虚拟主机均衡。

（6）实现 Nginx 均衡多个虚拟主机域名，分别对应后端不同 Tomcat。

（7）Shell 脚本支持删除 Nginx 均衡和删除 Tomcat 实例（单个和多个）。

Shell 编程实现 Tomcat 实例和 Nginx 均衡脚本，相关代码如下。

```bash
#!/bin/bash
#auto config nginx virtual
#by author www.jfedu.net
#########################
NGX_CNF="nginx.conf"
NGX_DIR="/usr/local/nginx"
NGX_YUM="yum install -y"
NGX_URL="http://nginx.org/download"
NGX_ARG="--user=www --group=www --with-http_stub_status_module"
function nginx_help(){
    echo -e "\033[33mNginx VIrtual Manager SHELL Scripts\033[0m"
    echo -e "\033[33m-------------------------------\033[0m"
        echo -e "\033[33m1)-I New Install Nginx WEB Server.\033[0m"
        echo -e "\033[33m2)-U Update Install Nginx WEB Server.\033[0m"
        echo -e "\033[33m3)-A v1.jfedu.net|v2.jfedu.net v3.jfedu.net\033[0m"
        echo -e "\033[33m4)-D v1.jfedu.net|v2.jfedu.net v3.jfedu.net\033[0m"
        echo -e "\033[33m5)-T v1.jfedu.net|v2.jfedu.net v3.jfedu.net\033[0m"
        echo -e "\033[35mUsage:{/bin/bash $0 -I(Install) | -U(Update)| -A(Add) | -D(Del) | -H(Help) -T(Tomcat)\033[0m"
        exit 0
}

function nginx_install(){
#Nginx Install Config
if [ $# -le 1 ];then
    nginx_help
fi
if [ ! -d ${NGX_DIR} ];then
   shift 1
   NGX_VER=$(echo $*)
   NGX_SOFT="nginx-${NGX_VER}.tar.gz"
   NGX_SRC=$(echo $NGX_SOFT|sed 's/.tar.*//g')
   NGX_CODE="src/core/nginx.h"
   echo -e "\033[33m-------------------------------\033[0m"
   echo -e "\033[33mStart Nginx install Proccess...\033[0m"
   $NGX_YUM wget make gzip tar gcc gcc-c++ >>/dev/null 2>&1
   $NGX_YUM pcre pcre-devel zlib zlib-devel >>/dev/null 2>&1
   wget -c $NGX_URL/$NGX_SOFT
   tar -xzf $NGX_SOFT
```

```
        cd $NGX_SRC
        sed -i "s/$NGX_VER//g" $NGX_CODE
        sed -i 's/nginx\//JWS/g' $NGX_CODE
        sed -i 's/"NGX"/"JWS"/g' $NGX_CODE
        useradd -s /sbin/nologin www -M
        ./configure --prefix=${NGX_DIR}/ $NGX_ARG
        make -j4
        make -j4 install
        ${NGX_DIR}/sbin/nginx
        ps -ef|grep -aiwE nginx
        netstat -tnlp|grep -aiwE 80
        setenforce 0
        sed -i '/SELINUX/s/enforcing/disabled/g' /etc/sysconfig/selinux
        if [ $(uname -r|awk -F"[-|.]" '{print $1}') -ge "3" ];then
            firewall-cmd --add-port=80/tcp --permanent
            systemctl reload firewalld.service
        else
            iptables -t filter -A INPUT -m tcp -p tcp --dport 80 -j ACCEPT
            service iptables save
        fi
else
        echo -e "\033[32mThe Nginx WEB Already Install,Please Exit.\033[0m"
        echo -e "\033[33m------------------------\033[0m"
        echo "ls -l $NGX_DIR/"
        ls -l $NGX_DIR/

        echo -e "\033[33m------------------------\033[0m"
        while true
        do
        echo -e -n "\033[33mPlease ensure to retry Nginx WEB service,yes or no ? \033[0m"
        read INPUT
        if [ -z $INPUT ];then
            continue
        fi
        if [ $INPUT == "yes" -o $INPUT == "YES" -o $INPUT == "y" ];then
            echo -e "-----------------------------"
            echo -e "Backup nginx to ${NGX_DIR}.bak,\mv $NGX_DIR ${NGX_DIR}.bak"
            \mv $NGX_DIR ${NGX_DIR}.bak
            shift 1
                NGX_VER=$(echo $*)
                NGX_SOFT="nginx-${NGX_VER}.tar.gz"
                NGX_SRC=$(echo $NGX_SOFT|sed 's/.tar.*//g')
                NGX_CODE="src/core/nginx.h"
                echo -e "\033[33m-----------------------------\033[0m"
                echo -e "\033[33mStart Nginx install Proccess...\033[0m"
```

```shell
                $NGX_YUM wget make gzip tar gcc gcc-c++ >>/dev/null 2>&1
                $NGX_YUM pcre pcre-devel zlib zlib-devel >>/dev/null 2>&1
                wget -c $NGX_URL/$NGX_SOFT
                tar -xzf $NGX_SOFT
                cd $NGX_SRC
                sed -i "s/$NGX_VER//g" $NGX_CODE
                sed -i 's/nginx\///JWS/g' $NGX_CODE
                sed -i 's/"NGX"/"JWS"/g' $NGX_CODE
                useradd -s /sbin/nologin www -M
                ./configure --prefix=${NGX_DIR}/ $NGX_ARG
                make -j4
                make -j4 install
                ${NGX_DIR}/sbin/nginx
                ps -ef|grep -aiwE nginx
                netstat -tnlp|grep -aiwE 80
                setenforce 0
                sed -i '/SELINUX/s/enforcing/disabled/g' /etc/sysconfig/selinux
                if [ $(uname -r|awk -F"[-|.]" '{print $1}') -ge "3" ];then
                    firewall-cmd --add-port=80/tcp --permanent
                    systemctl reload firewalld.service
                else
                    iptables -t filter -A INPUT -m tcp -p tcp --dport 80 -j ACCEPT
                    service iptables save
                fi
            break

        fi
    done
fi
}

function nginx_update(){
#Nginx Install Config
if [ $# -le 1 ];then
        nginx_help
fi
if [ ! -d ${NGX_DIR} ];then
    shift 1
    NGX_VER=$(echo $*)
    NGX_SOFT="nginx-${NGX_VER}.tar.gz"
    NGX_SRC=$(echo $NGX_SOFT|sed 's/.tar.*//g')
    NGX_CODE="src/core/nginx.h"
    echo -e "\033[33m-------------------------------\033[0m"
    echo -e "\033[33mStart Nginx install Proccess...\033[0m"
    $NGX_YUM wget make gzip tar gcc gcc-c++ >>/dev/null 2>&1
    $NGX_YUM pcre pcre-devel zlib zlib-devel >>/dev/null 2>&1
```

```bash
        wget -c $NGX_URL/$NGX_SOFT
        tar -xzf $NGX_SOFT
        cd $NGX_SRC
        sed -i "s/$NGX_VER//g" $NGX_CODE
        sed -i 's/nginx\///JWS/g' $NGX_CODE
        sed -i 's/"NGX"/"JWS"/g' $NGX_CODE
        useradd -s /sbin/nologin www -M
        ./configure --prefix=${NGX_DIR}/ $NGX_ARG
        make -j4
        make -j4 install
        ${NGX_DIR}/sbin/nginx
        ps -ef|grep -aiwE nginx
        netstat -tnlp|grep -aiwE 80
        setenforce 0
        sed -i '/SELINUX/s/enforcing/disabled/g' /etc/sysconfig/selinux
        if [ $(uname -r|awk -F"[-|.]" '{print $1}') -ge "3" ];then
            firewall-cmd --add-port=80/tcp --permanent
            systemctl reload firewalld.service
        else
            iptables -t filter -A INPUT -m tcp -p tcp --dport 80 -j ACCEPT
            service iptables save
        fi
else
        echo -e "\033[32mThe Nginx WEB Already Install,Please Exit.\033[0m"
        echo -e "\033[33m-----------------------\033[0m"
        echo "ls -l $NGX_DIR/"
        ls -l $NGX_DIR/

        echo -e "\033[33m-----------------------\033[0m"
        while true
        do
        echo -e -n "\033[33mPlease ensure to Update Nginx WEB service,yes or no ?\033[0m"
        read INPUT
        if [ -z $INPUT ];then
            continue
        fi
        if [ $INPUT == "yes" -o $INPUT == "YES" -o $INPUT == "y" ];then
            echo -e "-------------------------------"
            echo -e "Backup nginx to ${NGX_DIR}.bak,\mv $NGX_DIR ${NGX_DIR}.bak"
            \mv $NGX_DIR ${NGX_DIR}.bak
            shift 1
                NGX_VER=$(echo $*)
                NGX_SOFT="nginx-${NGX_VER}.tar.gz"
                NGX_SRC=$(echo $NGX_SOFT|sed 's/.tar.*//g')
                NGX_CODE="src/core/nginx.h"
```

```bash
            echo -e "\033[33m------------------------------\033[0m"
            echo -e "\033[33mStart Nginx install Proccess...\033[0m"
            $NGX_YUM wget make gzip tar gcc gcc-c++ >>/dev/null 2>&1
            $NGX_YUM pcre pcre-devel zlib zlib-devel >>/dev/null 2>&1
            wget -c $NGX_URL/$NGX_SOFT
            tar -xzf $NGX_SOFT
            cd $NGX_SRC
            sed -i "s/$NGX_VER//g" $NGX_CODE
            sed -i 's/nginx\//JWS/g' $NGX_CODE
            sed -i 's/"NGX"/"JWS"/g' $NGX_CODE
            useradd -s /sbin/nologin www -M
            ./configure --prefix=${NGX_DIR}/ $NGX_ARG
            make -j4
            \mv ${NGX_DIR}/sbin/nginx ${NGX_DIR}/sbin/nginx.old
        \cp objs/nginx ${NGX_DIR}/sbin/
            ${NGX_DIR}/sbin/nginx
            ps -ef|grep -aiwE nginx
            netstat -tnlp|grep -aiwE 80
            setenforce 0
            sed -i '/SELINUX/s/enforcing/disabled/g' /etc/sysconfig/selinux
            if [ $(uname -r|awk -F"[-|.]" '{print $1}') -ge "3" ];then
                firewall-cmd --add-port=80/tcp --permanent
                systemctl reload firewalld.service
            else
                iptables -t filter -A INPUT -m tcp -p tcp --dport 80 -j ACCEPT
                service iptables save
            fi
        break
        
    fi
    done
fi
}

function virtual_add(){
    #Nginx Config Virtual Host
    if [ $# -le 1 ];then
        nginx_help
    fi
    cd ${NGX_DIR}/conf/
    grep -aiE "include domains" ${NGX_CNF} >>/dev/null 2>&1
    if [ $? -ne 0 ];then
        grep -aiE -vE "^$|#" ${NGX_CNF} > ${NGX_CNF}.swp
        \cp ${NGX_CNF}.swp ${NGX_CNF}
        sed -i '/server/,$d' ${NGX_CNF}
        echo -e -e "    include domains/*;\n}" >>${NGX_CNF}
```

```bash
            ${NGX_DIR}/sbin/nginx -t
            mkdir domains -p
    fi
    shift 1
    for NGX_VHOSTS in $*
    do
        CHECK_NGX_NUM=`ls domains/|grep -aiE -c $NGX_VHOSTS`
        if [ $CHECK_NGX_NUM -eq 0 ];then
        cat>domains/$NGX_VHOSTS<<-EOF
        server {
                listen       80;
                server_name  $NGX_VHOSTS;
                location / {
                    root   html/$NGX_VHOSTS;
                    index  index.html index.htm;
                }
        }
        EOF
        mkdir -p ${NGX_DIR}/html/$NGX_VHOSTS
        cat>${NGX_DIR}/html/$NGX_VHOSTS/index.html<<-EOF
        <h1>$* Welcome to nginx!</h1>
        <hr color=red>
        EOF
        echo -e "\033[32m------------------------\033[0m"
        echo -e "\033[32mThe Nginx $NGX_VHOSTS ADD Success.\033[0m"
        cat domains/$NGX_VHOSTS
        echo -e "\033[32m------------------------\033[0m"
        $NGX_DIR/sbin/nginx -t
        $NGX_DIR/sbin/nginx -s reload
        echo
        else
            echo -e "\033[32m----------------------\033[0m"
            echo -e "\033[32mThe Nginx $NGX_VHOSTS Already Exist,Please Exit.\033[0m"
            cat domains/$NGX_VHOSTS
        fi
    done
}

function virtual_del(){
    if [ $# -le 1 ];then
        nginx_help
    fi
    shift 1
    for NGX_VHOSTS in $*
    do
```

```
            cd ${NGX_DIR}/conf/domains/ >/dev/null 2>&1
        if [ $? -eq 0 ];then
            ls -l|grep -aiE "$NGX_VHOSTS" >/dev/null 2>&1
            if [ $? -eq 0 ];then
                cat $NGX_VHOSTS
                if [ $? -eq 0 ];then
                    mkdir -p /data/backup/'date +%F'
                    \cp -a $NGX_VHOSTS /data/backup/'date +%F'
                    rm -rf $NGX_VHOSTS
                    $NGX_DIR/sbin/nginx -s reload
                    echo -e "\033[32m-----------------------\033[0m"
                    echo -e "\033[32mThe Nginx $NGX_VHOSTS Already remove,
reload nginx...\033[0m"
                fi
            else
                shift 1
                    echo -e "\033[31m-----------------------\033[0m"
                    echo -e "\033[31mNginx $NGX_VHOSTS Virtual hosts does
not exist.please check.\033[0m"
                    ls -l $NGX_DIR/conf/ |head -10
            fi
        else
            shift 1
            echo -e "\033[31m-----------------------\033[0m"
            echo -e "\033[31mNginx $NGX_VHOSTS Virtual hosts does not exist.
please check.\033[0m"
            ls -l $NGX_DIR/conf/ |head -10
        fi
    done
}

function tomcat_install(){
    #auto config tomcat web
    #change tomcat port :6001 7001 8001
    #upload jdk and tomcat for shell dir
    TOMCAT_VER="8.0.50"
    JAVA_VER="1.8.0_131"
    JAVA_DIR="/usr/java"
    TOMCAT_DIR="/usr/local"
    JAVA_SOFT="jdk${JAVA_VER}.tar.gz"
    TOMCAT_SOFT="apache-tomcat-${TOMCAT_VER}.tar.gz"
    if [ $# -le 1 ];then
            nginx_help
    fi
    shift 1
    #Install JAVA JDK
```

```
            grep -ai "^export" /etc/profile|grep -ai "JAVA_HOME" >/dev/null
            if [ $? -ne 0 ];then
                ls -l $JAVA_SOFT
                tar -xzvf $JAVA_SOFT
                mkdir -p $JAVA_DIR/
                \mv jdk$JAVA_VER $JAVA_DIR/
                ls -l $JAVA_DIR/jdk$JAVA_VER/
                $JAVA_DIR/jdk$JAVA_VER/bin/java -version
                cat>>/etc/profile<<-EOF
                export JAVA_HOME=$JAVA_DIR/jdk$JAVA_VER
                export CLASSPATH=\$CLASSPATH:\$JAVA_HOME/lib:\$JAVA_HOME/jre/lib
                export PATH=\$PATH:\$JAVA_HOME/bin/
                EOF
                java -version
                source /etc/profile
            fi
        source /etc/profile
        #Install Tomcat WEB
        MAX_PORT=$(for i in $(find /usr/local/ -name "server.xml");do grep -ai
"port=" $i;done|awk -F"=" '{print $2}'|awk '{print $1}'|sed 's/\"//g'|grep
-aivE "8443"|sort -nr|head -1)
            if [ -z $MAX_PORT ];then
                for TOMCAT_DOMAINS in $*
                do
                    MAX_PORT=$(for i in $(find /usr/local/ -name "server.xml");do grep
-ai "port=" $i;done|awk -F"=" '{print $2}'|awk '{print $1}'|sed 's/\"//g'
|grep -aivE "8443"|sort -nr|head -1)
                    if [ -z $MAX_PORT ];then
                        #Install Tomcat WEB
                        ls -l $TOMCAT_SOFT
                        tar -xzvf $TOMCAT_SOFT >/dev/null
                        mkdir -p $TOMCAT_DIR/tomcat_$TOMCAT_DOMAINS/
                        \mv apache-tomcat-$TOMCAT_VER/* $TOMCAT_DIR/tomcat_$TOMCAT_
DOMAINS/  >/dev/null 2>&1
                        $TOMCAT_DIR/tomcat_$TOMCAT_DOMAINS/bin/startup.sh
                        echo -e "\033[32m----------------------\033[0m"
                            echo -e "\033[32mThe Nginx $TOMCAT_DOMAINS ADD Success.\
033[0m"
                        sleep 5
                        ps -ef|grep "$TOMCAT_DOMAINS"|grep -v "$0"
                        netstat -tnlp|grep -aiwE "6001|7001|8001"
                        setenforce 0
                        systemctl stop firewalld.service
                        service iptables stop
                    else
                        ls -l $TOMCAT_DIR/ |grep "$TOMCAT_DOMAINS" >>/dev/null 2>&1
```

```
                    if [ $? -ne 0 ];then
                        #Install Tomcat WEB
                            PORT1=$(expr $MAX_PORT - 2000 + 1)
                            PORT2=$(expr $MAX_PORT - 1000 + 1)
                            PORT3=$(expr $MAX_PORT + 1)
                            ls -l $TOMCAT_SOFT
                            tar -xzvf $TOMCAT_SOFT >/dev/null
                            mkdir -p $TOMCAT_DIR/tomcat_$TOMCAT_DOMAINS/
                            \mv apache-tomcat-$TOMCAT_VER/* $TOMCAT_DIR/
tomcat_$TOMCAT_DOMAINS/ >/dev/null 2>&1
                            sed -i "s/6001/$PORT1/g" $TOMCAT_DIR/tomcat_
$TOMCAT_DOMAINS/conf/server.xml
                            sed -i "s/7001/$PORT2/g" $TOMCAT_DIR/tomcat_
$TOMCAT_DOMAINS/conf/server.xml
                            sed -i "s/8001/$PORT3/g" $TOMCAT_DIR/tomcat_
$TOMCAT_DOMAINS/conf/server.xml
                            $TOMCAT_DIR/tomcat_$TOMCAT_DOMAINS/bin/startup.sh
                        echo -e "\033[32m----------------------\033[0m"
                            echo -e "\033[32mThe Nginx $TOMCAT_DOMAINS ADD
Success.\033[0m"
                            sleep 5
                    ps -ef|grep "$TOMCAT_DOMAINS"|grep -v "$0"
                        netstat -tnlp|grep -aiwE "$PORT1|$PORT2|$PORT3"
                        setenforce 0
                        systemctl stop firewalld.service
                        service iptables stop
                else
                    echo -e "\033[32m----------------------\033[0m"
                            echo -e "\033[32mThe Tomcat $TOMCAT_DOMAINS
Already Exist,Please Exit.\033[0m"
                            echo "ls -l $TOMCAT_DIR/tomcat_$TOMCAT_DOMAINS/"
                                ls -l $TOMCAT_DIR/tomcat_$TOMCAT_DOMAINS/
                fi
            fi
        done
    else
        for TOMCAT_DOMAINS in $*
        do
            ls -l $TOMCAT_DIR/ |grep "$TOMCAT_DOMAINS" >>/dev/null 2>&1
        if [ $? -ne 0 ];then
            MAX_PORT=$(for i in $(find /usr/local/ -name "server.xml");do
grep -ai "port=" $i;done|awk -F"=" '{print $2}'|awk '{print $1}'|sed
's/\"//g'|grep -aivE "8443"|sort -nr|head -1)
            #Install Tomcat WEB
            PORT1=$(expr $MAX_PORT - 2000 + 1)
            PORT2=$(expr $MAX_PORT - 1000 + 1)
```

```
                PORT3=$(expr $MAX_PORT + 1)
                ls -l $TOMCAT_SOFT
                tar -xzvf $TOMCAT_SOFT >/dev/null
                mkdir -p $TOMCAT_DIR/tomcat_$TOMCAT_DOMAINS/
                \mv apache-tomcat-$TOMCAT_VER/* $TOMCAT_DIR/tomcat_$TOMCAT_DOMAINS/ >/dev/null 2>&1
                sed -i "s/6001/$PORT1/g" $TOMCAT_DIR/tomcat_$TOMCAT_DOMAINS/conf/server.xml
                sed -i "s/7001/$PORT2/g" $TOMCAT_DIR/tomcat_$TOMCAT_DOMAINS/conf/server.xml
                sed -i "s/8001/$PORT3/g" $TOMCAT_DIR/tomcat_$TOMCAT_DOMAINS/conf/server.xml
                $TOMCAT_DIR/tomcat_$TOMCAT_DOMAINS/bin/startup.sh
                echo -e "\033[32m-----------------------\033[0m"
                    echo -e "\033[32mThe Tomcat $TOMCAT_DOMAINS ADD Success.\033[0m"
                sleep 5
                ps -ef|grep "$TOMCAT_DOMAINS"|grep -v "$0"
                netstat -tnlp|grep -aiwE "$PORT1|$PORT2|$PORT3"
                setenforce 0
                systemctl stop firewalld.service
                service iptables stop
            else
                    echo -e "\033[32m-----------------------\033[0m"
                    echo -e "\033[32mThe Tomcat $TOMCAT_DOMAINS Already Exist,Please Exit.\033[0m"
                echo "ls -l $TOMCAT_DIR/tomcat_$TOMCAT_DOMAINS/"
                    ls -l $TOMCAT_DIR/tomcat_$TOMCAT_DOMAINS/
            fi
        done
    fi
}
case $1 in
    -i|-I)
    nginx_install $*
    ;;
    -u|-U)
    nginx_update $*
    ;;
    -a|-A)
    virtual_add $*
    ;;
    -d|-D)
    virtual_del $*
    ;;
    -t|-T)
```

```
    tomcat_install $*
    ;;
    * )
    nginx_help
    ;;
esac
```

2.23　Shell 编程密码远程执行命令脚本

Shell 编程实现密码远程执行命令脚本，编写思路如下。

（1）自动将服务器 IP 地址和用户名、密码保存至文件 list.txt。

（2）自动安装 Expect 远程交互工具。

（3）自动编写 Expect 远程执行命令文件。

（4）支持任意命令远程执行。

（5）支持列表循环操作多台服务器。

Shell 编程实现密码远程执行命令脚本，相关代码如下。

```
#!/bin/sh
#auto exec expect shell scripts
#by author www.jfedu.net 2021
if
    [ ! -e /usr/bin/expect ];then
    yum  install expect -y
fi
#Judge passwd.txt exist
if
    [ ! -e ./passwd.txt ];then
    echo -e "The passwd.txt is not exist......Please touch ./passwd.txt ,
Content Example:\n192.168.1.11 passwd1\n192.168.1.12 passwd2"
    sleep 2 &&exit 0
fi
#Auto Touch login.exp File
cat>login.exp <<EOF
#!/usr/bin/expect -f
set ip [lindex \$argv 0 ]
set passwd [lindex \$argv 1 ]
set command [lindex \$argv 2 ]
set timeout -1
spawn ssh root@\$ip
expect {
"yes/no" { send "yes\r";exp_continue }
"password:" { send "\$passwd\r" }
}
```

```
expect "*#*" { send "\$command\r" }
expect  "#*" { send "exit\r" }
expect eof
EOF
##Auto exec shell scripts
CMD="$*"
if
    [ "$1" == "" ];then
    echo ==============================================================
    echo "Please insert your command ,Example {/bin/sh $0 'mkdir -p
/tmp'} ,waiting exit .......... "
    sleep 2
    exit 1
fi
for i in `awk '{print $1}' passwd.txt`
do
    j=`awk -v I="$i" '{if(I==$1)print $2}' passwd.txt`
    expect ./login.exp $i $j "$CMD"
done
```

2.24　Shell 编程密码远程复制文件脚本

Shell 编程实现密码远程复制文件脚本，编写思路如下。

（1）自动将服务器 IP 地址和用户名、密码保存至文件 list.txt。

（2）自动安装 Expect 远程交互工具。

（3）自动编写 Expect 远程执行复制文件。

（4）支持任意文件远程复制的操作。

（5）支持列表循环操作多台服务器。

Shell 编程实现密码远程复制文件脚本，相关代码如下。

```
#!/bin/sh
#auto exec expect shell scripts
#by author www.jfedu.net 2021
if
    [ ! -e /usr/bin/expect ];then
    yum  install expect -y
fi
#Judge passwd.txt exist
if
    [ ! -e ./passwd.txt ];then
    echo -e "The passwd.txt is not exist......Please touch ./passwd.txt ,
Content Example:\n192.168.1.11 passwd1\n192.168.1.12 passwd2"
    sleep 2 &&exit 0
```

```
fi
#Auto Touch login.exp File
cat>login.exp <<EOF
#!/usr/bin/expect -f
set ip [lindex \$argv 0]
set passwd [lindex \$argv 1]
set src_file [lindex \$argv 2]
set des_dir [lindex \$argv 3]
set timeout -1
spawn scp -r \$src_file root@\$ip:\$des_dir
expect {
"yes/no"    { send "yes\r"; exp_continue }
"password:"  { send "\$passwd\r" }
}
expect "100%"
expect eof
EOF
##Auto exec shell scripts
if
    [ "$1" == "" ];then
    echo ==========================================================
    echo "Please insert your are command ,Example {/bin/sh $0 /src /des } ,waiting exit .......... "
    sleep 2
    exit 1
fi
for i in 'awk '{print $1}' passwd.txt'
do
    j='awk -v I="$i" '{if(I==$1)print $2}' passwd.txt'
    expect ./login.exp $i $j $1 $2
done
```

2.25　Shell 编程 Bind DNS 管理脚本

Bind 主要应用于企业 DNS 构建平台，而 DNS 用于解析域名与 IP 地址，用户在浏览器只需输入域名，即可访问服务器对应 IP 地址的虚拟主机网站。

Bind 难点在于创建各种记录，如 A 记录、mail 记录、反向记录、资源记录等，Shell 脚本可以减轻人工的操作，节省大量的时间成本。

Shell 编程实现 Bind 自动安装、初始化 Bind 环境、自动添加 A 记录、反向记录和批量添加 A 记录脚本，编写思路如下。

（1）YUM 方式自动安装 Bind。
（2）自动初始化 Bind 配置。

（3）创建安装，初始化，添加记录函数。

（4）自动添加单个 A 记录及批量添加 A 记录和反向记录。

Shell 编程实现 Bind 自动安装、初始化 Bind 环境、自动添加 A 记录、反向记录和批量添加 A 记录脚本，相关代码如下。

```bash
#!/bin/bash
#Auto install config bind server
#By author jfedu.net 2021
#Define Path variables
BND_ETC=/var/named/chroot/etc
BND_VAR=/var/named/chroot/var/named
BAK_DIR=/data/backup/dns_`date +%Y%m%d-%H%M`
##Backup named server
if
     [ ! -d $BAK_DIR ];then
     echo "Please waiting  Backup Named Config ............"
     mkdir  -p $BAK_DIR
     cp -a  /var/named/chroot/{etc,var}  $BAK_DIR
     cp -a  /etc/named.* $BAK_DIR
fi
##Define Shell Install Function
Install ()
{
  if
     [ ! -e /etc/init.d/named ];then
     yum install bind* -y
else
     echo ----------------------------------------------------
     echo "The Named Server is exists ,Please exit ........."
     sleep 1
 fi
}
##Define Shell Init Function
Init_Config ()
{
     sed  -i -e 's/localhost;/any;/g' -e '/port/s/127.0.0.1/any/g' /etc/named.conf
     echo ----------------------------------------------------
     sleep 2
     echo "The named.conf config Init success !"
}
##Define Shell Add Name Function
Add_named ()
{
##DNS name
```

```
        read -p "Please  Insert Into Your Add Name ,Example 51cto.com :" NAME
        echo $NAME |grep -E "com|cn|net|org"
      while
       [ "$?" -ne 0 ]
         do
        read -p "Please  reInsert Into Your Add Name ,Example 51cto.com :" NAME
        echo $NAME |grep -E "com|cn|net|org"
      done
## IP address
        read -p "Please  Insert Into Your Name Server IP ADDress:" IP
        echo $IP |egrep -o "([0-9]{1,3}\.){3}[0-9]{1,3}"
      while
       [ "$?" -ne "0" ]
         do
        read -p "Please  reInsert Into Your Name Server IP ADDress:" IP
        echo $IP |egrep -o "([0-9]{1,3}\.){3}[0-9]{1,3}"
      done
      ARPA_IP=`echo $IP|awk -F. '{print $3"."$2"."$1}'`
      ARPA_IP1=`echo $IP|awk -F. '{print $4}'`
      cd $BND_ETC
      grep  "$NAME" named.rfc1912.zones
if
      [ $? -eq 0 ];then
       echo "The $NAME IS exist named.rfc1912.zones conf ,please exit ..."
       exit
else
       read -p "Please  Insert Into SLAVE Name Server IP ADDress:" SLAVE
       echo $SLAVE |egrep -o "([0-9]{1,3}\.){3}[0-9]{1,3}"
      while
       [ "$?" -ne "0" ]
         do
           read -p "Please  Insert Into SLAVE Name Server IP ADDress:" SLAVE
           echo $SLAVE |egrep -o "([0-9]{1,3}\.){3}[0-9]{1,3}"
      done
           grep  "rev" named.rfc1912.zones
       if
        [ $? -ne 0 ];then
           cat >>named.rfc1912.zones <<EOF
#'date +%Y-%m-%d' Add $NAME CONFIG
zone "$NAME" IN {
    type master;
    file "$NAME.zone";
    allow-update { none; };
};
zone "$ARPA_IP.in-addr.arpa" IN {
```

```
                type master;
                file "$ARPA_IP.rev";
                allow-update { none; };
        };
EOF
        else
            cat >>named.rfc1912.zones <<EOF

#'date +%Y-%m-%d' Add $NAME CONFIG
zone "$NAME" IN {
                type master;
                file "$NAME.zone";
                allow-update { none; };
        };
EOF
        fi
fi
        [ $? -eq 0 ]&& echo "The $NAME config name.rfc1912.zones success !"
        sleep 3 ;echo "Please waiting config $NAME zone File ............"
        cd $BND_VAR
        read -p "Please insert Name DNS A HOST ,EXample  www or mail :" HOST
        read -p "Please insert Name DNS A NS IP ADDR ,EXample 192.168.111.130 :" IP_HOST
        echo $IP_HOST |egrep -o "([0-9]{1,3}\.){3}[0-9]{1,3}"
        ARPA_IP2='echo $IP_HOST|awk -F. '{print $3"."$2"."$1}''
        ARPA_IP3='echo $IP_HOST|awk -F. '{print $4}''
        while
        [ "$?" -ne "0" ]
do
        read -p "Please Reinsert Name DNS A IPADDRESS ,EXample 192.168.111.130 :" IP_HOST
        echo $IP_HOST |egrep -o "([0-9]{1,3}\.){3}[0-9]{1,3}"
done
        cat >$NAME.zone <<EOF
\$TTL    86400
@               IN SOA  localhost.      root.localhost. (
                                        43              ; serial (d. adams)
                                        1H              ; refresh
                                        15M             ; retry
                                        1W              ; expiry
                                        1D )            ; minimum

                IN NS           $NAME.
EOF
        REV='ls *.rev'
        ls *.rev >>/dev/null
```

```
if
      [ $? -ne 0 ];then
      cat >>$ARPA_IP.rev <<EOF
\$TTL    86400
@       IN      SOA     localhost.      root.localhost. (
                                1997022703 ; Serial
                                28800      ; Refresh
                                14400      ; Retry
                                3600000    ; Expire
                                86400 )    ; Minimum


          IN   NS   $NAME.
EOF
      echo "$HOST             IN   A       $IP_HOST" >>$NAME.zone
      echo "$ARPA_IP3          IN   PTR     $HOST.$NAME." >>$ARPA_IP.rev
      [ $? -eq 0 ]&& echo -e "The $NAME config success:\n$HOST       IN    A
$IP_HOST\n$ARPA_IP3          IN   PTR     $HOST.$NAME."
else
      sed -i "9a IN  NS  $NAME." $REV
      echo "$HOST             IN   A       $IP_HOST" >>$NAME.zone
      echo "$ARPA_IP3          IN   PTR     $HOST.$NAME." >>$REV
      [ $? -eq 0 ]&& echo -e "The $NAME config success1:\n$HOST       IN    A
$IP_HOST\n$ARPA_IP3          IN   PTR     $HOST.$NAME."
fi
}
##Define Shell List A Function
Add_A_List ()
{
if
      cd  $BND_VAR
      REV=`ls  *.rev`
      read -p "Please  Insert Into Your Add Name ,Example 51cto.com :" NAME
      [ ! -e "$NAME.zone" ];then
      echo "The $NAME.zone File is not exist ,Please ADD $NAME.zone File :"
      Add_named ;
else
      read -p "Please Enter List Name A NS File ,Example /tmp/name_list.txt:
" FILE
    if
      [ -e $FILE ];then
      for i in  `cat $FILE|awk '{print $2}'|sed "s/$NAME//g"|sed 's/\.$//g'`
      #for i in   `cat $FILE|awk '{print $1}'|sed "s/$NAME//g"|sed 's/\.$//g'`
do
      j=`awk -v I="$i.$NAME" '{if(I==$2)print $1}' $FILE`
      echo -------------------------------------------------------
      echo "The $NAME.zone File is exist ,Please Enter insert NAME HOST ...."
```

```
            sleep 1
            ARPA_IP=`echo $j|awk -F. '{print $3"."$2"."$1}'`
            ARPA_IP2=`echo $j|awk -F. '{print $4}'`
            echo  "$i            IN  A            $j" >>$NAME.zone
            echo  "$ARPA_IP2     IN  PTR          $i.$NAME." >>$REV
            [ $? -eq 0 ]&& echo -e "The $NAME config success:\n$i        IN   A
$j\n$ARPA_IP2         IN   PTR          $i.$NAME."
done
    else
        echo "The $FILE List File IS Not Exist .......,Please exit ..."
    fi
fi
}
##Define Shell Select Menu
PS3="Please select Menu Name Config: "
select i in "自动安装Bind服务"  "自动初始化Bind配置"  "添加解析域名"   "批量添加A
记录"
do
case  $i   in
    "自动安装Bind服务")
        Install
;;
    "自动初始化Bind配置")
        Init_Config
;;
    "添加解析域名")
        Add_named
;;
    "批量添加A记录")
        Add_A_List
;;
    * )
        echo ------------------------------------------------------
        sleep 1
        echo "Please exec: sh $0 { Install(1)   or Init_Config(2) or
Add_named(3) or Add_config_A(4) }"
;;
esac
done
```

2.26　Shell 编程 Docker 虚拟化管理脚本

Docker 虚拟化是目前主流的虚拟化解决方案,越来越多的企业在使用 Docker 轻量级虚拟化,构建、维护和管理 Docker 虚拟化平台都是非常重要的环节,开发 Docker Shell 脚本可以在命令行界面快速管理和维护 Docker 虚拟机。

Shell 编程实现 Docker 自动安装、自动导入镜像、创建虚拟机、指定 IP 地址和将创建的 Docker 虚拟机加入 Excel 存档或加入 MySQL 数据库脚本，编写思路如下。

（1）基于 CentOS 6.5+操作系统或 CentOS 7.x 操作系统使用 YUM 安装 Docker 环境。

（2）Docker 脚本参数指定 CPU、内存和硬盘容量。

（3）Docker 自动检测局域网 IP 地址并赋予 Docker 虚拟机。

（4）Docker 基于 Pipework 指定 IP 地址。

（5）将创建的 Docker 虚拟机加入 Excel 存档或 MySQL 数据库。

Shell 编程实现 Docker 自动安装、自动导入镜像、创建虚拟机、指定 IP 地址和将创建的 Docker 虚拟机加入 Excel 存档或加入 MySQL 数据库脚本，相关代码如下。

```bash
#!/bin/bash
#Auto install docker and Create VM
#By author jfedu.net 2021
#Define Path variables
IPADDR=`ifconfig|grep -E "\<inet\>"|awk '{print $2}'|grep "192.168"|head -1`
GATEWAY=`route -n|grep "UG"|awk '{print $2}'|grep "192.168"|head -1`
IPADDR_NET=`ifconfig|grep -E "\<inet\>"|awk '{print $2}'|grep "192.168"\
|head -1|awk -F. '{print $1"."$2"."$3"."}'`
LIST="/root/docker_vmlist.csv"
if [ ! -f /usr/sbin/ifconfig ];then
    yum install net-tools* -y
fi
for i in `seq 1 253`;do ping -c 1 ${IPADDR_NET}${i} ;[ $? -ne 0 ]&&
DOCKER_IPADDR="${IPADDR_NET}${i}" &&break;done >>/dev/null 2>&1
echo "###################"
echo -e "Dynamic get docker IP,The Docker IP address\n\n$DOCKER_IPADDR"
NETWORK=(
    HWADDR=`ifconfig eth0|grep ether|awk '{print $2}'`
    IPADDR=`ifconfig eth0|grep -E "\<inet\>"|awk '{print $2}'`
    NETMASK=`ifconfig eth0|grep -E "\<inet\>"|awk '{print $4}'`
    GATEWAY=`route -n|grep "UG"|awk '{print $2}'`
)
if [ -z "$1" -o -z "$2" ];then
    echo -e "\033[32m--------------------------------\033[0m"
    echo -e "\033[32mPlease exec $0 CPU(C) MEM(G),example $0 4 8\033[0m"
    exit 0
fi
#CPU=`expr $2 - 1`
if [ ! -e /usr/bin/bc ];then
    yum install bc -y >>/dev/null 2>&1
fi
CPU_ALL=`cat /proc/cpuinfo |grep processor|wc -l`
```

```
if [ ! -f $LIST ];then
    CPU_COUNT=$1
    CPU_1="0"
    CPU1=`expr $CPU_1 + 0`
    CPU2=`expr $CPU1 + $CPU_COUNT - 1`
    if [ $CPU2 -gt $CPU_ALL ];then
        echo -e "\033[32mThe System CPU count is $CPU_ALL,not more than it.\033[0m"
        exit
    fi
else
    CPU_COUNT=$1
    CPU_1=`cat $LIST|tail -1|awk -F"," '{print $4}'|awk -F"-" '{print $2}'`
    CPU1=`expr $CPU_1 + 1`
    CPU2=`expr $CPU1 + $CPU_COUNT - 1`
    if [ $CPU2 -gt $CPU_ALL ];then
        echo -e "\033[32mThe System CPU count is $CPU_ALL,not more than it.\033[0m"
        exit
    fi
fi
MEM_F=`echo $2 \* 1024|bc`
MEM=`printf "%.0f\n" $MEM_F`
DISK=20
USER=$3
REMARK=$4
ping $DOCKER_IPADDR -c 1 >>/dev/null 2>&1
if [ $? -eq 0 ];then
    echo -e "\033[32m----------------------------------\033[0m"
    echo -e "\033[32mThe IP address to be used,Please change other IP,exit.\033[0m"
    exit 0
fi
if [ ! -e /usr/bin/docker ];then
    yum install docker* device-mapper* -y
    mkdir -p /export/docker/
    cd /var/lib/ ;rm -rf docker ;ln -s /export/docker/ .
    mkdir -p /var/lib/docker/devicemapper/devicemapper
    dd if=/dev/zero of=/var/lib/docker/devicemapper/devicemapper/data bs=1G count=0 seek=2000
    service docker start
    if [ $? -ne 0 ];then
        echo "Docker install error ,please check."
        exit
    fi
```

```
    fi
cd  /etc/sysconfig/network-scripts/
    mkdir -p /data/backup/'date +%Y%m%d-%H%M'
    yes|cp ifcfg-eth* /data/backup/'date +%Y%m%d-%H%M'/
if
    [ -e /etc/sysconfig/network-scripts/ifcfg-br0 ];then
    echo
else
    cat >ifcfg-eth0 <<EOF
    DEVICE=eth0
    BOOTPROTO=none
    ${NETWORK[0]}
    NM_CONTROLLED=no
    ONBOOT=yes
    TYPE=Ethernet
    BRIDGE="br0"
    ${NETWORK[1]}
    ${NETWORK[2]}
    ${NETWORK[3]}
    USERCTL=no
EOF
    cat >ifcfg-br0 <<EOF
    DEVICE="br0"
    BOOTPROTO=none
    ${NETWORK[0]}
    IPV6INIT=no
    NM_CONTROLLED=no
    ONBOOT=yes
    TYPE="Bridge"
    ${NETWORK[1]}
    ${NETWORK[2]}
    ${NETWORK[3]}
    USERCTL=no
EOF
    /etc/init.d/network restart
fi
echo 'Your can restart Ethernet Service: /etc/init.d/network restart !'
echo '-----------------------------------------------------------'

cd -
#######create docker container
service docker status >>/dev/null
if [ $? -ne 0 ];then
    service docker restart
```

```bash
fi
NAME="Docker_`echo $DOCKER_IPADDR|awk -F"." '{print $(NF-1)"_"$NF}'`"
IMAGES=`docker images|grep -v "REPOSITORY"|grep -v "none"|grep "jfedu"|head -1|awk '{print $1}'`
if [ -z $IMAGES ];then
    echo "Plesae Download Docker Centos Images,you can to be use docker search centos,and docker pull centos6.5-ssh,exit 0"
        if [ ! -f jfedu_centos68.tar ];then
            echo "Please upload jfedu_centos68.tar for docker server."
            exit
        fi
        cat jfedu_centos68.tar|docker import - jfedu_centos6.8
fi
IMAGES=`docker images|grep -v "REPOSITORY"|grep -v "none"|grep "jfedu"|head -1|awk '{print $1}'`
CID=$(docker run -itd --privileged --cpuset-cpus=${CPU1}-${CPU2} -m ${MEM}m --net=none --name=$NAME $IMAGES /bin/bash)
echo $CID
docker ps -a |grep "$NAME"
pipework br0 $NAME $DOCKER_IPADDR/24@$IPADDR
docker exec $NAME /etc/init.d/sshd start
if [ ! -e $LIST ];then
    echo "编号,容器ID,容器名称,CPU,内存,硬盘,容器IP,宿主机IP,使用人,备注" >$LIST
fi
####################
NUM=`cat $LIST |grep -v CPU|tail -1|awk -F, '{print $1}'`
if [[ $NUM -eq "" ]];then
        NUM="1"
else
        NUM=`expr $NUM + 1`
fi
###################
echo -e "\033[32mCreate virtual client Successfully.\n$NUM `echo $CID|cut -b 1-12`,$NAME,$CPU1-$CPU2,${MEM}M,${DISK}G,$DOCKER_IPADDR,$IPADDR,$USER,$REMARK\033[0m"
if [ -z $USER ];then
    USER="NULL"
    REMARK="NULL"
fi
echo $NUM, `echo $CID|cut -b 1-12`,$NAME,$CPU1-$CPU2,${MEM}M,${DISK}G,$DOCKER_IPADDR,$IPADDR,$USER,$REMARK >>$LIST
rm -rf /root/docker_vmlist_*
iconv -c -f utf-8 -t gb2312 $LIST -o /root/docker_vmlist_`date +%H%M`.csv
```

2.27 Shell 编程脚本

2.27.1 Shell 编程采集服务器硬件信息脚本

Shell 编程实现采集服务器硬件信息脚本，编写思路如下。
（1）创建数据库和表用来存储服务器信息。
（2）获取服务器 CPU、内存、硬盘、网卡等相关硬件信息。
（3）将获取的信息写成 SQL 语句，并插入数据库。
（4）定期对 SQL 数据进行备份。
（5）将脚本加入 Crontab 实现自动备份。

2.27.2 Shell 编程 Linux 操作系统初始化脚本

Shell 编程实现 Linux 操作系统初始化脚本，编写思路如下。
（1）Linux 操作系统安装完成，自动初始化系统。
（2）关闭不必要的端口。
（3）关闭不必要的服务。
（4）添加同步时间任务计划。
（5）优化相关 Linux 内核参数。

2.27.3 Shell 编程 Xtrabackup 数据库自动备份脚本

Shell 编程实现 Xtrabackup 数据库自动备份脚本，编写思路如下。
（1）支持 MySQL 单个库备份。
（2）支持 MySQL 多个库备份。
（3）支持 MySQL 全数据库备份。
（4）支持 MySQL 指定数据库增量备份。
（5）支持 MySQL 指定多个数据库增量备份。
（6）支持 MySQL 数据库定期删除。

2.27.4 Shell 编程 Linux 服务器免密钥分发脚本

Shell 编程实现 Linux 服务器免密钥分发脚本，编写思路如下。
（1）基于 ssh-keygen 自动生成公钥和私钥。
（2）给定所有客户端的 IP 地址、用户名和密码信息。

(3)基于命令工具将公钥自动复制到远程机器。

给定的客户端 ip.txt 文件信息如下。

```
#ip              user      password
192.168.1.100    root      123456
192.168.1.101    root      1qaz@WSX
192.168.1.102    root      123
192.168.1.103    root      456
```

2.27.5　Shell 编程 Nginx 多版本软件安装脚本

Shell 编程实现 Nginx 多版本软件安装脚本,编写思路如下。

(1)安装不同版本的 Nginx。

(2)检测系统是否已经存在,是否可以覆盖版本。

(3)启动 Nginx 并测试访问。

(4)支持多个版本指定目录安装和启动。

2.27.6　Shell 编程自动收集软件、端口、进程脚本

Shell 编程实现自动收集软件、端口、进程脚本,编写思路如下。

(1)收集服务器所有端口和对应的服务。

(2)收集服务器对应服务 Nginx 发布目录和虚拟主机。

(3)收集服务器对应服务 MySQL 数据目录和配置文件。

2.27.7　Shell 编程 LVS 负载均衡管理脚本

Shell 编程实现 LVS 负载均衡管理脚本,编写思路如下。

(1)Shell 脚本自动安装 LVS 负载均衡。

(2)支持添加 VIP 地址。

(3)支持在 VIP 地址上添加后端 Realserver IP。

(4)支持删除后端真实 IP 地址和 VIP 地址。

2.27.8　Shell 编程 Keepalived 管理脚本

Shell 编程实现 Keepalived 管理脚本,编写思路如下。

(1)Shell 脚本支持自动配置 Keepalived 服务。

(2)自动添加负载均衡 VIP 地址。

(3)能够在已有的 VIP 地址均衡中添加 Realserver IP。

（4）支持删除某个 VIP 地址中指定的 Realserver IP。

2.27.9　Shell 编程 Discuz 门户网站自动部署脚本

Shell 编程实现 Discuz 门户网站自动部署脚本，编写思路如下。
（1）Shell 脚本支持从 SVN 获取网站代码。
（2）实现创建备份目录和备份源网站。
（3）将新的文件更新至 Discuz 对应的目录。

2.27.10　Shell 编程监控 Linux 磁盘分区容量脚本

Shell 编程实现监控 Linux 磁盘分区容量脚本，编写思路如下。
（1）Shell 脚本实现 Linux 磁盘多个分区监控。
（2）打印磁盘使用率超过 85% 的分区。
（3）将分区使用率超过 85% 的磁盘发送到邮件报警。
（4）将分区使用率超过 85% 的磁盘发送到微信报警。

第 3 章　自动化运维发展

随着企业服务器数量越来越多,服务器日常管理也逐渐繁杂,如果每天通过人工频繁地更新、部署及管理这些服务器,势必会浪费大量的时间,且有可能出现某些疏忽或遗漏。

本章将介绍如何构建企业自动化运维之路、传统运维方式存在的问题、自动化运维的具体内容、建立高效的 IT 自动化运维管理的步骤及工厂自动化运维工具、体系等内容。

3.1　传统运维方式简介

传统的 IT 运维是等到 IT 故障出现后再由运维人员采取相应的补救措施。这种被动的、孤立的、半自动式的 IT 运维管理模式经常让 IT 部门疲惫不堪,主要表现在以下 3 个方面。

(1) 运维人员被动、效率低。

在 IT 运维过程中,只有当事件已经发生并造成业务影响时才能被发现和着手处理,这种被动"救火"不但使 IT 运维人员终日忙碌,而且使 IT 运维质量很难提高,导致 IT 部门和业务部门对 IT 运维服务的满意度都不高。

(2) 缺乏一套高效的 IT 运维机制。

许多企业在 IT 运维管理过程中缺少自动化的运维管理模式,也没有明确的角色定义和责任划分,使问题出现后很难快速、准确地找到根本原因,无法及时联系相应的人员进行修复和处理,或者在问题找到后缺乏流程化的故障处理机制,并且在处理问题时不仅欠缺规范化的解决方案,还缺乏全面的跟踪记录。

(3) 缺乏高效的 IT 运维技术工具。

随着信息化建设的深入,企业 IT 系统日趋复杂,林林总总的网络设备、服务器、中间件、业务系统等让 IT 运维人员难以从容应对,即使加班加点地维护、部署、管理,还是经常会因设备出现故障而导致业务中断,严重影响企业的正常运转。

出现这些问题的部分原因是企业缺乏事件监控和诊断工具等 IT 运维技术工具。没有高效的技术工具的支持，故障事件就很难得到主动、快速的处理。

3.2 自动化运维简介

IT 运维已经在风风雨雨中走过了十几个春秋，如今它正以一种全新的姿态出现在众人面前。IT 系统的复杂性已经在客观上要求 IT 运维能够实现数字化、自动化维护，自动化运维是 IT 技术发展的必然结果。

自动化运维是指将 IT 运维中日常的、大量的重复性工作自动化，把过去的手动执行转为自动化操作。自动化是 IT 运维工作的升华，IT 自动化运维不只是一个维护过程，更是一个管理的提升过程，是 IT 运维的最高层次，也是未来的发展趋势。

3.3 自动化运维的具体内容

日常 IT 运维中大量的重复性工作（小到简单的日常检查、配置变更和软件安装，大到整个流程变更的组织调度）由过去的手动执行转为自动化操作，从而减少乃至消除运维中的延迟，实现"零延时"的 IT 运维。

简单地说，IT 自动化运维是指基于流程化的框架，将事件与 IT 流程相关联，一旦被监控系统监测到性能超标或宕机，就会触发相关事件和事先定义好的流程，可以自动启动故障响应和恢复机制。

3.4 建立高效的 IT 自动化运维管理

建立高效的 IT 自动化运维管理主要有以下几个步骤。

（1）建立自动化运维管理平台。

IT 运维自动化管理建设的第一步是要建立 IT 运维的自动化监控和管理平台。通过监控工具实现对用户操作规范的约束和对 IT 资源的实时监控，包括服务器、数据库、中间件、存储备份、网络、安全、机房、业务应用和客户端等，通过自动监控管理平台实现故障或问题的综合处理和集中管理。

（2）建立故障事件自动触发流程，提高故障处理效率。

所有 IT 设备在遇到问题时都要报警，无论是系统自动报警还是人员报警。报警信息应以红色标识显示在运维屏幕上。IT 运维人员只需要按照相关知识库的数据，一步一步操作即可。

（3）建立规范的事件跟踪流程，强化运维执行力度。

需要建立故障和事件处理跟踪流程，利用表格工具等记录故障及其处理情况，以建立运维日志,并定期回顾,从中发现问题的线索和根源。

(4)设立IT运维关键流程,引入优先处理原则。

设置自动化流程时还需要引入优先处理原则,例行的事件按常规处理,特别事件要按优先级次序处理,也就是把事件细分为例行事件和例外关键事件。

3.5 IT自动化运维工具

随着IT运维的飞速发展,市场上涌现了大量的自动化配置维护工具,如PSSH、Puppet、Chef、SaltStack、Ansible等。自动配置工具存在的初衷就是为了更方便、快捷地进行配置管理,它易于安装和使用,语法也非常简单易学。

对于企业来说,要特别关注两类自动化工具:一是IT运维监控和诊断优化工具;二是运维流程自动化工具。这两类工具主要有以下应用。

(1)监控自动化:对重要的IT设备实施主动式监控,如路由器、交换机、防火墙等。

(2)配置变更检测自动化:IT设备配置参数一旦发生变化,将触发变更流程并转发给相关技术人员进行确认,通过自动检测协助IT运维人员发现和维护配置变更。

(3)维护事件提醒自动化:对IT设备和应用活动进行实时监控,当发生异常事件时系统自动启动报警和响应机制,第一时间通知相关责任人。

(4)系统健康检测自动化:定期自动对IT设备硬件和应用系统进行健康巡检,配合IT运维团队实施对系统的健康检查和监控。

(5)维护报告生成自动化:定期自动对系统做日志的收集分析,记录系统运行状况,并通过阶段性的监控、分析和总结,定时提供IT运维的可用性性能和系统资源利用状况的分析报告。

3.6 IT自动化运维体系

一个完善的自动化运维体系包括系统预备、配置管理和监控报警3个环节,每个环节实现的功能也各不相同,具体功能如下。

(1)系统预备类主要功能包括以下几点。

① 自动化安装操作系统。

② 自动初始化系统。

③ 自动安装各种软件包。

(2)配置管理类主要功能包括以下几点。

① 自动化部署业务系统软件包并完成配置。

② 远程管理服务器。

③ 配置文件、自动部署 Jenkins、网站代码变更回滚。

（3）监控报警类主要功能包括以下几点。

① 服务器可用性、性能和安全监控。

② 向管理员发送报警信息。

根据提供的功能不同，自动化运维工具可分为 4 类，如表 3-1 所示。

表 3-1　自动化运维工具分类

编　号	系统预备类工具	配置管理类工具	监控报警类工具
1	Kickstart	Puppet	Nagios
2	Cobbler	SaltStack	Cacti
3	openQRM	Func	Ganglia
4	Spacewalk	Ansible	Zabbix

第 4 章 Puppet 自动运维企业实战

Puppet 是目前互联网三大主流自动化运维工具（Puppet、Ansible、SaltStack）之一，是一种 Linux 平台和 UNIX 平台的集中配置管理系统。配置管理系统用于管理文件、用户、进程和软件包等资源，设计目标是简化对这些资源的管理及妥善处理资源间的依赖关系。

本章将介绍 Puppet 工作原理、Puppet 安装配置、企业资源案例、Puppet 高可用集群配置、Puppet 批量更新部署网站、Puppet+SVN 实现代码自动部署等内容。

4.1 Puppet 入门

Puppet 使用一种描述性语言定义配置项，其中配置项称为资源，描述性语言可以声明配置的状态，比如，声明一个软件包应该被安装，或者一个服务应该被启用。

Puppet 可以运行一台服务器，每个客户端通过 SSL 证书连接服务器，得到本机器的配置列表，然后根据列表完成配置工作。如果硬件性能比较高，维护和管理成千上万台机器是非常轻松的，前提是客户端机器的配置、服务器路径、软件都保持一致。

在企业级的大规模生成环境中，如果只有一台 Puppet Master，压力会非常大，因为 Puppet 是用 Ruby 语言编写的，Ruby 是解析型语言，每个客户端来访问都要解析一次，当客户端很多时，服务器压力将会很大，所以需要扩展成一个服务器集群组。

Puppet Master 可以看作一个 Web 服务器，实际上也是基于 Ruby 提供的 Web 服务器模块。因此，可以利用 Web 代理软件配合 Puppet Master 做集群设置，一般使用 Nginx+Puppet Master 整合构建大型企业自动化运维管理工具，Puppet 遵循 GPLv2 版权协议，该项目主要开发者是 Luke Kanies。

Kanies 从 1997 年开始参与 UNIX 的系统管理工作，Puppet 的开发源于他在这份工作中积累的经验。因为对已有的配置工具不甚满意，2001—2005 年，Kanies 开始在 Reductive 实验室从事工具的开发。很快，Reductive 实验室发布了他们新的旗舰产品。

Puppet 是开源的基于 Ruby 的系统配置管理工具。Puppet 是 C/S 结构,所有的 Puppet 客户端同一个服务器的 Puppet 通信,每个 Puppet 客户端每 30min(可以设置)连接一次服务器,下载最新的配置文件,并严格按照配置文件配置。配置完成以后 Puppet 客户端可以反馈给服务器一个消息,如果报错也会给服务器反馈一个消息。

4.2 Puppet 工作原理

要熟练掌握 Puppet 在企业生产环境中的应用,需要深入理解 Puppet 服务器与客户端详细的工作流程及原理。图 4-1 所示为 Puppet 和 SVN 结构图。

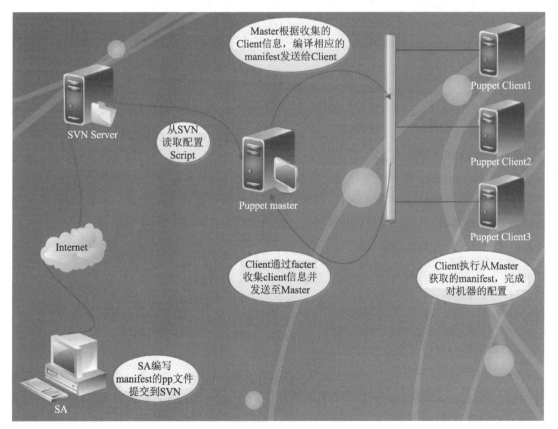

图 4-1 Puppet 和 SVN 结构图

Puppet 工作原理详解如下。

(1)客户端 Puppet 调用本地 facter,facter 会探测出该主机的常用变量,如主机名、内存大小、IP 地址等,然后 Puppet 把这些信息发送给 Puppet 服务器。

（2）Puppet 服务器检测到客户端的主机名，然后会检测 manifest 中对应的 node 配置，并对这段内容进行解析，facter 发送过来的信息可以作为变量进行处理。

（3）Puppet 服务器匹配 Puppet 客户端相关联的代码才进行解析，其他代码将不解析。解析分以下几个过程：语法检查；生成一个中间的伪代码；把伪代码发送给 Puppet 客户端。

（4）Puppet 客户端接收到伪代码之后就会执行，执行完后会将执行的结果发送给 Puppet 服务器。

（5）Puppet 服务器把客户端的执行结果写入日志。

4.3 Puppet 安装配置

由于 Puppet 是 C/S 结构，构建 Puppet 平台需安装 Puppet 服务器和客户端，安装之前需准备好系统环境。

```
操作系统版本：CentOS 6.5 x64
服务器 ip 192.168.149.128  hostname: 192-168-149-128-jfedu.net
客户端 ip 192.168.149.130  hostname: 192-168-149-130-jfedu.net
```

（1）Puppet 服务器安装。

由于 Puppet 主要是基于 hostname 检测的，所以 Puppet 服务器需修改主机名称为 192-168-149-128-jfedu.net，并在 hosts 文件中添加主机名和本机 IP 地址的对应关系，如果本地局域网有 DNS 服务器，则无须修改 hosts 文件。修改主机名及配置的 hosts 代码如下。

```
hostname `ifconfig eth0 |grep Bcast|awk '{print $2}'|cut -d: -f 2 |sed 's/\./\-/g'`-jfedu.net
cat >>/etc/hosts<<EOF
192.168.149.128 192-168-149-128-jfedu.net
192.168.149.130 192-168-149-130-jfedu.net
EOF
```

Puppet 服务器除了需要安装 Puppet 外，还需要 Ruby 语言的支持，所以需要安装 Ruby 相关软件包。默认 YUM 安装 Puppet 时，会自动下载并安装 Ruby 相关软件。相关代码如下，运行结果如图 4-2 所示。

```
rpm -Uvh http://yum.puppetlabs.com/el/6/products/x86_64/puppetlabs-release-6-1.noarch.rpm
yum install puppet-server -y
/etc/init.d/puppetmaster  start
/etc/init.d/iptables      stop
```

```
sed -i '/SELINUX/S/enforce/disabled/' /etc/selinux/config
setenforce 0
```

```
[root@192-168-149-128-jfedu ~]# rpm -Uvh http://yum.puppetlabs.com/
Retrieving http://yum.puppetlabs.com/el/6/products/x86_64/puppetlab
warning: /var/tmp/rpm-tmp.03j7NA: Header V4 RSA/SHA1 Signature, key
Preparing...                ################################
   1:puppetlabs-release      ################################
[root@192-168-149-128-jfedu ~]# yum install puppet-server -y
Loaded plugins: fastestmirror
Determining fastest mirrors
epel/metalink
 * base: mirrors.btte.net
 * epel: mirrors.tuna.tsinghua.edu.cn
 * extras: mirrors.tuna.tsinghua.edu.cn
 * updates: mirror.bit.edu.cn
base
epel
```

图 4-2　Puppet 服务器安装

（2）Puppet 客户端安装。

Puppet 主要是基于 hostname 检测的，所以 Puppet 客户端也需要修改主机名称为 192-168-149-130-jfedu.net，并在 hosts 文件中添加主机名和本机 IP 地址的对应关系，如果本地局域网有 DNS 服务器，则无须修改 hosts 文件。修改主机名及配置的 hosts 代码如下。

```
hostname 'ifconfig eth0 |grep Bcast|awk '{print $2}'|cut -d: -f 2 |sed 's/\./\-/g''-jfedu.net
cat >>/etc/hosts<<EOF
192.168.149.128 192-168-149-128-jfedu.net
192.168.149.130 192-168-149-130-jfedu.net
EOF
```

Puppet 客户端除了需要安装 Puppet 外，还需要 Ruby 语言的支持，所以需要安装 Ruby 相关软件包。默认 YUM 安装 Puppet 时，会自动下载并安装 Ruby 相关软件。相关代码如下，运行结果如图 4-3 所示。

```
rpm -Uvh http://yum.puppetlabs.com/el/6/products/x86_64/puppetlabs-release-6-1.noarch.rpm
yum install puppet -y
/etc/init.d/puppetmaster start
/etc/init.d/iptables stop
sed -i '/SELINUX/S/enforce/disabled/' /etc/selinux/config
setenforce 0
```

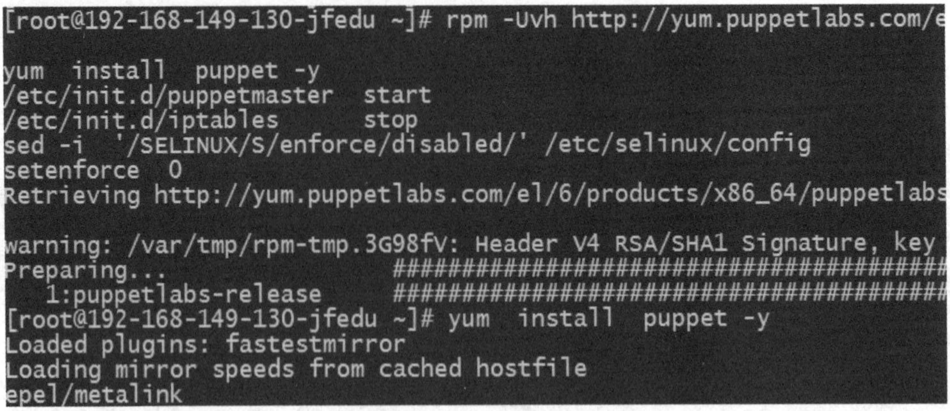

```
[root@192-168-149-130-jfedu ~]# rpm -Uvh http://yum.puppetlabs.com/e
yum install puppet -y
/etc/init.d/puppetmaster start
/etc/init.d/iptables stop
sed -i '/SELINUX/S/enforce/disabled/' /etc/selinux/config
setenforce 0
Retrieving http://yum.puppetlabs.com/el/6/products/x86_64/puppetlabs
warning: /var/tmp/rpm-tmp.3G98fV: Header V4 RSA/SHA1 Signature, key
Preparing...                ###########################################
   1:puppetlabs-release      ###########################################
[root@192-168-149-130-jfedu ~]# yum install puppet -y
Loaded plugins: fastestmirror
Loading mirror speeds from cached hostfile
epel/metalink
```

图 4-3 Puppet 客户端安装

（3）Puppet 客户端申请证书。

由于 Puppet 客户端与 Puppet 服务器是通过 SSL 隧道通信的，客户端安装完成后，首次使用需向服务器申请 Puppet 通信证书。Puppet 客户端在第一次连接服务器时会发起证书申请，在 Puppet 客户端执行的命令如下，运行结果如图 4-4 所示。

```
puppet agent --server 192-168-149-128-jfedu.net --test
```

```
-jfedu ~]#
-jfedu ~]# puppet agent --server 192-168-149-128-jfedu.net --test
cate for ca
file loading from /etc/puppet/csr_attributes.yaml
SSL certificate request for 192-168-149-130-jfedu.net
est fingerprint (SHA256): EC:89:AE:7C:42:50:5A:66:45:D1:0F:1D:4B:5F
cate for ca
e found and waitforcert is disabled
-jfedu ~]#
```

图 4-4 Puppet 客户端发起证书申请

（4）Puppet 服务器颁发证书。

Puppet 客户端向服务器发起证书申请，服务器必须审核证书，如果不审核，客户端与服务器将无法进行后续正常通信。Puppet 服务器颁发证书命令代码如下，运行结果如图 4-5 所示。

```
puppet cert --list                                    #查看申请证书的客户端主机名
puppet cert -s 192-168-149-130-jfedu.net              #颁发证书给客户端
puppet cert -s                                        #为特定的主机颁发证书
puppet cert -s and -a                                 #给所有的主机颁发证书
puppet cert --list --all                              #查看已经颁发的所有证书
```

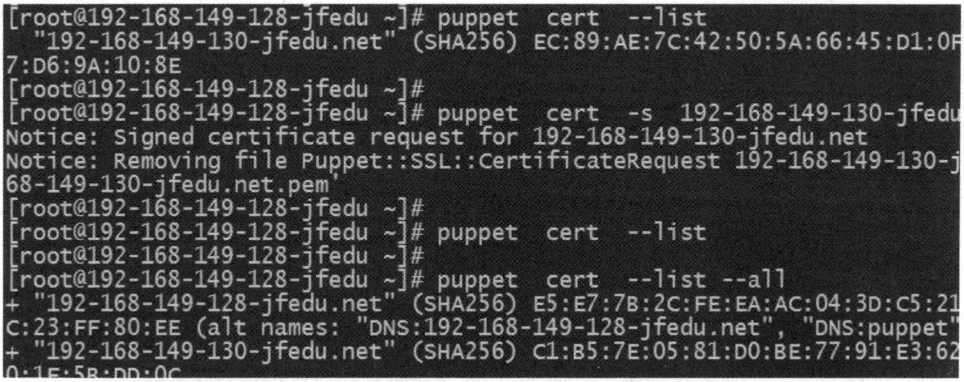

图 4-5　Puppet 服务器颁发证书

4.4　Puppet 企业案例演示

Puppet 基于 C/S 架构，服务器保存着所有对客户端的配置代码，在 Puppet 服务器该配置文件为 manifest。客户端下载文件 manifest 之后，可以根据文件内容对客户端进行配置，如软件包管理、用户管理、文件管理、命令管理、脚本管理等，Puppet 主要基于各种资源或模块管理客户端。

默认 Puppet 服务器文件 manifest 在/etc/puppet/manifests/目录下，只需要在该目录下创建一个 site.pp 文件，然后写入相应的配置代码，Puppet 客户端与 Puppet 服务器同步时，会检查客户端 node 配置文件，匹配之后会将该代码下载至客户端，对代码进行解析，然后在客户端执行。

在 Puppet 客户端创建 test.txt 文件，并在该文件中写入测试内容，操作方法如下。

（1）Puppet 服务器创建 node 代码，创建或编辑/etc/puppet/manifests/site.pp 文件，在文件中加入以下代码。

```
node  default {
file  {
"/tmp/test.txt":
 content => "Hello World,jfedu.net  2021";
   }
}
```

site.pp 配置文件代码详解如下。

```
node  default              #新建node节点,default表示所有主机,可修改为特定主机名
file                       #基于file资源模块管理客户端文件或者目录操作
"/tmp/test.txt":           #需在客户端创建文件的文件名
 content                   #客户端文件内容
```

（2）客户端执行同步命令，获取 Puppet 服务器 node 配置，其代码如下，运行结果如图 4-6 所示。

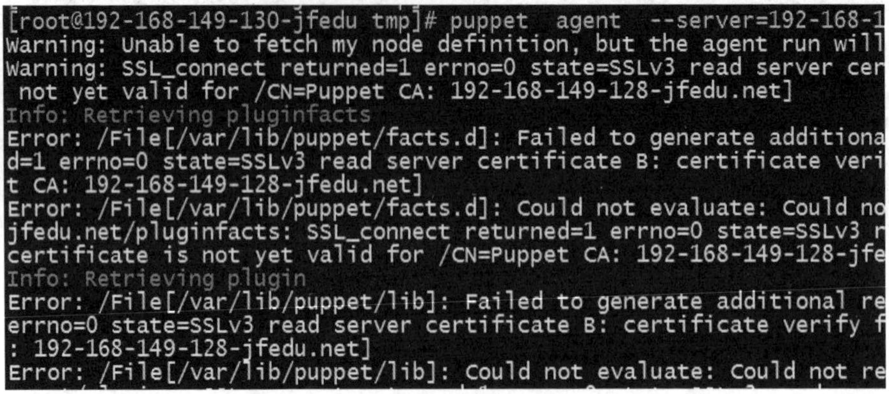

```
puppet agent --server=192-168-149-128-jfedu.net --test
```

图 4-6 Puppet 客户端同步服务器配置

报错原因是服务器与客户端时间不同步，需要同步时间，其代码如下，然后再次执行 puppet agent 命令，如图 4-7 所示。

```
ntpdate pool.ntp.org
puppet agent --server=192-168-149-128-jfedu.net --test
```

图 4-7 Puppet 客户端获取服务器 node 配置

Puppet 客户端执行同步命令，执行日志如下，会在/tmp/目录下创建 test.txt 文件，内容为"Hello World, jfedu.net"，即证明 Puppet 客户端成功获取了服务器 node 配置。

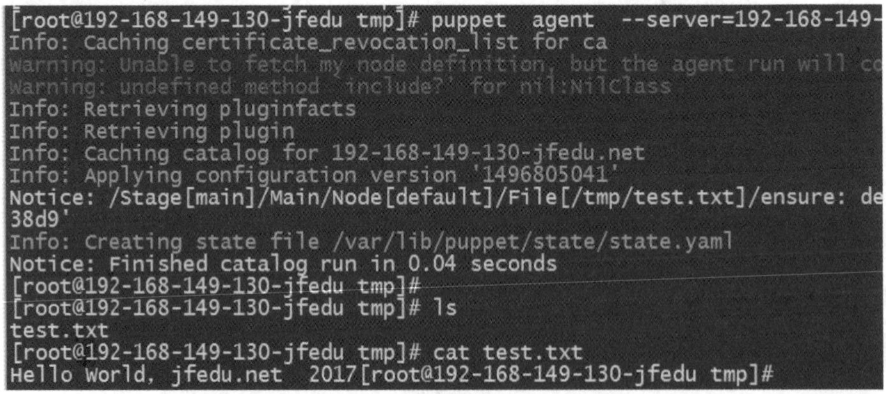

```
Info: Caching certificate_revocation_list for ca
Warning: Unable to fetch my node definition, but the agent run will continue:
Warning: undefined method 'include?' for nil:NilClass
Info: Retrieving pluginfacts
```

```
Info: Retrieving plugin
Info: Caching catalog for 192-168-149-130-jfedu.net
Info: Applying configuration version '1496805041'
Notice: /Stage[main]/Main/Node[default]/File[/tmp/test.txt]/ensure:
defined content as '{md5}d1c2906ad0b249a330e936e3bc1d38d9'
Info: Creating state file /var/lib/puppet/state/state.yaml
Notice: Finished catalog run in 0.04 seconds
```

4.5 Puppet 常见资源及模块

Puppet 主要是基于各种资源模块管理客户端，目前企业主流的 Puppet 管理客户端资源模块如下。

```
file             #主要负责管理文件
package          #软件包的安装管理
service          #系统服务的管理
cron             #配置自动任务计划
exec             #远程执行运行命令
```

通过命令 puppet describe -l 可以查看 Puppet 支持的所有资源和模块，如图 4-8 所示。

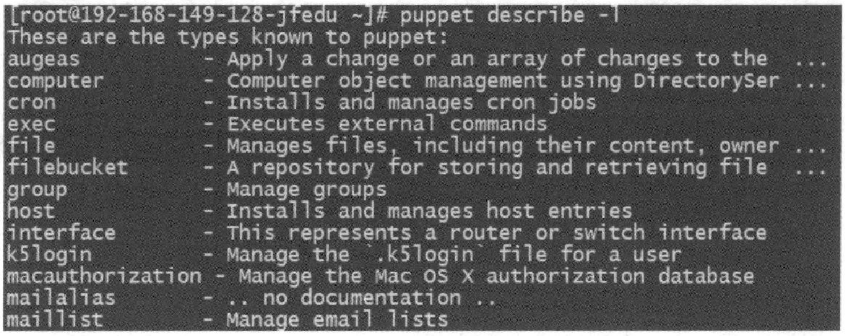

（a）

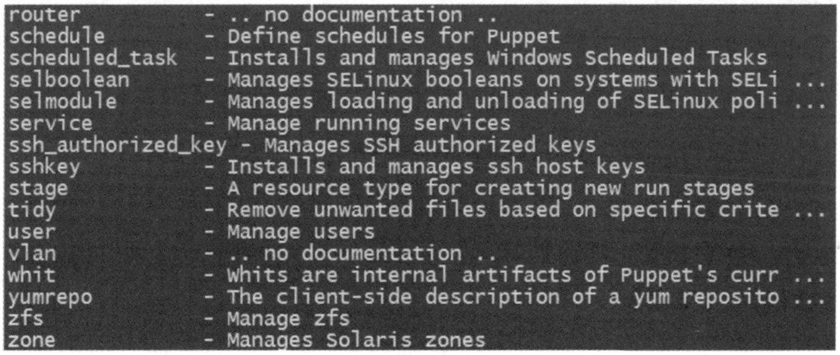

（b）

图 4-8 Puppet 支持的资源及模块

（a）puppet describe 查看资源；（b）puppet describe 查看模块

通过命令 puppet describe -s file 可以查看 Puppet file 资源所有的帮助信息，如图 4-9 所示。

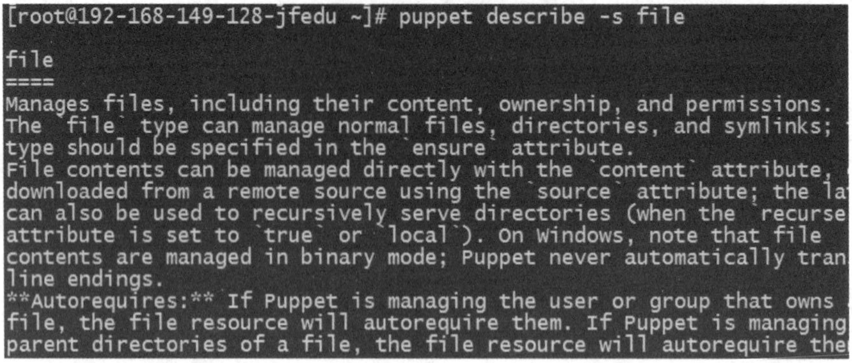

（a）

（b）

图 4-9　Puppet file 资源模块详情

（a）puppet describe 查看 Puppet file 资源；（b）puppet describe 查看 Puppet file 模块

4.6　Puppet file 资源案例

Puppet file 资源主要用于管理客户端文件，包括文件的内容、所有权和权限，可管理的文件类型包括普通文件、目录以及符号链接等。

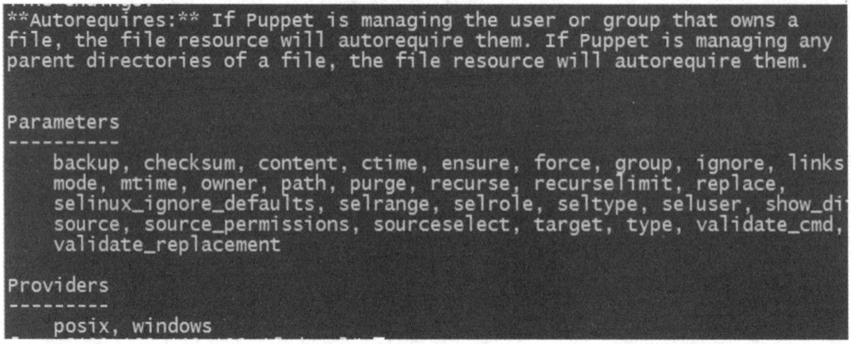

类型应在"确保"属性中指定。如果是文件内容，既可以直接用 content 属性管理，又可以使用 source 属性从远程源下载，还可以用 recurse 服务目录（当 recurse 属性设置为 true 或 local 时）。Puppet file 资源支持参数如下。

```
ensure                          #默认为文件或目录
backup                          #通过 filebucket 备份文件
checksum                        #检查文件是否被修改
ctime                           #只读属性,文件的更新时间
```

```
    mtime                           #只读属性,文件的修改时间
    content                         #文件的内容,与source和target互斥
    force                           #强制执行文件、软链接目录的删除操作
    owner                           #用户名或用户ID
    group                           #指定加年的用户组或组ID
    link                            #软链接
    mode                            #文件权限配置,通常采用数字符号
    path                            #文件路径
    Parameters
        backup, checksum, content, ctime, ensure, force, group, ignore, links,
        mode, mtime, owner, path, purge, recurse, recurselimit, replace,
        selinux_ignore_defaults, selrange, selrole, seltype, seluser, show_diff,
        source, source_permissions, sourceselect, target, type, validate_cmd,
        validate_replacement
    Providers:
        posix, windows
```

（1）从 Puppet 服务器下载 nginx.conf 文件到客户端/tmp 目录，首先需要将 nginx.conf 文件复制到/etc/puppet/files 目录，然后在/etc/puppet/fileserver.conf 文件中添加如下 3 行代码，并重启 puppet master。

```
[files]
path /etc/puppet/files/
allow *
```

创建 site.pp 文件，文件代码如下。

```
node default {
file {
    '/tmp/nginx.conf':
    mode  => '644',
    owner => 'root',
    group => 'root',
    source => 'puppet://192-168-149-128-jfedu.net/files/nginx.conf',
    }
}
```

客户端同步配置，运行结果如图 4-10 所示。

（2）从 Puppet 服务器下载 sysctl.conf 文件，如果客户端该文件已存在，那么备份为 sysctl.conf.bak 文件，然后再覆盖原文件。site.pp 文件代码如下，运行结果如图 4-11 所示。

```
node default {
file {
    "/etc/sysctl.conf":
    source => "puppet://192-168-149-128-jfedu.net/files/sysctl.conf",
    backup => ".bak_$uptime_seconds",
    }
}
```

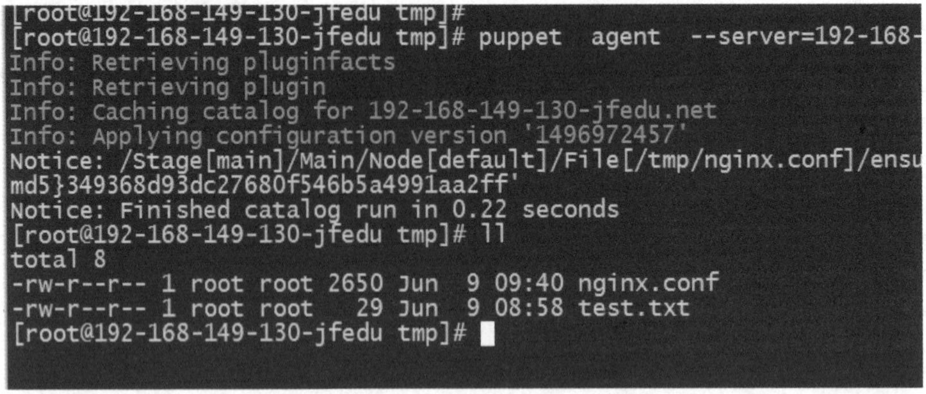

```
[root@192-168-149-130-jfedu tmp]#
[root@192-168-149-130-jfedu tmp]# puppet  agent  --server=192-168-
Info: Retrieving pluginfacts
Info: Retrieving plugin
Info: Caching catalog for 192-168-149-130-jfedu.net
Info: Applying configuration version '1496972457'
Notice: /Stage[main]/Main/Node[default]/File[/tmp/nginx.conf]/ensu
md5}349368d93dc27680f546b5a4991aa2ff'
Notice: Finished catalog run in 0.22 seconds
[root@192-168-149-130-jfedu tmp]# ll
total 8
-rw-r--r-- 1 root root 2650 Jun  9 09:40 nginx.conf
-rw-r--r-- 1 root root   29 Jun  9 08:58 test.txt
[root@192-168-149-130-jfedu tmp]#
```

图 4-10　Puppet file 资源远程下载文件

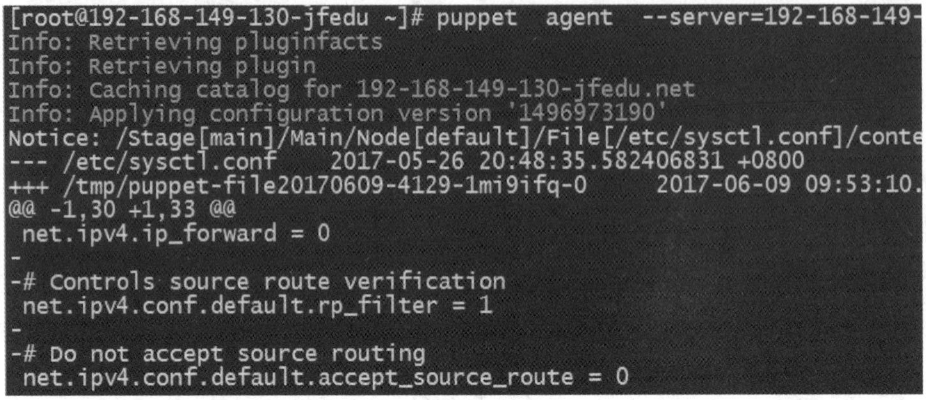

```
[root@192-168-149-130-jfedu ~]# puppet  agent  --server=192-168-149-
Info: Retrieving pluginfacts
Info: Retrieving plugin
Info: Caching catalog for 192-168-149-130-jfedu.net
Info: Applying configuration version '1496973190'
Notice: /Stage[main]/Main/Node[default]/File[/etc/sysctl.conf]/conte
--- /etc/sysctl.conf    2017-05-26 20:48:35.582406831 +0800
+++ /tmp/puppet-file20170609-4129-1mi9ifq-0     2017-06-09 09:53:10.
@@ -1,30 +1,33 @@
 net.ipv4.ip_forward = 0
-
-# Controls source route verification
 net.ipv4.conf.default.rp_filter = 1
-
-# Do not accept source routing
 net.ipv4.conf.default.accept_source_route = 0
```

图 4-11　Puppet file 资源备份文件（1）

（3）在 Agent 上创建/export/docker 目录的软链接为/var/lib/docker/，site.pp 文件代码如下，运行结果如图 4-12 所示。

```
node  default {
file {
       "/var/lib/docker":
       ensure => link,
       target => "/export/docker",
 }
```

（4）在 Agent 上创建/tmp/20501212 目录，site.pp 文件代码如下，运行结果如图 4-13 所示。

```
node  default {
file {
       "/tmp/20501212":
       ensure => directory;
       }
}
```

图 4-12 Puppet file 资源备份文件（2）

图 4-13 Puppet file 创建目录

4.7 Puppet package 资源案例

Puppet package 资源主要用于管理客户端的软件包，通过 Puppet 基于 YUM 源自动安装软件包，所以需要先配置 YUM 源。

可以对软件包进行安装、卸载以及升级操作。Puppet package 资源支持参数如下。

```
Parameters
    adminfile, allow_virtual, allowcdrom, category, configfiles,
    description, ensure, flavor, install_options, instance, name,
    package_settings, platform, responsefile, root, source, status,
    uninstall_options, vendor
Providers
    aix, appdmg, apple, apt, aptitude, aptrpm, blastwave, dpkg, fink,
    freebsd, gem, hpux, macports, msi, nim, openbsd, opkg, pacman, pip, pkg,
    pkgdmg, pkgin, pkgutil, portage, ports, portupgrade, rpm, rug, sun,
    sunfreeware, up2date, urpmi, windows, yum, zipper
ensure => {installed|absent|pureged|latest}
present                    #检查软件是否存在,不存在则安装
installed                  #表示安装软件
```

absent	#删除（无依赖）。当别的软件包依赖时，不可删除
pureged	#删除所有配置文件和依赖包，有潜在风险，慎用
latest	#升级到最新版本
version	#指定安装具体的某个版本号

（1）客户端安装 ntpdate 软件及 screen 软件，代码如下，运行结果如图 4-14 所示。

```
node default {
package {
 ["screen","ntpdate"]:
 ensure => "installed";
}
```

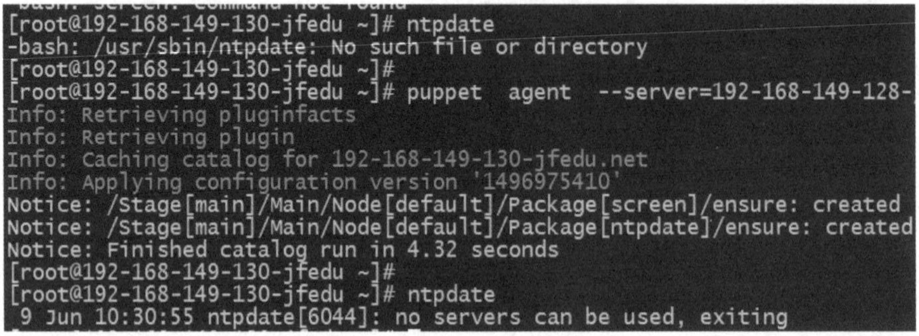

图 4-14 Puppet package 安装软件

（2）客户端卸载 ntpdate 软件及 screen 软件，代码如下，运行结果如图 4-15 所示。

```
node default {
package {
 ["screen","ntp"]:
 ensure => "absent";
}
```

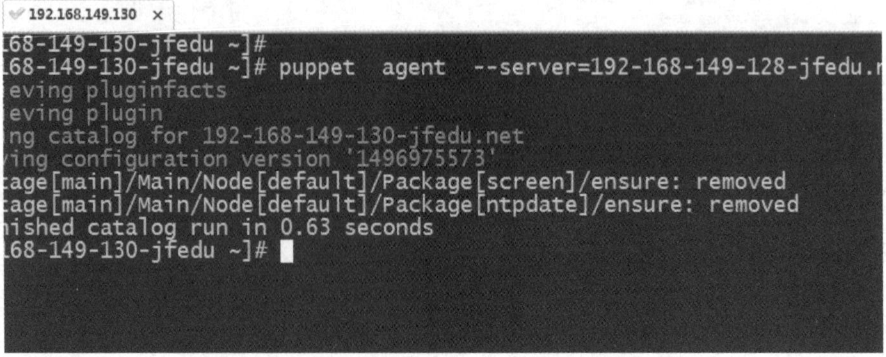

图 4-15 Puppet package 卸载软件

4.8 Puppet service 资源案例

Puppet service 资源主要用于启动、重启和关闭客户端的守护进程，同时可以监控进程的状态，还可以将守护进程加入开机自动启动列表。Puppet service 资源支持参数如下。

```
Parameters
    binary, control, enable, ensure, flags, hasrestart, hasstatus, manifest,
    name, path, pattern, restart, start, status, stop

Providers
    base, bsd, daemontools, debian, freebsd, gentoo, init, launchd, openbsd,
    openrc, openwrt, redhat, runit, service, smf, src, systemd, upstart,
    windows
enable                  #指定服务在开机的时候是否启动,可以设置 true 和 false
ensure                  #是否运行服务,running 表示运行,stopped 表示停止服务
name                    #守护进程的名称
path                    #启动脚本搜索路径
provider                #默认为 init
hasrestart              #管理脚本是否支持 restart 参数,如果不支持,就用 stop 和
                        #start 实现 restart 效果
hasstatus               #管理脚本是否支持 status 参数,Puppet 用 status 参数判断服
                        #务是否已经在运行,如果不支持 status 参数,Puppet 将利用查
                        #找运行进程列表里面是否有服务名的方法来判断服务是否在运行
```

（1）启动 Agent httpd 服务，停止 nfs 服务，代码如下，运行结果如图 4-16 所示。

```
node default {
service {
        "httpd":
        ensure => running;
        "nfs":
        ensure => stopped;
    }
}
```

（2）启动 Agent httpd 服务并设置为开机启动；停止 nfs 服务，设置为开机不启动，代码如下，运行结果如图 4-17 所示。

```
node default {
service {
        "httpd":
        ensure => running,
        enable => true;
        "nfs":
        ensure => stopped,
        enable => false;
    }
}
```

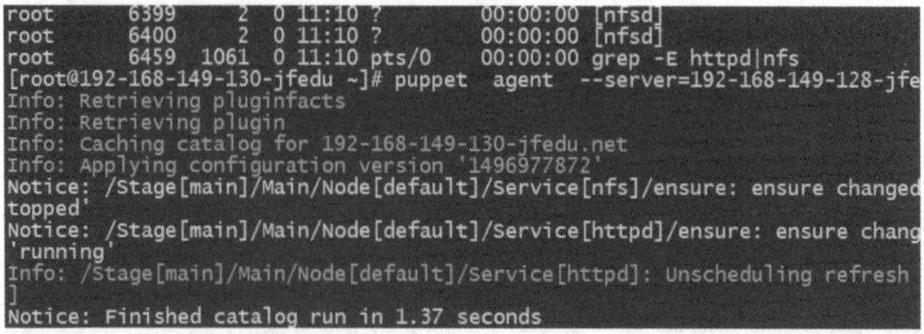

（a）

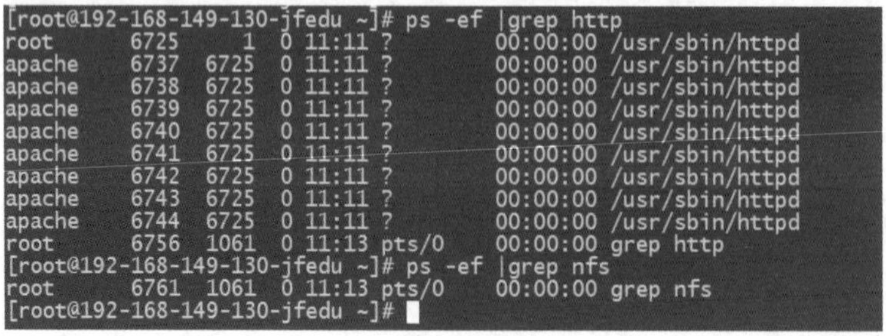

（b）

图 4-16　Puppet service 重启服务

（a）puppet 同步 server 配置；（b）puppet 启动 httpd 服务

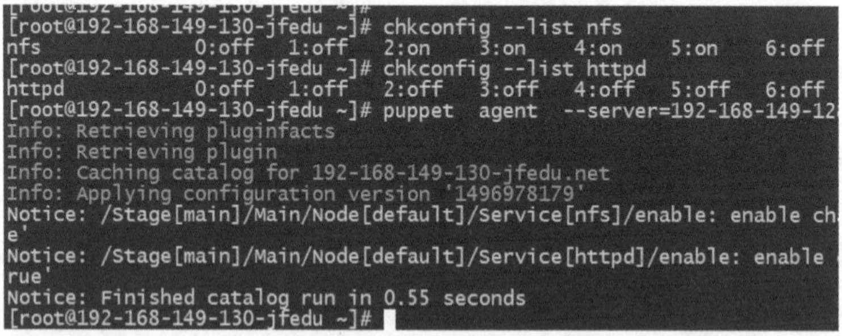

（a）

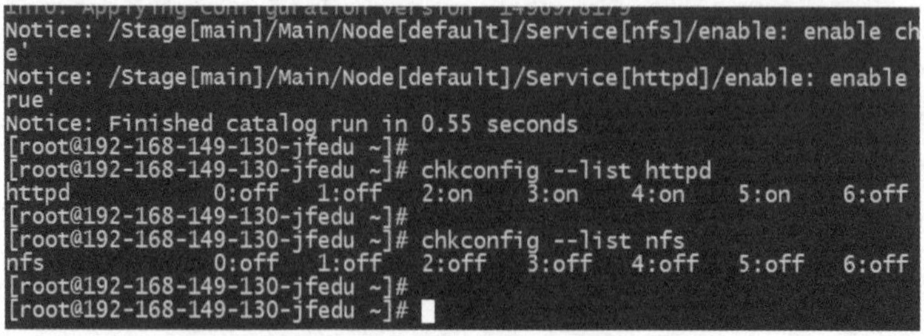

（b）

图 4-17　Puppet Service 开机启动服务

（a）puppet 同步 server 配置；（b）puppet 启动 httpd 服务

4.9 Puppet exec 资源案例

Puppet exec 资源主要用于客户端远程执行命令或安装软件等，相当于 Shell 的调用。exec 是一次性执行资源，在不同类里面 exec 名称可以相同。Puppet exec 资源支持参数如下。

```
Parameters
    command, creates, cwd, environment, group, logoutput, onlyif, path,
    refresh, refreshonly, returns, timeout, tries, try_sleep, umask, unless,
    user
Providers
    posix, shell, windows
command                          #指定要执行的系统命令
creates                          #指定命令所生成的文件
cwd                              #指定命令执行目录,如果目录不存在,则命令执行失败
group                            #执行命令运行的账户组
logoutput                        #是否输出记录
onlyif                           #exec 在 onlyif 设定的命令返回 0 时才执行
path                             #命令执行的搜索路径
refresh =>true|false             #刷新命令执行状态
refreshonly =>true|false         #该属性可以使命令变成仅刷新触发
returns                          #指定返回的代码
timeout                          #命令运行的最长时间
tries                            #命令执行重试次数,默认为 1
try_sleep                        #设置命令重试的间隔时间,单位为 s
user                             #指定执行命令的账户
provider                         #Shell 和 Windows
environment                      #为命令设定额外的环境变量;要注意的是,如果设定过 path,
                                 #则 path 的属性会被覆盖
```

（1）Agent 服务器执行命令 tar 解压 nginx 软件包，其代码如下，运行结果如图 4-18 所示。

```
node default {
exec {
    'Agent tar xzf nginx-1.12.0.tar.gz':
    path => ["/usr/bin","/bin"],
    user => 'root',
    group => 'root',
    timeout => '10',
    command => 'tar -xzf /tmp/nginx-1.12.0.tar.gz',
}
}
```

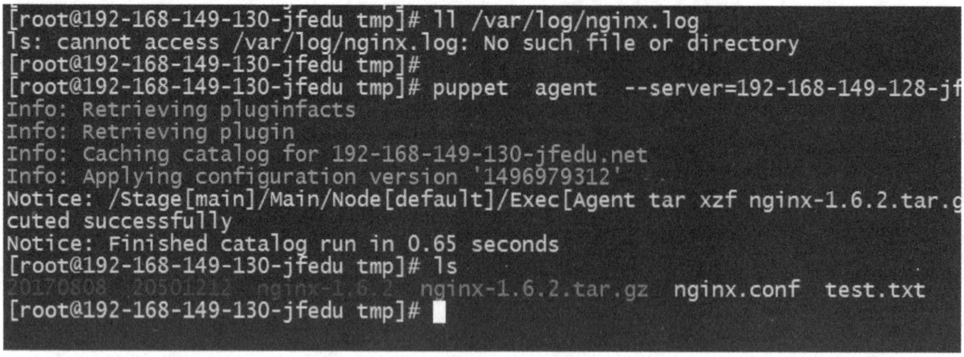

```
[root@192-168-149-130-jfedu tmp]# ll /var/log/nginx.log
ls: cannot access /var/log/nginx.log: No such file or directory
[root@192-168-149-130-jfedu tmp]#
[root@192-168-149-130-jfedu tmp]# puppet  agent  --server=192-168-149-128-jf
Info: Retrieving pluginfacts
Info: Retrieving plugin
Info: Caching catalog for 192-168-149-130-jfedu.net
Info: Applying configuration version '1496979312'
Notice: /Stage[main]/Main/Node[default]/Exec[Agent tar xzf nginx-1.6.2.tar.g
cuted successfully
Notice: Finished catalog run in 0.65 seconds
[root@192-168-149-130-jfedu tmp]# ls
20170808__20501212    nginx-1.6.2   nginx-1.6.2.tar.gz   nginx.conf  test.txt
[root@192-168-149-130-jfedu tmp]#
```

图 4-18　Puppet exec 远程执行命令

（2）Agent 服务器远程执行 auto_install_nginx.sh 脚本，其代码如下，运行结果如图 4-19 所示。

```
node  default {
file {
      "/tmp/auto_install_nginx.sh":
      source =>"puppet://192-168-149-128-jfedu.net/files/auto_install_
nginx.sh",
      owner => "root",
      group => "root",
      mode => 755,
   }
exec {
      "/tmp/auto_install_nginx.sh":
      cwd => "/tmp",
      user => root,
      path => ["/usr/bin","/usr/sbin","/bin","/bin/sh"],
}
```

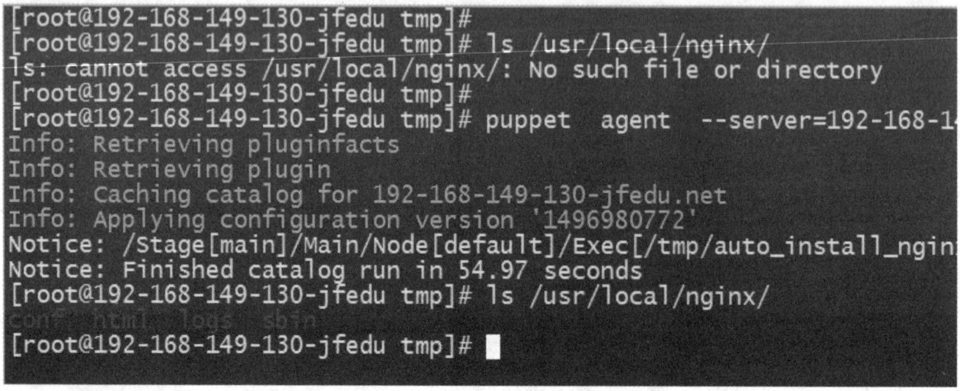

```
[root@192-168-149-130-jfedu tmp]#
[root@192-168-149-130-jfedu tmp]# ls /usr/local/nginx/
ls: cannot access /usr/local/nginx/: No such file or directory
[root@192-168-149-130-jfedu tmp]#
[root@192-168-149-130-jfedu tmp]# puppet  agent  --server=192-168-1
Info: Retrieving pluginfacts
Info: Retrieving plugin
Info: Caching catalog for 192-168-149-130-jfedu.net
Info: Applying configuration version '1496980772'
Notice: /Stage[main]/Main/Node[default]/Exec[/tmp/auto_install_ngin
Notice: Finished catalog run in 54.97 seconds
[root@192-168-149-130-jfedu tmp]# ls /usr/local/nginx/
conf  html  logs  sbin
[root@192-168-149-130-jfedu tmp]#
```

图 4-19　Puppet exec 执行 Nginx 安装脚本

（3）Agent 服务器更新 sysctl.conf 文件，如果该文件发生改变，则执行命令 sysctl -p，其代码如下，运行结果如图 4-20 所示。

```
node default {
file {
        "/etc/sysctl.conf":
        source =>"puppet://192-168-149-128-jfedu.net/files/sysctl.conf",
        owner => "root",
        group => "root",
        mode => 644,
    }
exec {
        "sysctl refresh kernel config":
        path => ["/usr/bin", "/usr/sbin", "/bin", "/sbin"],
        command => "/sbin/sysctl -p",
        subscribe => File["/etc/sysctl.conf"],
        refreshonly => true
    }
}
```

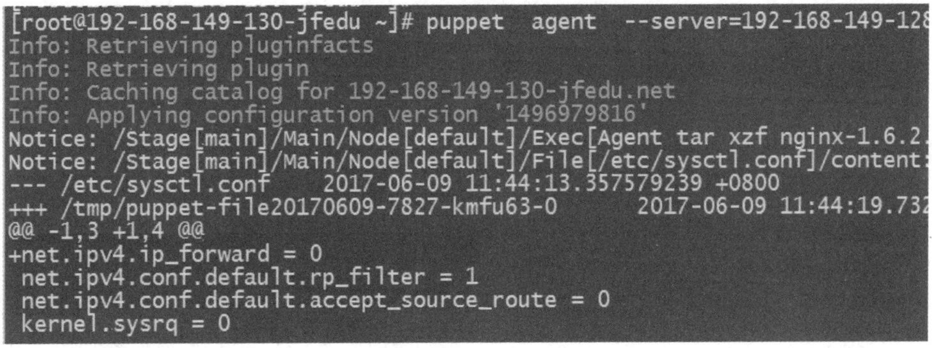

（a）

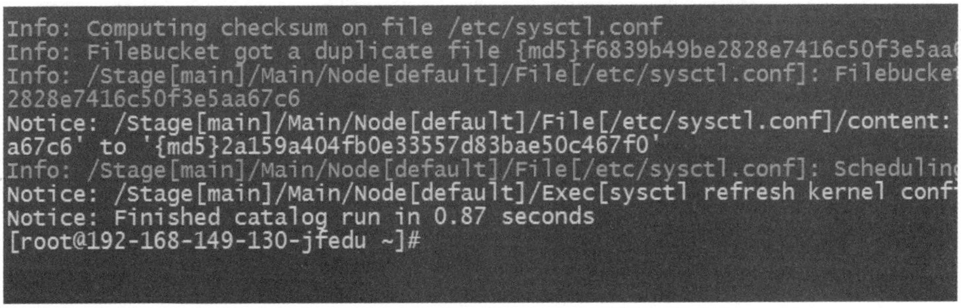

（b）

图 4-20　Puppet exec 更新执行触发命令

（a）puppet 同步 server 配置；（b）puppet 更新 sysctl.conf 文件

4.10　Puppet cron 资源案例

Puppet cron 资源主要用于安装和管理 crontab 计划任务，每一个 cron 资源需要一个 command 属性、一个 user 属性和至少一个周期属性（hour、minute、month、monthday、weekday）。

crontab 计划任务的名称不是计划任务的一部分，它是 Puppet 用来存储和检索该资源的线索。假如指定了一个计划任务，除了名称外其他内容都和另一个已经存在的计划任务相同，那么这两个计划任务被认为是等效的，且新名称将永久地与该计划任务相关联。Puppet cron 资源支持参数如下。

```
Parameters
    command, ensure, environment, hour, minute, month, monthday, name,
    special, target, user, weekday
Providers
    crontab
user                    #加某个用户的 crontab 任务,默认是运行 Puppet 的用户
command                 #要执行的命令或脚本路径,可不写,默认是 title(名称)
ensure                  #确定该资源是否启用,可设置成 true 或 false
environment             #crontab 环境中指定环境变量
hour                    #设置 crontab 的小时数,可设置成 0~23
minute                  #指定 crontab 的分钟数,可设置成 0~59
month                   #指定 crontab 运行的月份,可设置成 1~12
monthday                #指定月的天数,可设置成 1~31
name                    #crontab 的名字,区分不同的 crontab
provider                #可用的 provider 有 crontab 默认的 crontab 程序
target                  #crontab 作业存放的位置
weekday                 #行 crontab 的星期数,可设置成 0~7,周日为 0
```

（1）Agent 服务器添加 ntpdate 时间同步任务的代码如下，运行结果如图 4-21 所示。

```
node default {
cron{
        "ntpdate":
        command => "/usr/sbin/ntpdate  pool.ntp.org",
        user => root,
        hour => 0,
        minute => 0,
    }
}
```

（2）Agent 服务器删除 ntpdate 时间同步任务的代码如下，运行结果如图 4-22 所示。

```
node default {
cron{
```

```
        "ntpdate":
        command => "/usr/sbin/ntpdate  pool.ntp.org",
        user => root,
        hour => 0,
        minute => 0,
        ensure => absent,
    }
}
```

```
[root@192-168-149-130-jfedu ~]# puppet  agent  --server=192-168-149-1
Info: Retrieving pluginfacts
Info: Retrieving plugin
Info: Caching catalog for 192-168-149-130-jfedu.net
Info: Applying configuration version '1496985201'
Notice: /Stage[main]/Main/Node[default]/Cron[ntpdate]/ensure: created
Notice: Finished catalog run in 0.46 seconds
[root@192-168-149-130-jfedu ~]#
[root@192-168-149-130-jfedu ~]# crontab  -l
# HEADER: This file was autogenerated at Fri Jun 09 13:13:22 +0800 20
# HEADER: While it can still be managed manually, it is definitely no
# HEADER: Note particularly that the comments starting with 'Puppet N
# HEADER: not be deleted, as doing so could cause duplicate cron jobs
# Puppet Name: ntpdate
0 0 * * * /usr/sbin/ntpdate pool.ntp.org
[root@192-168-149-130-jfedu ~]#
```

图 4-21　Puppet cron 添加任务计划

```
# HEADER: not be deleted, as doing so could cause duplicate cron jobs.
# Puppet Name: ntpdate
0 0 * * * /usr/sbin/ntpdate pool.ntp.org
[root@192-168-149-130-jfedu ~]# puppet  agent  --server=192-168-149-128-j
Info: Retrieving pluginfacts
Info: Retrieving plugin
Info: Caching catalog for 192-168-149-130-jfedu.net
Info: Applying configuration version '1496988132'
Notice: /Stage[main]/Main/Node[default]/Cron[ntpdate]/ensure: removed
Notice: Finished catalog run in 0.53 seconds
[root@192-168-149-130-jfedu ~]# crontab  -l
# HEADER: This file was autogenerated at Fri Jun 09 14:02:13 +0800 2017 b
# HEADER: While it can still be managed manually, it is definitely not re
# HEADER: Note particularly that the comments starting with 'Puppet Name'
# HEADER: not be deleted, as doing so could cause duplicate cron jobs.

[root@192-168-149-130-jfedu ~]#
```

图 4-22　Puppet cron 删除任务计划

4.11　Puppet 日常管理与配置

Puppet 平台构建完毕，即可使用 Puppet 管理客户端对文件、服务、脚本及各种配置的变更，如果要管理批量服务器，还需要进行一些配置。

4.11.1　Puppet 自动认证

企业新服务器通过 Kickstart 自动安装 Linux 操作系统，安装完毕后可以自动安装 Puppet 相关软件包。Puppet 客户端安装完毕，需向 Puppet 服务器请求证书，然后 Puppet 服务器颁发证书给客户端。默认需要手动颁发，也可以通过配置让 Puppet 服务器自动颁发证书。

自动颁发证书的前提是服务器与客户端能 ping 通彼此的主机名，配置自动颁发证书需在 Puppet 服务器的 puppet.conf 配置文件的 main 段加入如下代码，运行结果如图 4-23 所示。

```
[main]
autosign = true
```

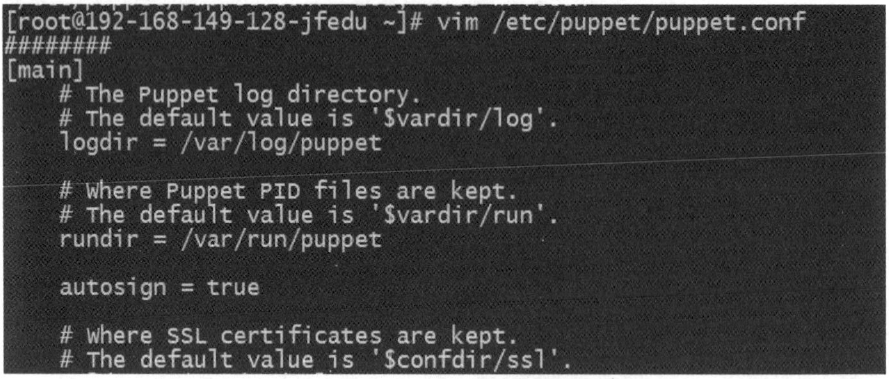

图 4-23 Puppet 服务器添加自动颁发证书

重启 puppetmaster 服务，并删除 192.168.149.130 的证书。

```
/etc/init.d/puppetmaster  restart
puppet  cert  --clean  192-168-149-130-jfedu.net
```

删除 Puppet 客户端 SSL 文件，再重新生成 SSL 文件，执行如下命令自动申请证书。

```
rm  -rf  /var/lib/puppet/ssl/
puppet  agent  --server=192-168-149-128-jfedu.net  --test
```

Puppet 服务器会自动认证，即服务器不必手动颁发证书，这样可减轻人工的干预和操作，运行结果如图 4-24 所示。

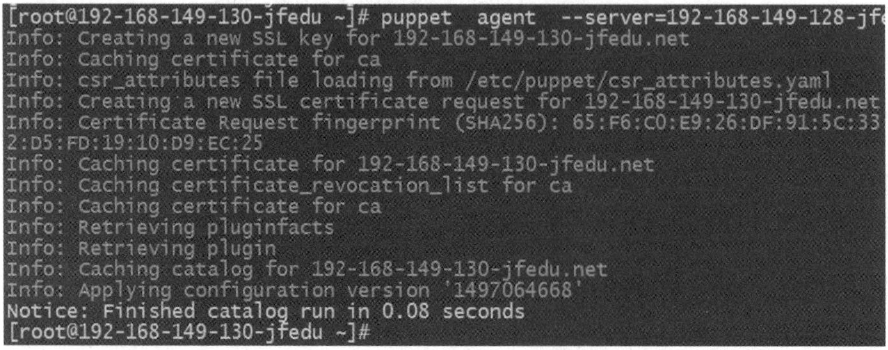

图 4-24 Puppet 客户端自动获取证书

4.11.2　Puppet 客户端自动同步

Puppet 客户端安装并认证完毕之后，如果在 Puppet 服务器配置了 node 信息，那么客户端启动服务，默认 30 min 自动与服务器同步信息。如何修改同步的时间频率呢？修改 Puppet 客户端配置信息即可。

Puppet 客户端配置相关参数和同步时间，修改/etc/sysconfig/puppet 配置文件，最终代码如下。

```
# The puppetmaster server
PUPPET_SERVER=192-168-149-128-jfedu.net
# If you wish to specify the port to connect to do so here
PUPPET_PORT=8140
# Where to log to. Specify syslog to send log messages to the system log.
PUPPET_LOG=/var/log/puppet/puppet.log
# You may specify other parameters to the puppet client here
PUPPET_EXTRA_OPTS=--waitforcert=500
```

/etc/sysconfig/puppet 配置文件参数详解如下。

```
PUPPET_SERVER=192-168-149-128-jfedu.net       #指定 Puppet Master 主机名
PUPPET_PORT=8140                              #指定 Puppet Master 端口
PUPPET_LOG=/var/log/puppet/puppet.log         #Puppet 客户端日志路径
PUPPET_EXTRA_OPTS=--waitforcert=500           #获取 Puppet Master 证书返回等待时间
```

重启 Puppet 客户端服务，客户端会每 30min 与服务器同步一次配置信息。

```
/etc/init.d/puppet restart
```

可以修改与服务器同步配置信息的时间，修改/etc/puppet/puppet 配置文件，在[agent]段加入如下语句，表示 60 s 与服务器同步一次配置信息。重启 Puppet，同步结果如图 4-25 所示。

```
[agent]
runinterval = 60
```

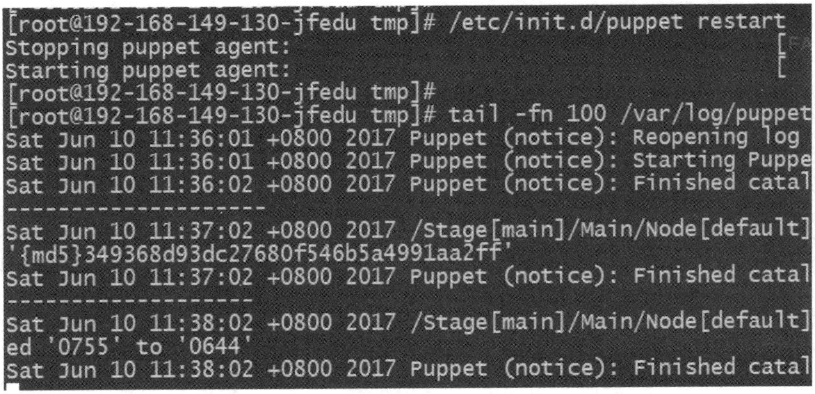

图 4-25　Puppet 客户端自动同步服务器配置

4.11.3 Puppet 服务器主动推送

4.11.2 节中 Puppet 客户端配置每 60 s 与服务器同步配置信息，如果服务器更新了配置信息，想立刻让客户端同步，如何通知客户端获取最新的配置信息呢？可以使用 Puppet Master 主动推送的方式进行配置。

Puppet 服务器使用 puppet run 命令可以给客户端发送一段信号，告诉客户端立刻与服务器同步配置信息。配置方法如下。

修改 Puppet 客户端配置文件/etc/puppet/puppet.conf，在 agent 段加入如下代码。

```
[agent]
listen = true
```

修改 Puppet 客户端配置文件/etc/sysconfig/puppet.conf，指定 Puppet Master 端主机名。

```
PUPPET_SERVER=192-168-149-128-jfedu.net
```

创建 Puppet 客户端配置文件 namespaceauth.conf，写入如下代码。

```
[puppetrunner]
allow  *
```

在 auth.conf 源文件的 path /前添加如下三行代码。

```
path  /run
method  save
allow  *
```

重启 Puppet 客户端。

```
/etc/init.d/puppet  restart
```

Puppet 服务器执行如下命令，通知客户端同步配置，也可以批量通知其他客户端，只需将客户端的主机名写入 host.txt 文件即可，运行结果如图 4-26 所示。

```
puppet  kick  -d  192-168-149-130-jfedu.net
#puppet  kick  -d  'cat host.txt'
```

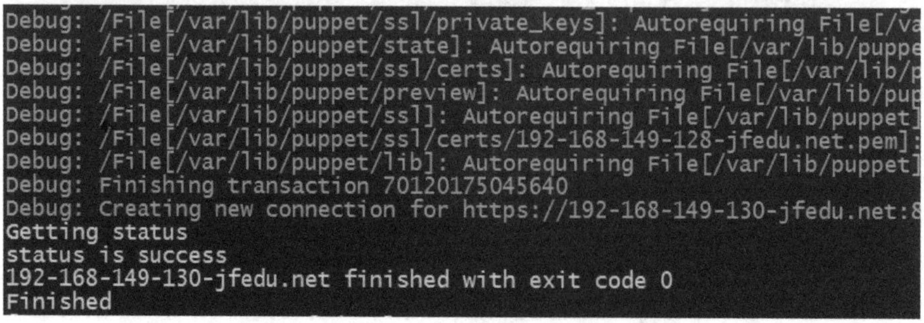

```
Debug: /File[/var/lib/puppet/ssl/private_keys]: Autorequiring File[/va
Debug: /File[/var/lib/puppet/state]: Autorequiring File[/var/lib/puppe
Debug: /File[/var/lib/puppet/ssl/certs]: Autorequiring File[/var/lib/p
Debug: /File[/var/lib/puppet/preview]: Autorequiring File[/var/lib/pup
Debug: /File[/var/lib/puppet/ssl]: Autorequiring File[/var/lib/puppet]
Debug: /File[/var/lib/puppet/ssl/certs/192-168-149-128-jfedu.net.pem]:
Debug: /File[/var/lib/puppet/lib]: Autorequiring File[/var/lib/puppet]
Debug: Finishing transaction 70120175045640
Debug: Creating new connection for https://192-168-149-130-jfedu.net:8
Getting status
status is success
192-168-149-130-jfedu.net finished with exit code 0
Finished
```

图 4-26　Puppet 主动通知客户端同步配置

4.12 Puppet 批量部署案例

随着 IT 行业的迅猛发展，传统靠大量人力的运维方式比较吃力，而近几年自动化运维管理快速的发展，得到了很多 IT 运维人员的青睐。一个完整的自动化运维体系包括系统安装、配置管理、服务监控 3 方面。以下为 Puppet 在生产环境中的应用案例。

某互联网公司新到 100 台硬件服务器，要求统一安装 Linux 操作系统、部署上线以及后期的管理配置。对于 Linux 操作系统安装，需采用批量安装的方式，批量安装系统的主流工具为 Kickstart 和 Cobbler，任选其一即可。

如果采用自动安装，可以自动初始化系统、内核简单优化以及常见服务和软件客户端等的安装。当然，Puppet 客户端也可以放在 Kickstart 中安装并配置。

当 Linux 操作系统安装完成后，需要对服务器进行相应的配置，才能应对高并发网站，例如，修改动态 IP 地址为静态 IP 地址、安装及创建 crontab 任务计划、同步操作系统时间、安装 Zabbix 客户端软件、优化内核参数等，可以基于 Puppet 做统一调整。

4.12.1 Puppet 批量修改静态 IP 地址案例

现需要修改 100 台 Linux 服务器原 DHCP 动态获取的 IP 地址为 Static IP 地址。首先需要修改 IP 地址脚本，将该脚本推送到客户端，然后执行脚本并重启网卡。

（1）修改 IP 地址为静态 IP 地址的 Shell 脚本代码如下。

```
#!/bin/bash
#auto Change ip netmask gateway scripts
#By author jfedu.net 2021
#Define Path variables
ETHCONF=/etc/sysconfig/network-scripts/ifcfg-eth0
DIR=/data/backup/`date +%Y%m%d`
IPADDR=`ifconfig|grep inet|grep 192|head -1|cut -d: -f2|awk '{print $1}'`
NETMASK=255.255.255.0
grep dhcp $ETHCONF
if [ $? -eq 0 ];then
    sed -i 's/dhcp/static/g' $ETHCONF
    echo -e "IPADDR=$IPADDR\nNETMASK=$NETMASK\nGATEWAY=`echo $IPADDR|awk -F. '{print $1"."$2"."$3}'`.2" >>$ETHCONF
    echo "The IP configuration success. !"
    service network restart
fi
```

（2）Puppet Master 执行 kick 命令推送配置至 Agent 服务器远程，Puppet 客户端修改 IP 地址脚本代码如下，结果如图 4-27 所示。

```
node  default {
file {
     "/tmp/auto_change_ip.sh":
     source =>"puppet://192-168-149-128-jfedu.net/files/auto_change_ip.sh",
     owner => "root",
     group => "root",
     mode => 755,
   }
exec {
     "/tmp/auto_change_ip.sh":
     cwd => "/tmp/",
     user => root,
     path => ["/usr/bin","/usr/sbin","/bin","/bin/sh"],
   }
}
```

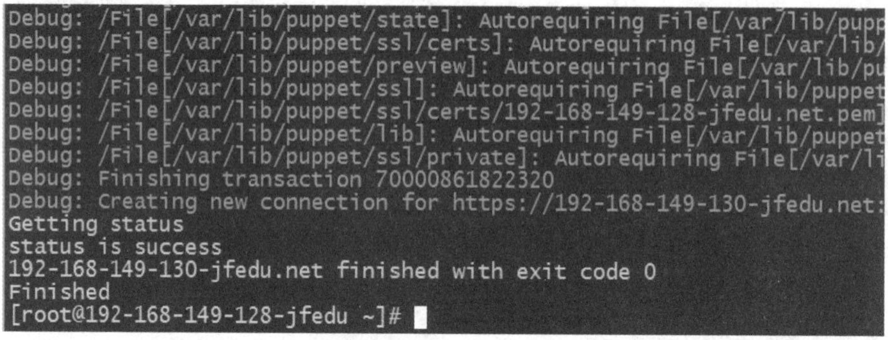

(a)

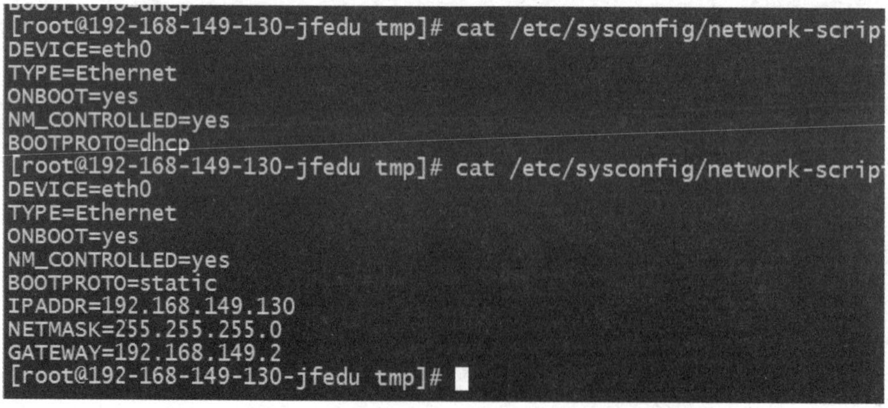

(b)

图 4-27 最终结果

（a）Puppet 主动通知客户端同步配置；（b）Puppet 客户端 IP 地址自动配置为 Static IP 地址方式

4.12.2　Puppet 批量配置 NTP 同步服务器

在 100 台 Linux 服务器上配置 crontab 任务，修改 ntpdate 与 NTP 服务器的同步时间。

（1）在 Puppet Master 上创建客户端 node 配置，可以编写 NTP 模块，使用 class 定义模块分组，对不同业务进行分组管理，/etc/puppet/modules/ntp/manifests/init.pp 配置文件代码如下，将原 ntpdate 同步时间从 0 min 改成每 5 min 同步一次时间，并修改原 pool.ntp.org 服务器为本地局域网 NTP 时间服务器的 IP 地址。

```
class ntp {
Exec { path =>"/bin:/sbin:/bin/sh:/usr/bin:/usr/sbin:/usr/local/bin:/usr/local/sbin"}
exec {
"auto change crontab ntp config":
command =>"sed -i -e '/ntpdate/s/0/*\/5 /2'  -e 's/pool.ntp.org/10.1.1.21/' /var/spool/cron/root",
 }
}
```

（2）在/etc/puppet/manifests 目录下创建两个文件，分别为 modules.pp 配置文件和 nodes.pp 配置文件，即模块入口文件和 node 配置段。

modules.pp 配置文件内容如下。

```
import "ntp"
```

nodes.pp 配置文件内容如下。

```
node default {
include ntp
}
```

（3）在 site.pp 文件中导入 modules.pp 配置文件和 nodes.pp 配置文件的名称，site.pp 文件代码如下。

```
import "modules.pp"
import "nodes.pp"
```

（4）Puppet Master 执行 kick 命令推送配置到 Agent 服务器远程，Puppet 客户端最终结果如图 4-28 所示。

当服务器分组之后，可以使用正则表达式定义 node，在定义一个 node 节点时，要指定节点的名称，并用单引号将名称引起来，然后在大括号中指定需要应用的配置。

客户端节点名称可以是主机名也可以是客户端的正式域名，目前 Puppet 版本还不能使用通配符来指定节点，例如，不能用*.jfedu.net 表达，但可以使用正则表达式。相关代码如下。

```
node /^Beijing-IDC-web0\d+\-jfedu\.net {
    include ntp
}
```

以上规则会匹配所有在 jfedu.net 域并且主机名以 Beijing-IDC 开头，紧跟 web01、web02、web03 等节点，由此可以进行批量服务器的分组管理。

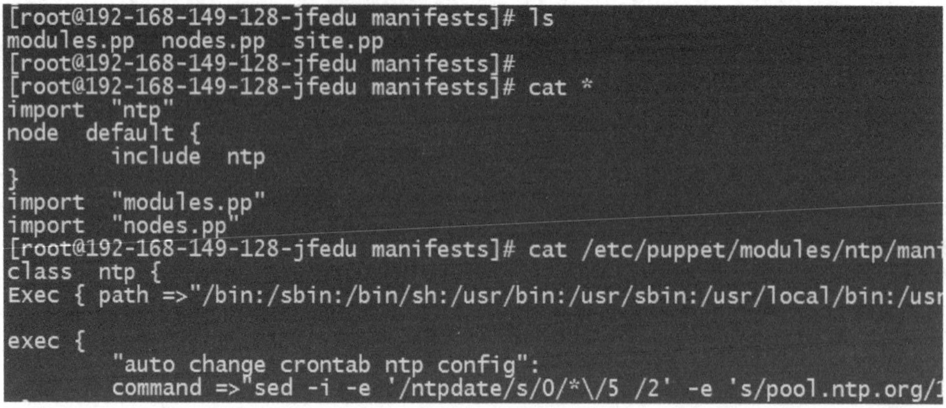

图 4-28 Puppet 客户端最终结果

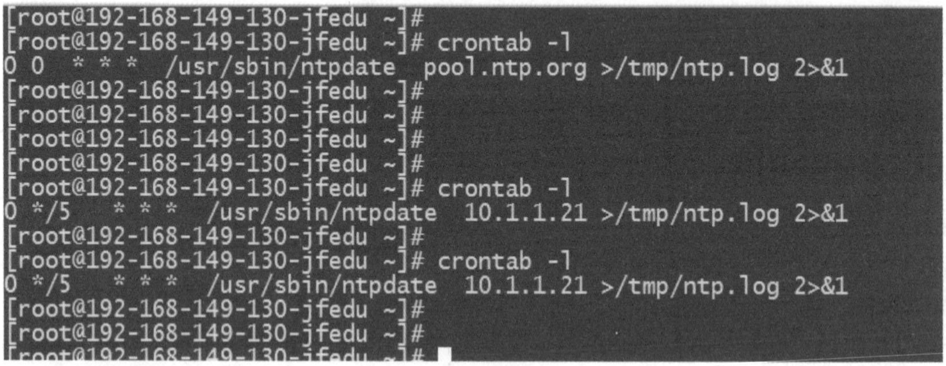

（a）Puppet 服务器 class 模块配置；（b）Puppet 主动通知客户端修改 NTP 同步配置

4.12.3 Puppet 自动部署及同步网站

企业生产环境下的所有服务器要求数据一致，可以采用 rsync 同步，配置 rsync 服务器，客户端执行脚本命令即可。同样，可以使用 Puppet+Shell 脚本同步，这样同步比较快捷，还可以使用 Puppet rsync 模块。

（1）Puppet 服务器配置，/etc/puppet/modules/www/manifests/init.pp 文件代码如下。

```
class www {
Exec { path =>"/bin:/sbin:/bin/sh:/usr/bin:/usr/sbin:/usr/local/bin:/usr/
local/sbin"}
file {
"/data/sh/rsync_www_client.sh":
source =>"puppet://192-9-11-162-tdt.com/files/www/rsync_www_client.sh",
owner =>"root",
group =>"root",
mode =>"755",
}
file {
"/etc/rsync.pas":
source =>"puppet://192-9-11-162-tdt.com/files/www/rsync.pas",
owner =>"root",
group =>"root",
mode =>"600",
}
exec {
"auto backup www data":
command =>"mkdir -p /data/backup/'date +%Y%m%d';mv /data/index /data/
backup/www/'date +%Y%m%d' ; /bin/sh /data/sh/rsync_www_client.sh ",
user =>"root",
subscribe =>File["/data/sh/rsync_www_client.sh"],
refreshonly =>"true",
 }
}
```

（2）在/etc/puppet/manifests 目录下创建两个文件，分别为 modules.pp 配置文件和 nodes.pp 配置文件，即模块入口文件和 node 配置段。

modules.pp 配置文件内容如下。

```
import "www"
```

nodes.pp 配置文件内容如下。

```
node /^Beijing-IDC-web0\d+\-jfedu\.net {
include www
}
```

（3）在 site.pp 文件中导入 modules.pp 配置文件和 nodes.pp 配置文件的名称，site.pp 文件代码如下。

```
import "modules.pp"
import "nodes.pp"
```

Puppet Master 端批量执行通知客户端同步配置,命令如下。

```
puppet kick -d --host 'cat hosts.txt'
```

(4) cat hosts.txt 文件内容为需要同步的客户端的主机名。

```
Beijing-IDC-web01-jfedu.net
Beijing-IDC-web02-jfedu.net
Beijing-IDC-web03-jfedu.net
Beijing-IDC-web04-jfedu.net
```

第 5 章 Ansible 自动运维企业实战

本章将介绍 Ansible 工作原理、Ansible 安装配置、生产环境模块讲解、Ansible 企业场景案例、PlayBook 剧本实战及 Ansible 性能调优等内容。

5.1 Ansible 工具特点

Ansible 与 SaltStack 均基于 Python 语言开发，Ansible 只需要在一台普通的服务器上运行，而不需要在客户端上安装客户端。因为 Ansible 是基于 SSH 远程管理，而 Linux 服务器大都离不开 SSH，所以 Ansible 不需要为配置工作添加额外的支持。

Ansible 的安装和使用非常简单，且基于上千个插件和模块实现了各种软件、平台和版本的管理，还支持虚拟容器多层级的部署。很多用户认为 Ansible 比 SaltStatck 执行效率慢，其实不是软件本身慢，而是由于 SSH 服务慢，可以通过优化 SSH 连接速度及使用 Ansible 加速模块，来满足企业上万台服务器的维护和管理。

5.2 Ansible 运维工具原理

Ansible 是一款极为灵活的开源工具套件，能够极大地简化 UNIX 操作系统管理员的自动化配置管理与流程控制方式。它利用推送方式对客户系统加以配置，这样所有工作都可在主服务器完成。它的命令行机制同样非常强大，允许用户利用商业许可 Web UI 实现授权管理与配置。

可以通过命令行或 GUI 使用 Ansible，运行 Ansible 的服务器，俗称管理节点；通过 Ansible 进行管理的服务器俗称受控节点。权威媒体报道，Ansible 于 2015 年被 Red Hat 公司以 1.5 亿美元的价格收购，新版 Red Hat 内置了 Ansible 软件。

本书以 Ansible 为案例，基于 Ansible 构建企业自动化运维平台，实现大规模服务器的快速管理和部署。Ansible 将平常复杂的配置工作变得简单、标准化且容易控制。

Ansible 自动运维管理工具的优点如下。

（1）轻量级，更新时只需要在操作机上进行一次更新即可。

（2）采用 SSH 协议。

（3）不需要去客户端安装 agent。

（4）执行批量任务可以写成脚本，且不用分发到远程就可以执行。

（5）使用 Python 语言编写，维护更简单。

（6）支持 sudo 普通用户命令。

（7）去中心化管理。

Ansible 自动运维管理工具工作原理如图 5-1 所示。

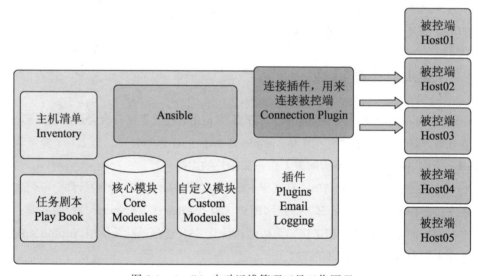

图 5-1　Ansible 自动运维管理工具工作原理

5.3　Ansible 管理工具安装配置

Ansible 工具可以工作在 Linux、BSD、macOS X 等操作系统上，对 Python 环境的版本最低要求为 Python 2.6 以上，如果操作系统的 Python 软件版本为 2.4,那么需要升级才能使用 Ansible 工具。

Red Hat、CentOS 操作系统可以直接基于 YUM 工具自动安装 Ansible 工具，CentOS 6.x 操作系统或 CentOS 7.x 操作系统安装前，需先安装 EPEL 扩展源，其代码如下。

```
rpm -Uvh http://mirrors.ustc.edu.cn/fedora/epel/6/x86_64/epel-release-
6-8.noarch.rpm
```

```
yum install epel-release -y
yum install ansible -y
```

Ansible 工具默认主目录为/etc/ansible/，其中 hosts 文件为被管理机 IP 地址或主机名列表，ansible.cfg 文件为 Ansible 的主配置文件，roles 为角色或者插件路径，默认该目录为空，如图 5-2 所示。

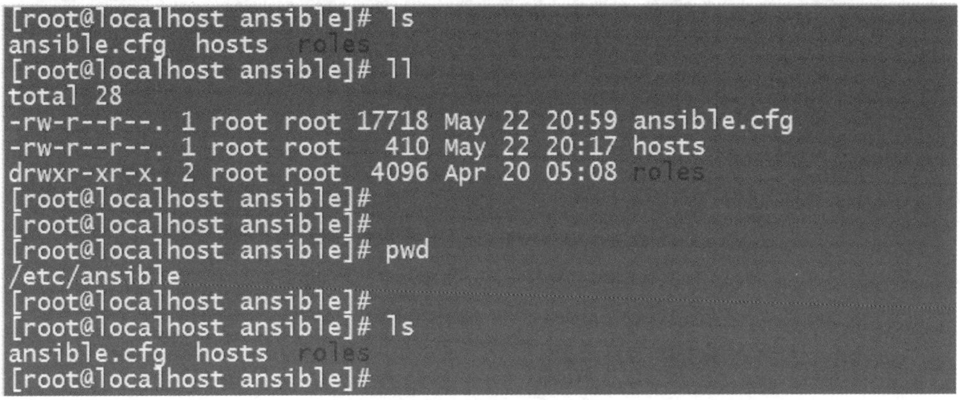

图 5-2 Ansible 主目录信息

Ansible 远程批量管理，其中执行命令通过 Ad-Hoc 完成，即点对点单条执行命令，能够快速执行，且不需要保存执行的命令。默认 hosts 文件配置主机列表，可以配置分组，可以定义各种 IP 地址及规则。Hosts 主机列表默认配置如图 5-3 所示。

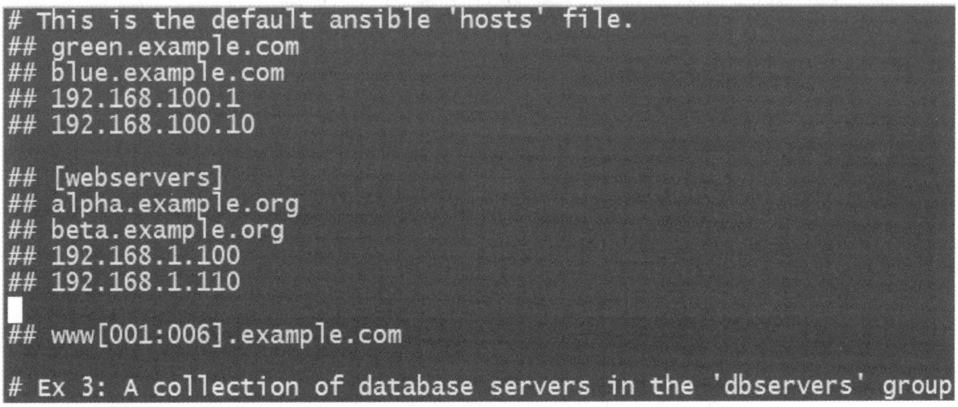

图 5-3 Hosts 主机列表默认配置

Ansible 是基于多模块管理，常用的 Ansible 工具管理模块包括 command、shell、script、yum、copy、file、async、docker、cron、mysql_user、ping、sysctl、user、acl、add_host、easy_install、haproxy 等。

可以使用 ansible-doc -l|more 命令来查看 Ansible 支持的模块，也可以查看每个模块的帮助文档，其命令为 ansible-doc module_name，如图 5-4 所示。

```
[root@www ~]# ansible-doc docker
> DOCKER    (/usr/lib/python2.7/site-packages/ansible/modules/cloud/doc

  This is the original Ansible module for managing the Docker container
  Additional and newer modules are available. For the latest on orchest
  Ansible visit our Getting Started with Docker Guide at
  https://github.com/ansible/ansible/blob/devel/docsite/rst/guide_docke

DEPRECATED:
In 2.2 use M(docker_container) and M(docker_image) instead.

Options (= is mandatory):

- cap_add
        Add capabilities for the container. Requires docker-py >= 0.5.0
        [Default: False]
```

图 5-4　Ansible-doc docker 帮助信息

5.4　Ansible 工具参数详解

在基于 Ansible 批量管理之前，需将被管理的服务器 IP 地址列表添加至/etc/ansible/hosts 文件中。图 5-5 所示为添加 4 台被管理的服务器 IP 地址，分成 Web 和 DB 两组，本机也可以是被管理机。

```
# This is the default ansible 'hosts' file.
#
# It should live in /etc/ansible/hosts
#
#   - Comments begin with the '#' character
#   - Blank lines are ignored
#   - Groups of hosts are delimited by [header] elements
#   - You can enter hostnames or ip addresses
#   - A hostname/ip can be a member of multiple groups
[web]
192.168.149.128
192.168.149.129
[db]
192.168.149.130
192.168.149.131
```

图 5-5　Ansible Hosts 主机列表

基于 Ansible 自动运维工具管理客户端案例的操作，由于 Ansible 管理远程服务器是基于 SSH，因此在登录远程服务器执行命令时需要输入远程服务器的用户名和密码，也可以加入-k 参数手动输入密码或基于 ssh-keygen 设置免密钥登录。

Ansible 自动化批量管理工具主要参数如下。

```
-v, -verbose                    #打印详细模式
-i PATH, -inventory=PATH        #指定host文件路径
-f NUM, -forks=NUM              #指定fork开启同步进程的个数,默认为5
-m NAME, -module-name=NAME      #指定module名称,默认模块command
-a MODULE_ARGS                  #Module模块的参数或命令
-k, -ask-pass                   #输入远程被管理端密码
 -sudo                          #基于sudo用户执行
-K, -ask-sudo-pass              #提示输入sudo密码,与sudo一起使用
-u USERNAME, -user=USERNAME     #指定移动端的执行用户
-C, -check                      #测试执行过程,不改变真实内容,相当于预演
-T TIMEOUT,                     #执行命令的超时时间,默认为10s
--version                       #查看Ansible软件版本信息
```

5.5 Ansible ping 模块实战

Ansible 最基础的模块为 ping 模块,主要用于判断远程客户端是否在线,用于 ping 本身服务器,返回值为 changed 和 ping。

Ansible ping 服务器状态的代码如下,运行结果如图 5-6 所示。

```
ansible -k all  -m ping
```

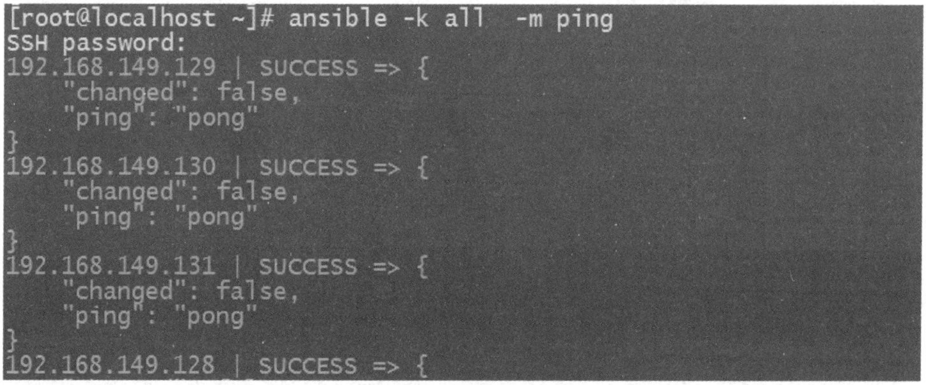

图 5-6 Ansible ping 服务器状态运行结果

5.6 Ansible command 模块实战

Ansible command 模块为 Ansible 默认模块,主要用于执行 Linux 基础命令,可以执行远程服务器命令和任务等。模块使用详解如下。

```
Chdir                           #执行命令前,切换到目录
Creates                         #当该文件存在时,不执行该步骤
```

Executable	#换用 Shell 环境执行命令
Free_form	#需要执行的脚本
Removes	#当该文件不存在时,不执行该步骤
Warn	#关闭或开启警告提示

Ansible command 模块的企业常用案例如下。

(1) Ansible command 模块远程执行 date 命令如下,执行结果如图 5-7 所示。

```
ansible -k -i /etc/ansible/hosts all -m command -a "date"
```

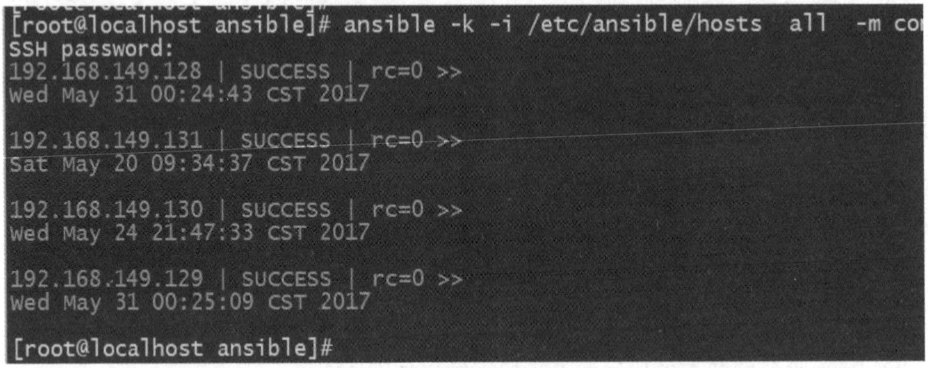

图 5-7　Ansible command 模块远程执行 date 命令结果

(2) Ansible command 模块远程执行 ping 命令如下,执行结果如图 5-8 所示。

```
ansible -k all -m command -a "ping -c 1 www.baidu.com"
```

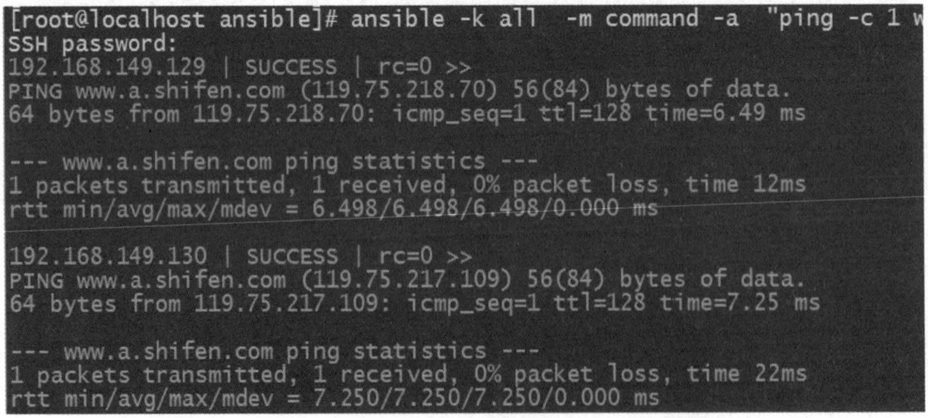

图 5-8　Ansible command 模块远程执行 ping 命令结果

(3) Ansible Hosts 正则模式远程执行 df -h 命令如下,执行结果如图 5-9 所示。

```
ansible -k 192.168.149.13* -m command -a "df -h"
```

```
[root@localhost ansible]# ansible -k 192.168.149.13* -m command
SSH password:
192.168.149.131 | SUCCESS | rc=0 >>
Filesystem      Size  Used Avail Use% Mounted on
/dev/sda3        30G   15G   14G  52% /
tmpfs           242M     0  242M   0% /dev/shm
/dev/sda1       194M  190M     0 100% /boot

192.168.149.130 | SUCCESS | rc=0 >>
Filesystem      Size  Used Avail Use% Mounted on
/dev/sda3        30G   16G   12G  57% /
tmpfs           242M     0  242M   0% /dev/shm
/dev/sda1       194M   27M  158M  15% /boot

[root@localhost ansible]#
```

图 5-9　Ansible command 正则模式远程执行 df -h 命令结果

5.7　Ansible copy 模块实战

Ansible copy 模块主要用于文件或目录的复制，支持文件、目录、权限和用户组功能。模块使用详解如下。

```
src                 #Ansible 端源文件或者目录,空文件夹不复制
content             #替代 src,用于将指定文件的内容复制到远程文件内
dest                #客户端目标目录或文件,需要绝对路径
backup              #复制之前,先备份远程节点上的原始文件
directory_mode      #用于复制文件夹,新建的文件会被复制,而老旧的不会被复制
follow              #支持 link 文件复制
force               #覆盖远程主机不一致的内容
group               #设定远程主机文件夹的组名
mode                #指定远程主机文件及文件的权限
owner               #设定远程主机文件夹的用户名
```

Ansible copy 模块的企业常用案例如下。

（1）Ansible copy 复制文件的操作代码如下，其中 src 表示源文件，dest 表示目标目录或者文件，owner 表示指定拥有者，执行结果如图 5-10 所示。

```
ansible -k all -m copy -a 'src=/etc/passwd dest=/tmp/ mode=755 owner=root'
```

（2）Ansible copy 追加内容的操作代码如下，其中 content 表示文件内容，dest 表示目标文件，owner 表示指定拥有者，执行结果如图 5-11 所示。

```
ansible -k all -m copy -a 'content="Hello World" dest=/tmp/jfedu.txt mode=755 owner=root'
```

```
[root@localhost ~]# ansible -k all -m copy -a 'src=/etc/passwd dest=
SSH password:
192.168.149.131 | SUCCESS => {
    "changed": true,
    "checksum": "6968f8053525cb8a821b3855d1860ad39f8f0e5d",
    "dest": "/tmp/passwd",
    "gid": 0,
    "group": "root",
    "md5sum": "d65ee7c1ff1b9868a6ee72eaf8666b03",
    "mode": "0755",
    "owner": "root",
    "size": 1791,
    "src": "/root/.ansible/tmp/ansible-tmp-1496163407.29-2790653592960
    "state": "file",
    "uid": 0
```

图 5-10　Ansible copy 复制文件执行结果

```
root'
SSH password:
192.168.149.129 | SUCCESS => {
    "changed": true,
    "checksum": "7b502c3a1f48c8609ae212cdfb639dee39673f5e",
    "dest": "/tmp/jfedu.txt",
    "gid": 0,
    "group": "root",
    "md5sum": "3e25960a79dbc69b674cd4ec67a72c62",
    "mode": "0755",
    "owner": "root",
    "size": 11,
    "src": "/root/.ansible/tmp/ansible-tmp-1496167669.85-1119688532
    "state": "file",
    "uid": 0
```

图 5-11　Ansible copy 追加内容执行结果

（3）Ansible copy 客户端备份结果的操作代码如下，其中 content 表示文件内容，dest 表示目标文件，owner 表示指定拥有者，backup=yes 表示开启备份，执行结果如图 5-12 所示。

```
ansible -k all -m copy -a 'content="Hello World" dest=/tmp/jfedu.txt
backup=yes mode=755 owner=root'
```

```
jfedu-net-129 tmp]# ls
ADO  ansible_YOWbvE   cacti.log  httpd-2.2.31.tar.gz  jfedu.txt  jfedu.
jfedu-net-129 tmp]#
jfedu-net-129 tmp]# ll
2 root root       4096 May 31 01:39 ansible_G2xADO
2 root root       4096 May 31 01:39 ansible_YOWbvE
1 root root       1064 May 31 01:35 cacti.log
1 root root    4711452 May 31 01:22 httpd-2.2.31.tar.gz
1 root root         16 May 31 02:03 jfedu.txt
1 root root         11 May 31 02:01 jfedu.txt.28886.2017-05-31@02:03:38~
3 root root       4096 May 31 01:36 test
jfedu-net-129 tmp]#
jfedu-net-129 tmp]# ll jfedu.txt.28886.2017-05-31\@02\:03\:38~
1 root root 11 May 31 02:01 jfedu.txt.28886.2017-05-31@02:03:38~
```

图 5-12　Ansible copy 客户端备份结果执行结果

5.8 Ansible YUM 模块实战

Ansible YUM 模块主要用于软件的安装、升级和卸载，支持红帽 .rpm 软件的管理，模块使用详解如下。

```
conf_file              #设定远程 YUM 执行时所依赖的 YUM 配置文件
disable_gpg_check      #安装软件包之前是否检查 gpg  key
name                   #需要安装的软件名称,支持软件组安装
update_cache           #安装软件前更新缓存
enablerepo             #指定 repo 源名称
skip_broken            #跳过异常软件节点
state          #软件包状态,包括 installed、present、latest、absent、removed
```

Ansible YUM 模块的企业常用案例如下。

（1）Ansible YUM 安装软件包的操作代码如下，其中 name 表示需安装的软件名称，state 表示软件包状态，state=installed 表示安装软件，执行结果如图 5-13 所示。

```
ansible all -k -m yum -a "name=sysstat,screen state=installed"
```

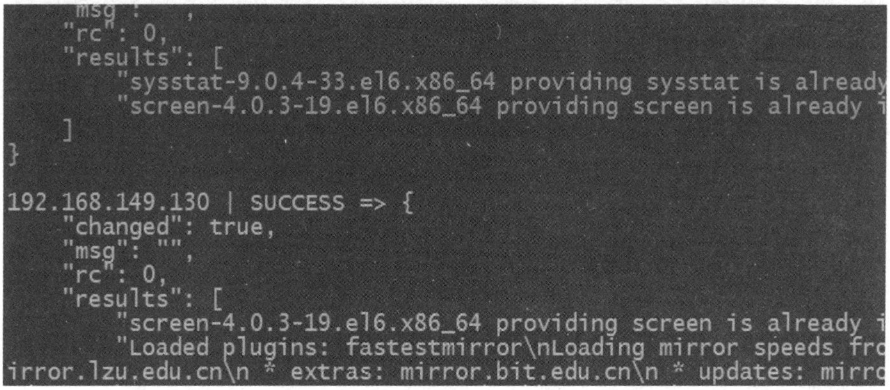

图 5-13 Ansible YUM 安装软件包执行结果

（2）Ansible YUM 卸载软件包的操作代码如下，其中 name 表示需卸载的软件名称，state 表示软件包状态，state=absent 表示卸载件，执行结果如图 5-14 所示。

```
ansible all -k -m yum -a "name=sysstat,screen state=absent"
```

（3）Ansible YUM 安装软件包，不检查 key 的操作代码如下，其中 name 表示需安装的软件名称，state 表示软件包状态，state=installed 表示安装软件，disable_gpg_check=no 表示不检查 key，执行结果如图 5-15 所示。

```
ansible 192.168.149.129 -k -m yum -a "name=sysstat,screen state=
installed disable_gpg_check=no"
```

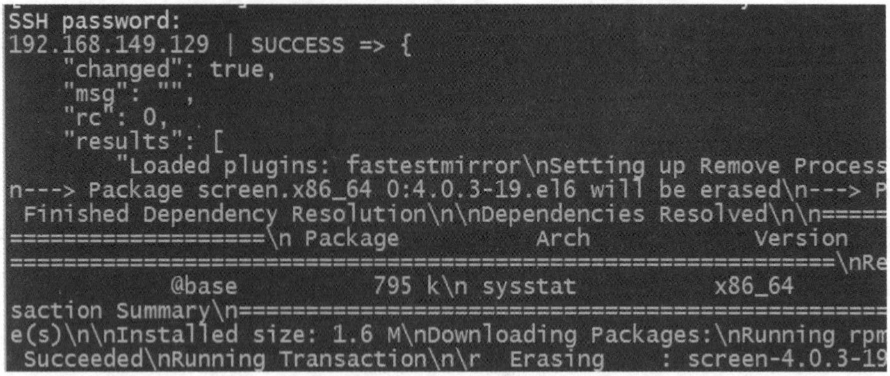

```
SSH password:
192.168.149.129 | SUCCESS => {
    "changed": true,
    "msg": "",
    "rc": 0,
    "results": [
        "Loaded plugins: fastestmirror\nSetting up Remove Process
n---> Package screen.x86_64 0:4.0.3-19.el6 will be erased\n---> P
Finished Dependency Resolution\n\nDependencies Resolved\n\n====
================================\n Package       Arch           Version
                 @base      795 k\n sysstat      x86_64
saction Summary\n=====================================
e(s)\n\nInstalled size: 1.6 M\nDownloading Packages:\n\nRunning rpm
Succeeded\nRunning Transaction\n\r  Erasing    : screen-4.0.3-19
```

图 5-14　Ansible YUM 卸载软件包执行结果

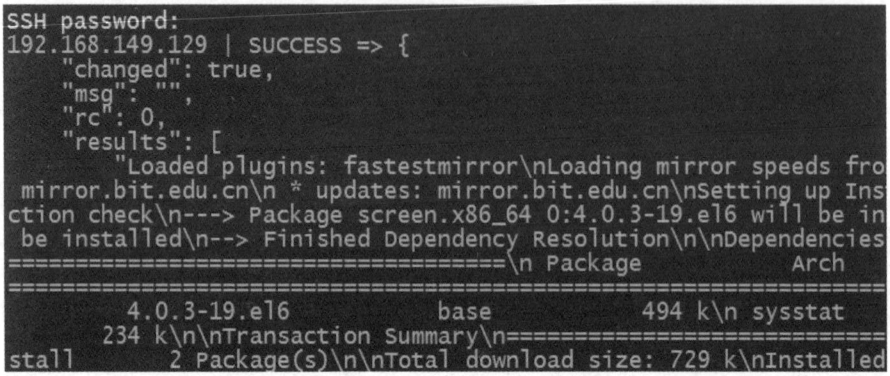

```
SSH password:
192.168.149.129 | SUCCESS => {
    "changed": true,
    "msg": "",
    "rc": 0,
    "results": [
        "Loaded plugins: fastestmirror\nLoading mirror speeds fro
mirror.bit.edu.cn\n * updates: mirror.bit.edu.cn\nSetting up Ins
ction check\n---> Package screen.x86_64 0:4.0.3-19.el6 will be in
be installed\n--> Finished Dependency Resolution\n\nDependencies
================================\n Package        Arch
         4.0.3-19.el6       base        494 k\n sysstat
         234 k\n\nTransaction Summary\n=====================
stall    2 Package(s)\n\nTotal download size: 729 k\nInstalled
```

图 5-15　Ansible YUM 安装软件包，不检查 key 执行结果

5.9　Ansible file 模块实战

Ansible file 模块主要用于对文件的创建、删除、修改、权限、属性的维护和管理等操作，模块使用详解如下。

```
src                    #Ansible 端源文件或目录
follow                 #支持链接文件复制
force                  #覆盖远程主机不一致的内容
group                  #设定远程主机文件夹的组名
mode                   #指定远程主机文件及文件夹的权限
owner                  #设定远程主机文件夹的用户名
path                   #目标路径，也可以用 dest,name 代替
state                  #状态包括 file、link、directory、hard、touch、absent
attributes             #文件或者目录特殊属性
```

Ansible file 模块的企业常用案例如下。

（1）Ansible file 创建目录的操作代码如下，其中 path 表示目标路径，state=directory 表示创建目录，执行结果如图 5-16 所示。

```
ansible -k 192.168.* -m file -a "path=/tmp/'date +%F' state=directory
mode=755"
```

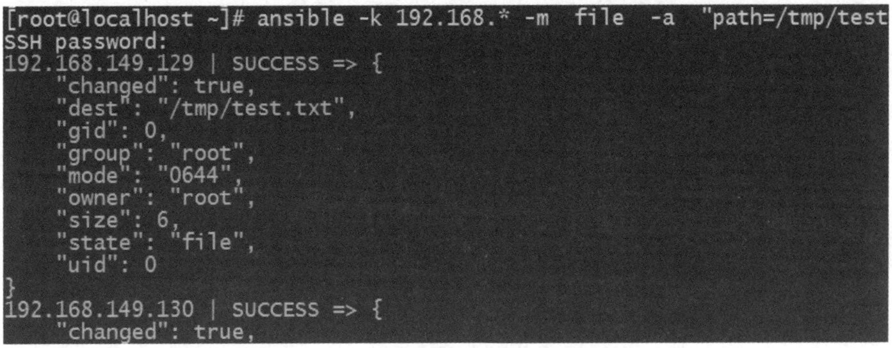

```
[root@localhost ~]# ansible -k 192.168.* -m file -a "path=/tmp/test
SSH password:
192.168.149.129 | SUCCESS => {
    "changed": true,
    "dest": "/tmp/test.txt",
    "gid": 0,
    "group": "root",
    "mode": "0644",
    "owner": "root",
    "size": 6,
    "state": "file",
    "uid": 0
}
192.168.149.130 | SUCCESS => {
    "changed": true,
```

图 5-16　Ansible file 创建目录执行结果

（2）Ansible file 创建文件的操作代码如下，其中 path 表示目标路径，state=touch 表示创建文件，执行结果如图 5-17 所示。

```
ansible -k 192.168.* -m file -a "path=/tmp/jfedu.txt state=touch
mode=755"
```

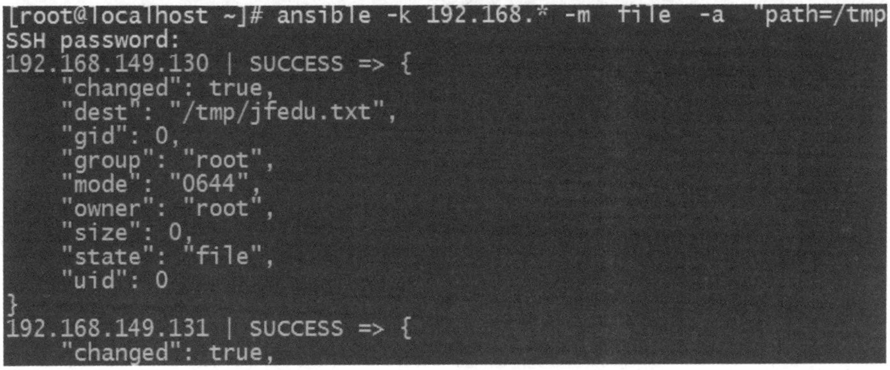

```
[root@localhost ~]# ansible -k 192.168.* -m file -a "path=/tmp
SSH password:
192.168.149.130 | SUCCESS => {
    "changed": true,
    "dest": "/tmp/jfedu.txt",
    "gid": 0,
    "group": "root",
    "mode": "0644",
    "owner": "root",
    "size": 0,
    "state": "file",
    "uid": 0
}
192.168.149.131 | SUCCESS => {
    "changed": true,
```

图 5-17　Ansible file 创建文件执行结果

5.10　Ansible user 模块实战

Ansible user 模块主要用于操作系统用户、组、权限、密码等操作，相关参数详解如下。

```
system                              #默认创建为普通用户，为 yes 则创建为系统用户
append                              #添加一个新的组
```

comment	#新增描述信息
createhome	#给用户创建主目录
force	#强制删除用户
group	#创建用户主组
groups	#将用户加入组或者附属组添加
home	#指定用户的主目录
name	#表示用户的名称
password	#指定用户的密码,此处为加密密码
remove	#删除用户
shell	#设置用户的Shell登录环境
uid	#设置用户ID
update_password	#修改用户密码
state	#用户状态,默认为present,表示新建用户

Ansible user 模块的企业常用案例如下。

（1）Ansible user 创建新用户的操作代码如下，其中 name 表示用户的名称，home 表示用户的主目录，执行结果如图 5-18 所示。

```
ansible -k 192.168.149.* -m user -a "name=jfedu home=/tmp/"
```

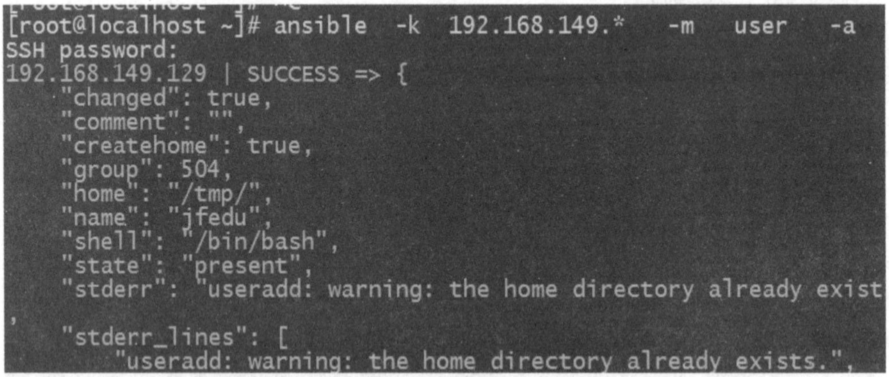

图 5-18　Ansible user 创建新用户执行结果

（2）Ansible user 指定 Shell 环境的操作代码如下，其中 name 表示用户的名称，home 表示用户的主目录，Shell 表示用户的 Shell 登录环境，执行结果如图 5-19 所示。

```
vansible -k 192.168.149.* -m user -a "name=jfedu home=/tmp/ shell=/sbin/nologin"
```

（3）Ansible user 删除用户的操作代码如下，其中 name 表示用户的名称，state=absent 表示删除用户，执行结果如图 5-20 所示。

```
ansible -k 192.168.149.* -m user -a "name=jfedu state=absent force=yes"
```

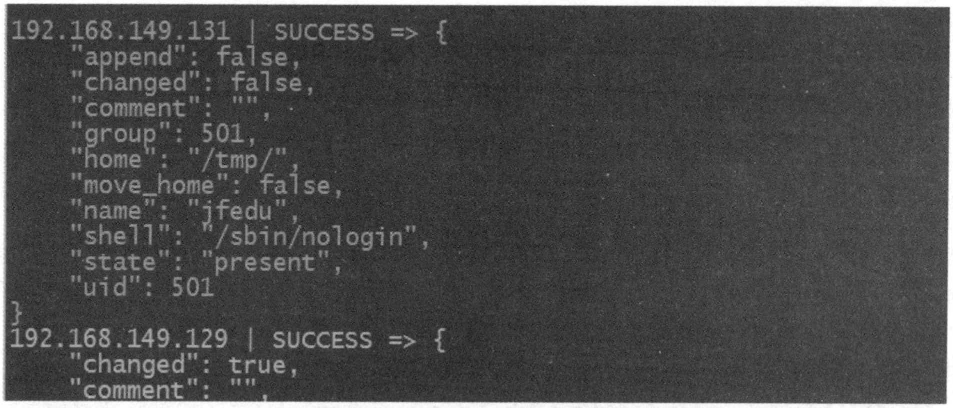

```
192.168.149.131 | SUCCESS => {
    "append": false,
    "changed": false,
    "comment": "",
    "group": 501,
    "home": "/tmp/",
    "move_home": false,
    "name": "jfedu",
    "shell": "/sbin/nologin",
    "state": "present",
    "uid": 501
}
192.168.149.129 | SUCCESS => {
    "changed": true,
    "comment": "",
```

图 5-19　Ansible user 指定 Shell 环境执行结果

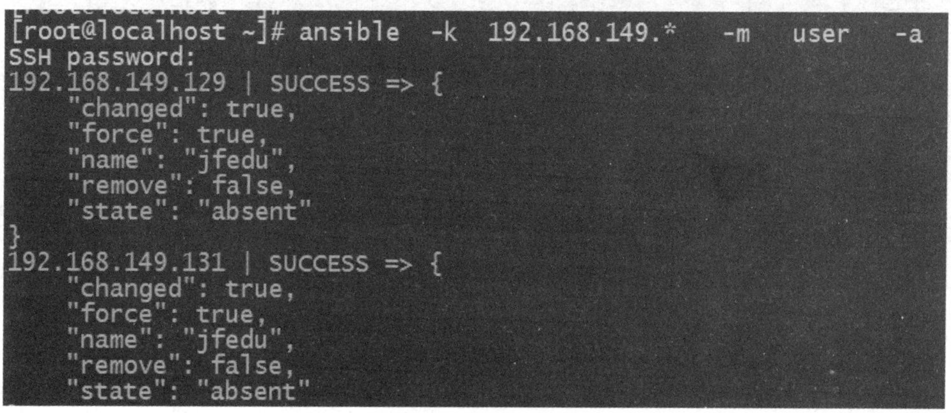

```
[root@localhost ~]# ansible -k 192.168.149.* -m user -a
SSH password:
192.168.149.129 | SUCCESS => {
    "changed": true,
    "force": true,
    "name": "jfedu",
    "remove": false,
    "state": "absent"
}
192.168.149.131 | SUCCESS => {
    "changed": true,
    "force": true,
    "name": "jfedu",
    "remove": false,
    "state": "absent"
}
```

图 5-20　Ansible user 删除用户执行结果

5.11　Ansible cron 模块实战

Ansible cron 模块主要用于添加、删除、更新操作系统 crontab 任务计划，模块使用详解如下。

```
name            #任务计划名称
cron_file       #替换客户端该用户的任务计划的文件
minute          #分（0~59 ,* ,*/2）
hour            #时（0~23 ,* ,*/2）
day             #日（1~31 ,* ,*/2）
month           #月（1~12 ,* ,*/2）
weekday         #周（0~6 或 1~7 ,*）
job             #任何计划执行的命令, state 要设为 present
backup          #是否备份之前的任务计划
```

```
user                              #新建任务计划的用户
state                             #指定任务计划状态,可设为present或absent
```

Ansible cron 模块的企业常用案例如下。

（1）Ansible cron 创建任务计划的代码如下，其中 name 表示任务计划名称，执行结果如图 5-21 所示。

```
ansible -k all -m cron -a "minute=0 hour=0 day=* month=* weekday=
* name='Ntpdate server for sync time'  job='/usr/sbin/ntpdate 139.224.
227.121'"
```

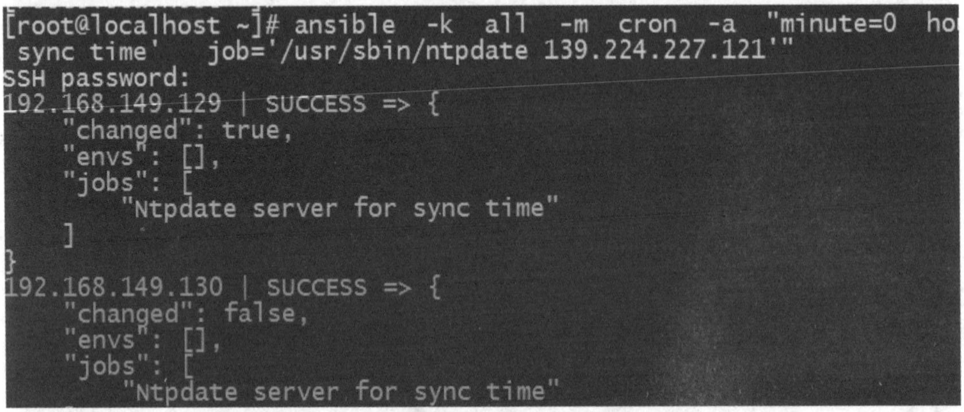

图 5-21 Ansible cron 创建任务计划执行结果

（2）Ansible cron 备份任务计划的代码如下，其中 backup=yes 表示开启备份，备份文件存放于客户端/tmp/目录下，执行结果如图 5-22 所示。

```
ansible -k all -m cron -a "minute=0 hour=0 day=* month=* weekday=
name='Ntpdate server for sync time' backup=yes job='/usr/sbin/ntpdate
pool.ntp.org'"
```

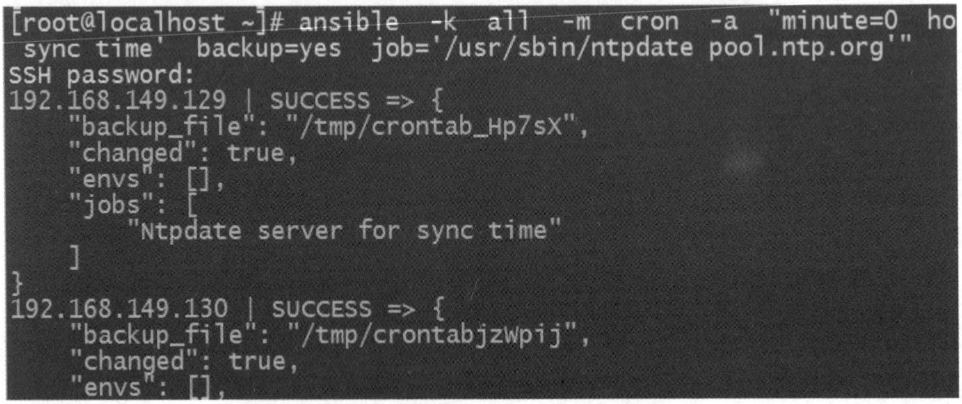

图 5-22 Ansible cron 备份任务计划执行结果

（3）Ansible cron 删除任务计划的代码如下，其中 state=absent 表示删除任务计划，执行结果如图 5-23 所示。

```
ansible -k all -m cron -a "name='Ntpdate server for sync time' state=absent"
```

```
[root@localhost ~]# ansible -k all -m cron -a "name='Ntpdat
SSH password:
192.168.149.130 | SUCCESS => {
    "changed": true,
    "envs": [],
    "jobs": []
}
192.168.149.129 | SUCCESS => {
    "changed": true,
    "envs": [],
    "jobs": []
}
192.168.149.131 | SUCCESS => {
    "changed": true,
    "envs": [],
```

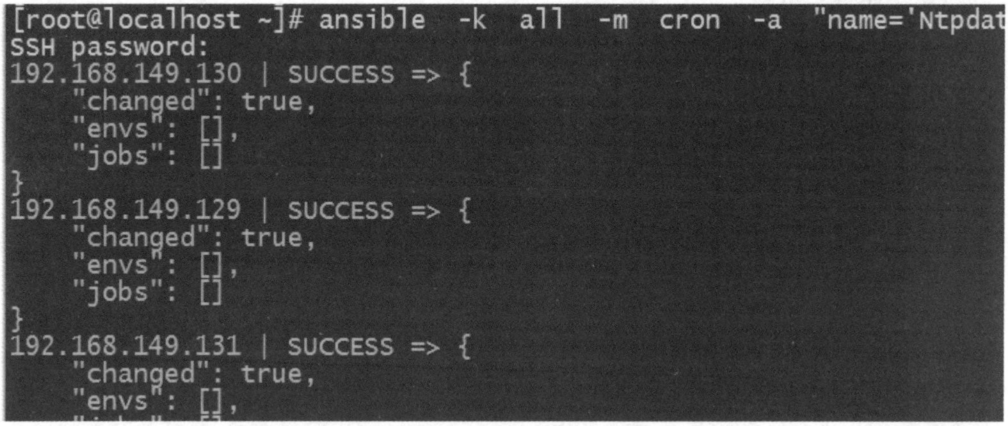

图 5-23　Ansible cron 删除任务计划执行结果

5.12　Ansible synchronize 模块实战

Ansible synchronize 模块主要用于目录同步和文件同步，是基于 rsync 命令同步目录。模块使用详解如下。

```
compress            #开启压缩,默认为开启
archive             #是否采用归档模式同步,保证源和目标文件属性一致
checksum            #是否校验
dirs                #以非递归的方式传输目录
links               #同步链接文件
recursive           #是否递归(yes/no)
rsync_opts          #使用 rsync 的参数
copy_links          #同步的时候是否复制链接
delete              #删除源中没有但目标中却存在的文件
src                 #源目录及文件
dest                #目标目录及文件
dest_port           #目标接收的端口
rsync_path          #服务的路径,指定 rsync 命令在远程服务器上运行
rsync_timeout       #指定 rsync 操作的 IP 地址的超时时间
set_remote_user     #设置远程用户名
```

```
--exclude=.log          #忽略以同步.log 结尾的文件
mode                    #同步的模式,rsync 同步的方式为 push、pull,默认都是推送 push
```

Ansible synchronize 模块的企业常用案例如下。

(1) Ansible synchronize 目录同步的操作代码如下,其中 src 表示源目录,dest 表示目标目录,执行结果如图 5-24 所示。

```
ansible -k all -m synchronize -a 'src=/tmp/ dest=/tmp/'
```

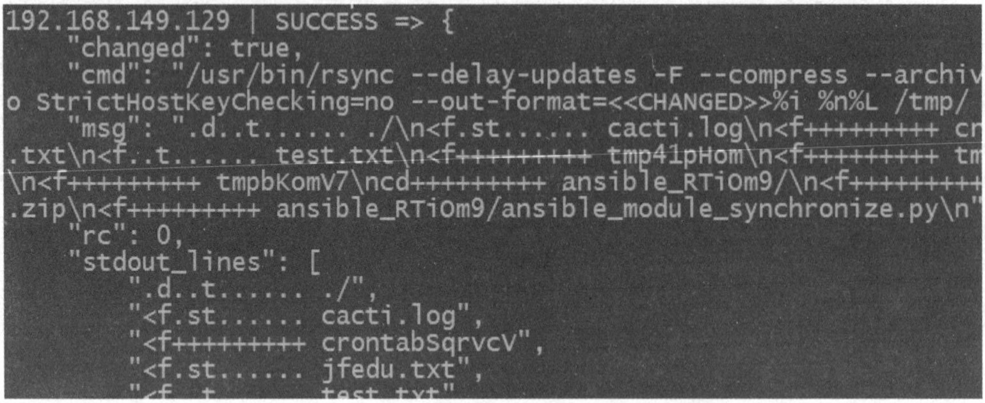

图 5-24　Ansible synchronize 目录同步执行结果

(2) Ansible synchronize 目录同步排除.txt 文件的操作代码如下,其中 src 表示源目录,dest 表示目标目录,compress=yes 表示开启压缩,delete=yes 表示源目录与目标目录中的数据一致,rsync_opts 表示同步参数,--exclude 表示排除文件,执行结果如图 5-25 所示。

```
ansible -k all -m synchronize -a 'src=/tmp/ dest=/tmp/ compress=yes
delete=yes rsync_opts=--no-motd,--exclude=.txt'
```

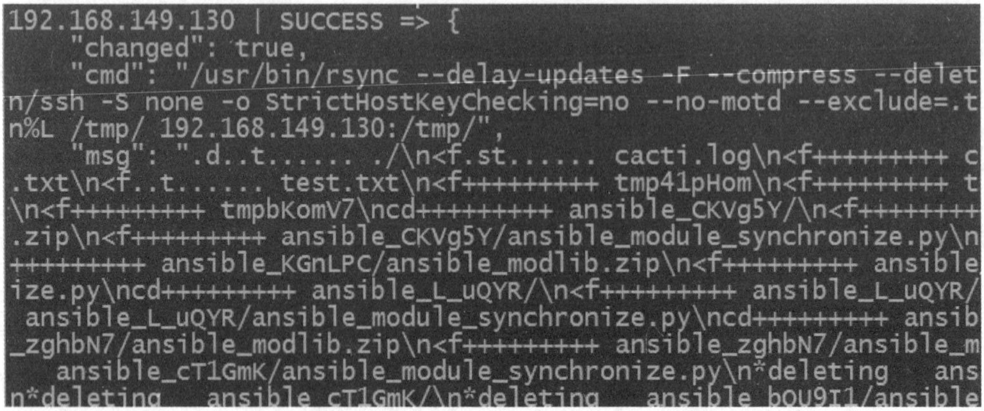

图 5-25　Ansible synchronize 目录同步排除.txt 文件执行结果

5.13 Ansible Shell 模块实战

Ansible Shell 模块主要用于在远程客户端上执行各种 Shell 命令或运行脚本，远程执行命令通过/bin/sh 环境执行，支持比 command 更多的指令。模块使用详解如下。

```
Chdir        #执行命令前,切换到目录
Creates      #当该文件存在时,不执行该步骤
Executable   #换用 Shell 环境执行命令
Free_form    #需要执行的脚本
Removes      #当该文件不存在时,不执行该步骤
Warn         #如果在 ansible.cfg 文件中存在警告,若设定了 False,则不会警告此行
```

Ansible Shell 模块的企业常用案例如下。

（1）Ansible Shell 远程执行脚本的操作代码如下，其中-m shell 表示指定模块为 Shell，远程执行 Shell 脚本。远程执行脚本也可采用 script 模块。把执行结果追加到客户端/tmp/var.log 文件中，执行结果如图 5-26 所示。

```
ansible -k all -m shell -a "/bin/sh /tmp/variables.sh >>/tmp/var.log"
```

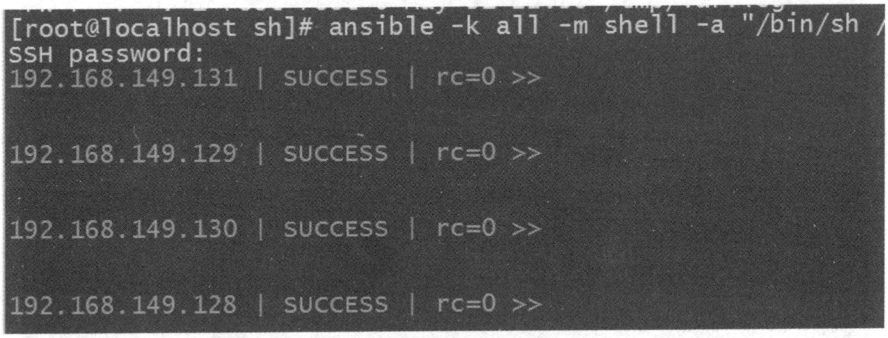

图 5-26 Ansible Shell 远程执行脚本执行结果

（2）Ansible Shell 远程创建目录，执行之前切换至/tmp 目录，屏蔽警告信息，其代码如下，执行结果如图 5-27 所示。

```
ansible -k all -m shell -a "mkdir -p 'date +%F' chdir=/tmp/ state=directory warn=no"
```

（3）Ansible Shell 远程查看进程的操作代码如下，其中-m shell 表示指定模块为 Shell，远程客户端查看 http 进程是否启动，执行结果如图 5-28 所示。

```
ansible -k all -m shell -a "ps -ef |grep http"
```

（4）Ansible Shell 远程查看任务计划的操作代码如下，其中-m shell 表示指定模块为 Shell，远程客户端查看 crontab 任务计划，执行结果如图 5-29 所示。

```
ansible  -k  all  -m  shell  -a  "crontab -l"
```

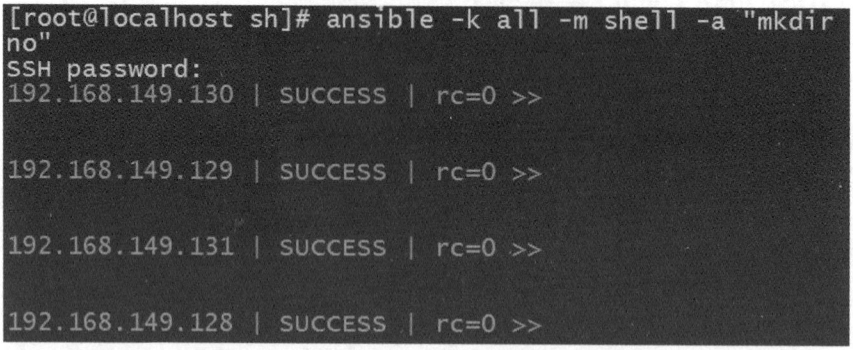

图 5-27　Ansible Shell 远程创建目录执行结果

图 5-28　Ansible Shell 远程查看进程执行结果

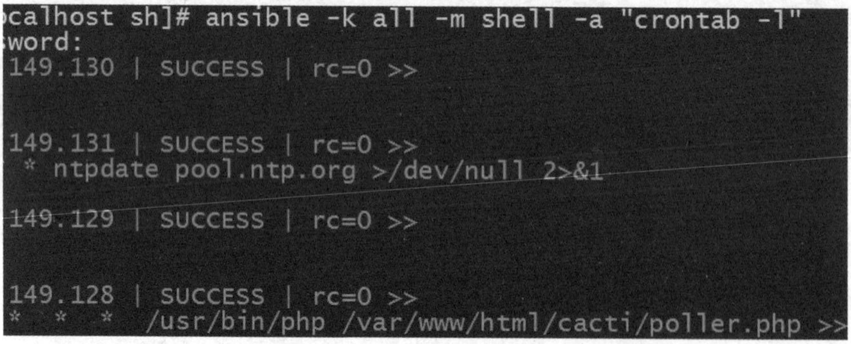

图 5-29　Ansible Shell 远程查看任务计划执行结果

5.14　Ansible service 模块实战

Ansible service 模块主要用于远程客户端各种服务管理，包括启动、停止、重启、重新加载等。模块使用详解如下。

```
enabled                 #是否启动服务
name                    #服务名称
runlevel                #服务启动级别
arguments               #服务命令行参数传递
state                   #服务操作状态,状态包括started、stopped、restarted、reloaded
```

Ansible service 模块的企业常用案例如下。

(1) Ansible service 重启 httpd 服务的代码如下,执行结果如图 5-30 所示。

```
ansible -k all -m service -a "name=httpd state=restarted"
```

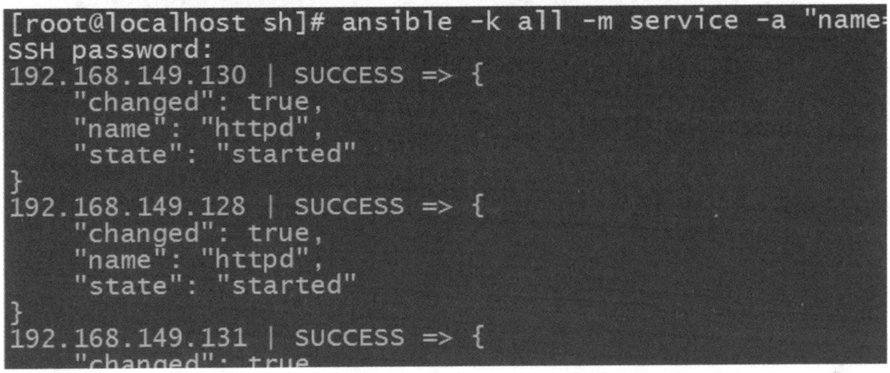

图 5-30　Ansible service 重启 httpd 服务执行结果

(2) Ansible service 重启 network 服务,指定参数为 eth0,其代码如下,执行结果如图 5-31 所示。

```
ansible -k all -m service -a "name=network args=eth0 state=restarted"
```

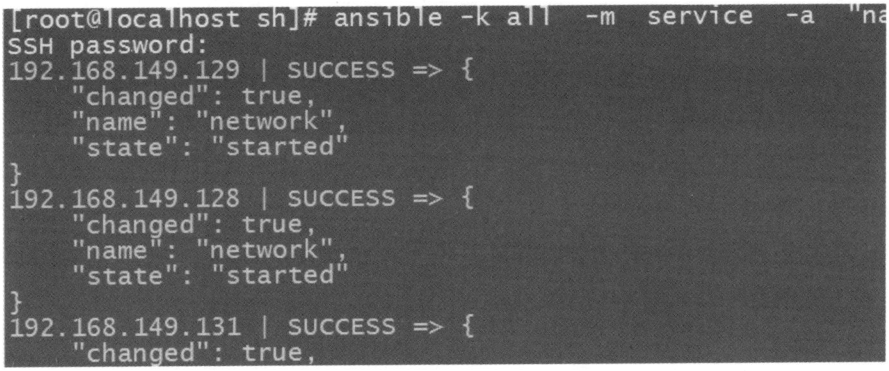

图 5-31　Ansible service 重启 network 服务执行结果

(3) Ansible service 开机启动 nfs 服务,设置 3 级别和 5 级别自动启动,代码如下,执行结果如图 5-32 所示。

```
ansible -k all -m service -a "name=nfs enabled=yes runlevel=3,5"
```

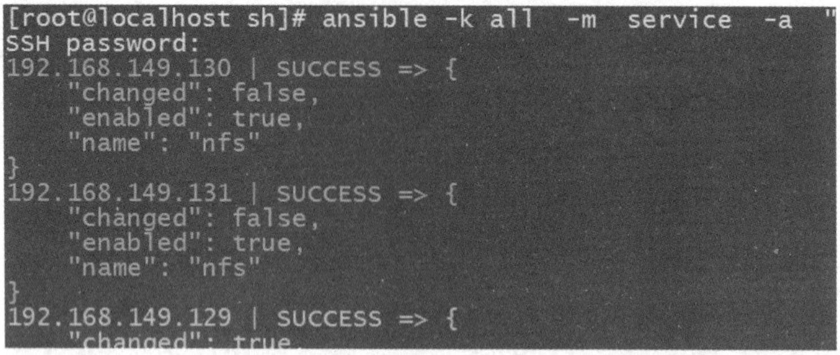

```
[root@localhost sh]# ansible -k all -m service -a "
SSH password:
192.168.149.130 | SUCCESS => {
    "changed": false,
    "enabled": true,
    "name": "nfs"
}
192.168.149.131 | SUCCESS => {
    "changed": false,
    "enabled": true,
    "name": "nfs"
}
192.168.149.129 | SUCCESS => {
    "changed": true,
```

图 5-32　Ansible service 开机启动 nfs 服务执行结果

5.15　Ansible Playbook 应用

服务器数量较少时，可以使用点对点的方式管理远程主机。如果服务器数量很多，配置信息比较多，还可以利用 Ansible Playbook 编写剧本，从而以非常简便的方式实现任务处理的自动化与流程化。

Playbook（剧本）是由一个或多个 play（角色）组成的列表，play 的主要功能是定义需要执行 task（任务）的主机或组，实现通过 Ansible 中的 task 定义好的 play，指定 Playbook 对应的服务器组。

task 调用 Ansible 各种模块（module），将多个 play 组织在一个 Playbook 中，然后组成一个非常完整的流程控制集合。

基于 Ansible Playbook 还可以收集命令、创建任务集，这样能够大大降低管理工作的复杂程度。Playbook 采用 YAML 语法结构，易于阅读、方便配置。

YAML（Yet Another Markup Language）是一种直观的能够被电脑识别的数据序列化格式，是一个可读性强、容易和脚本语言交互、用来表达资料序列的编程语言。YAML 参考了其他多种语言，包括 XML、C、Python、Perl 等，它类似于标准通用标记语言的子集 XML 的数据描述语言，但语法比 XML 语言简单很多。

YAML 使用空白字符和分行来分隔资料，适合用 GREP、Python、Perl、Ruby 操作。

（1）YAML 语言特性如下。

① 可读性强。

② 与脚本语言的交互性好。

③ 使用实现语言的数据类型。

④ 一致的信息模型。

⑤ 易于实现。

⑥ 可以基于流处理。

⑦ 可扩展性强。

（2）Playbooks 组件如下。

```
Target      #定义 Playbook 的远程主机组
Variable    #定义 Playbook 使用的变量
Task        #定义远程主机上执行的任务列表
Handler     #定义 task 执行完成以后需要调用的任务,如配置文件被改动,则启动 handler
            #任务重启相关联的服务
```

（3）Target 常用参数如下。

```
hosts             #定义远程主机组
user              #执行该任务的用户
sudo              #如设置为 yes,则执行任务的时候使用 root 权限
sudo_user         #指定 sudo 普通用户
connection        #默认基于 SSH 连接客户端
gather_facts      #获取远程主机 facts 基础信息
```

（4）Variable 常用参数如下。

```
vars              #定义格式,变量名:变量值
vars_files        #指定变量文件
vars_prompt       #用户交互模式自定义变量
setup             #模块设置为远程主机的值
```

（5）Task 常用参数如下。

```
name        #任务显示名称,即屏幕显示信息
action      #定义执行的动作
copy        #复制本地文件到远程主机
template    #复制本地文件到远程主机,可以引用本地变量
service     #定义服务的状态
```

Ansible Playbook 的案例演示如下。

（1）远程主机安装 Nginx Web 服务。其 Playbook 代码如下，执行结果如图 5-33 所示。

```
- hosts: all
  remote_user: root
  tasks:
  - name: Jfedu Pcre-devel and Zlib LIB Install.
    yum: name=pcre-devel,pcre,zlib-devel state=installed
  - name: Jfedu Nginx WEB Server Install Process.
    shell: cd /tmp;rm -rf nginx-1.12.0.tar.gz;wget http://nginx.org/download/nginx-1.12.0.tar.gz;tar xzf nginx-1.12.0.tar.gz;cd nginx-1.12.0;./configure --prefix=/usr/local/nginx;make;make install
```

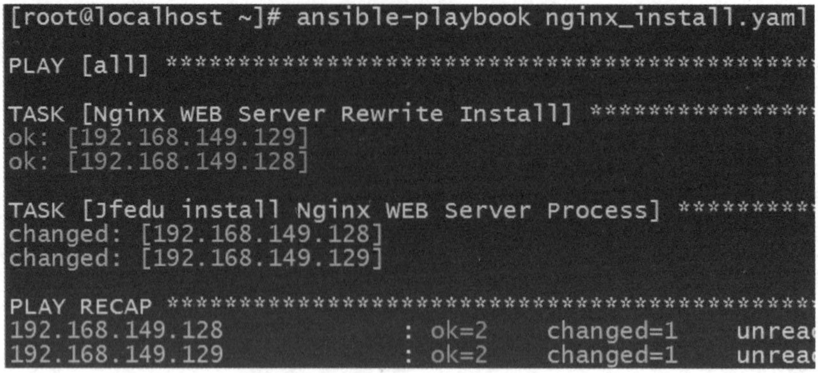

```
[root@localhost ~]# ansible-playbook nginx_install.yaml
PLAY [all] ****************************************************

TASK [Nginx WEB Server Rewrite Install] ***********************
ok: [192.168.149.129]
ok: [192.168.149.128]

TASK [Jfedu install Nginx WEB Server Process] *****************
changed: [192.168.149.128]
changed: [192.168.149.129]

PLAY RECAP ****************************************************
192.168.149.128            : ok=2    changed=1    unrea
192.168.149.129            : ok=2    changed=1    unrea
```

图 5-33　Ansible Playbook 远程 Nginx 安装

（2）检测远程主机 Nginx 目录是否存在，若不存在则安装 Nginx Web 服务，安装完成后启动 Nginx。Playbook 代码如下，执行结果如图 5-34 所示。

```
- hosts: all
  remote_user: root
  tasks:
    - name: Nginx server Install 2021
      file: path=/usr/local/nginx/ state=directory
      notify:
          - nginx install
          - nginx start
  handlers:
    - name: nginx install
      shell: cd /tmp;rm -rf nginx-1.12.0.tar.gz;wget http://nginx.org/download/nginx-1.12.0.tar.gz;tar xzf nginx-1.12.0
.tar.gz;cd nginx-1.12.0;./configure --prefix=/usr/local/nginx;make;make install
    - name: nginx start
      shell: /usr/local/nginx/sbin/nginx
```

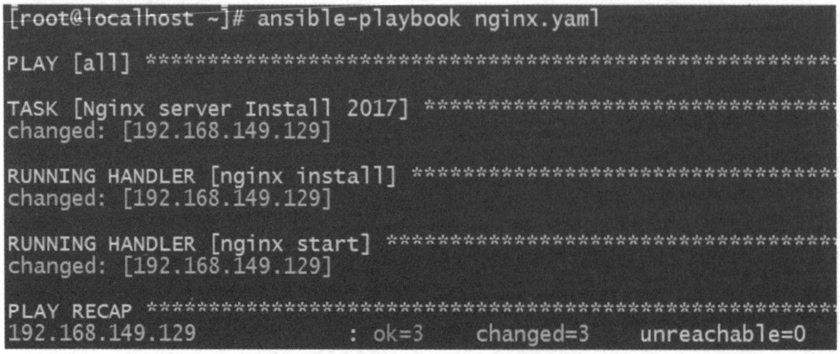

```
[root@localhost ~]# ansible-playbook nginx.yaml
PLAY [all] ****************************************************

TASK [Nginx server Install 2017] ******************************
changed: [192.168.149.129]

RUNNING HANDLER [nginx install] *******************************
changed: [192.168.149.129]

RUNNING HANDLER [nginx start] *********************************
changed: [192.168.149.129]

PLAY RECAP ****************************************************
192.168.149.129            : ok=3    changed=3    unreachable=0
```

图 5-34　Ansible Playbook Nginx 触发安装

（3）检测远程主机内核参数配置文件是否更新，如果更新，则执行命令 sysctl -p 使内核参数生效。Playbook 代码如下，执行结果如图 5-35 所示。

```
- hosts: all
  remote_user: root
  tasks:
    - name: Linux kernel config 2021
      copy: src=/data/sh/sysctl.conf dest=/etc/
      notify:
          - source sysctl
  handlers:
    - name: source sysctl
      shell: sysctl -p
```

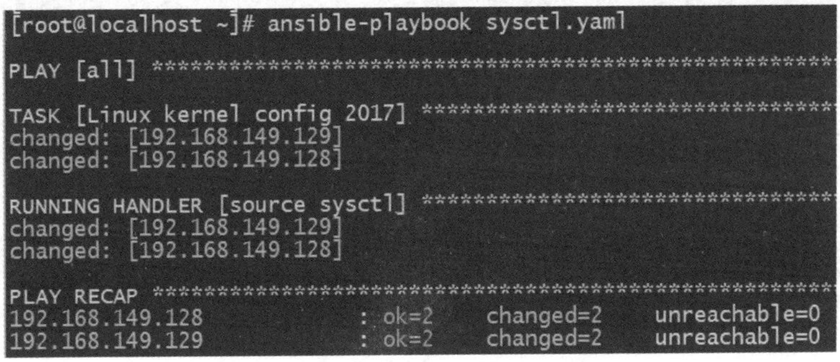

图 5-35　Ansible Playbook 内核参数优化

（4）基于列表 items 的多个值创建用户的代码如下，通过{{}}定义列表变量，with_items 选项传入变量的值，执行结果如图 5-36 所示。

```
- hosts: all
  remote_user: root
  tasks:
  - name: Linux system Add User list.
    user: name={{ item }} state=present
    with_items:
      - jfedu1
      - jfedu2
      - jfedu3
      - jfedu4
```

（5）Ansible Playbook 可以自定义模板文件（template），模板文件主要用于服务器需求不一致的情况，需要独立定义，如两台服务器都安装了 Nginx，安装完毕之后需将服务器 A 的 HTTP 端口改成 80，服务器 B 的 HTTP 端口改成 81。

① Ansible hosts 文件指定不同服务器的 httpd_port 端口不同，代码如下。

```
[web]
192.168.149.128 httpd_port=80
192.168.149.129 httpd_port=81
```

```
[root@localhost ~]# ansible-playbook user.yaml
PLAY [all] ********************************************************
TASK [Linux system Add User list.] *********************************
changed: [192.168.149.129] => (item=jfedu1)
changed: [192.168.149.129] => (item=jfedu2)
changed: [192.168.149.129] => (item=jfedu3)
changed: [192.168.149.129] => (item=jfedu4)

PLAY RECAP *********************************************************
192.168.149.129            : ok=1    changed=1    unreachable=0

[root@localhost ~]#
```

(a)

```
nfsnobody:x:65534:65534:Anonymous NFS User:/var/lib/nfs:/sbin/no
mysql:x:27:27:MySQL Server:/var/lib/mysql:/bin/bash
apache:x:48:48:Apache:/var/www/:/sbin/nologin
exim:x:93:93::/var/spool/exim:/sbin/nologin
sdfjsdklfskl
jfedu001:x:501:501::/home/jfedu001:/bin/bash
redis:x:496:496:Redis Server:/var/lib/redis:/sbin/nologin
a:x:502:502::/home/a:/bin/bash
zabbix:x:503:503::/home/zabbix:/sbin/nologin
tcpdump:x:72:72::/:/sbin/nologin
jfedu1:x:504:504::/home/jfedu1:/bin/bash
jfedu2:x:505:505::/home/jfedu2:/bin/bash
jfedu3:x:506:506::/home/jfedu3:/bin/bash
jfedu4:x:507:507::/home/jfedu4:/bin/bash
```

(b)

图 5-36 Ansible Playbook item 变量创建用户

(a) Ansible 创建用户; (b) 查看系统用户已创建

② Ansible 创建 nginx.conf jinja2 模板文件, 复制文件 nginx.conf nginx.conf.j2, 修改 listen 80 为 listen {{httpd_port}}, Nginx 其他配置项不变, 其代码如下。

```
cp nginx.conf nginx.conf.j2
listen  {{httpd_port}};
```

③ Ansible PlaybookYAML 文件创建的代码如下。

```
- hosts: all
  remote_user: root
  tasks:
    - name: Nginx server Install 2021
      file: path=/usr/local/nginx/ state=directory
      notify:
         - nginx install
         - nginx config
  handlers:
    - name: nginx install
      shell: cd /tmp;rm -rf nginx-1.12.0.tar.gz;wget http://nginx.org/
download/nginx-1.12.0.tar.gz;tar xzf nginx-1.12.0
.tar.gz;cd nginx-1.12.0;./configure --prefix=/usr/local/nginx;make;make
```

```
install
    - name: nginx config
      template: src=/data/sh/nginx.conf.j2 dest=/usr/local/nginx/conf/
nginx.conf
```

④ Ansible Playbook 执行剧本文件的执行结果如图 5-37 所示。

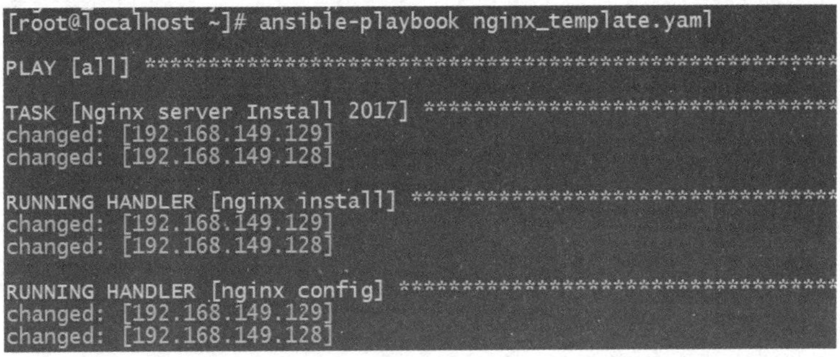

```
[root@localhost ~]# ansible-playbook nginx_template.yaml

PLAY [all] ************************************************************

TASK [Nginx server Install 2017] **************************************
changed: [192.168.149.129]
changed: [192.168.149.128]

RUNNING HANDLER [nginx install] ***************************************
changed: [192.168.149.129]
changed: [192.168.149.128]

RUNNING HANDLER [nginx config] ****************************************
changed: [192.168.149.129]
changed: [192.168.149.128]
```

（a）

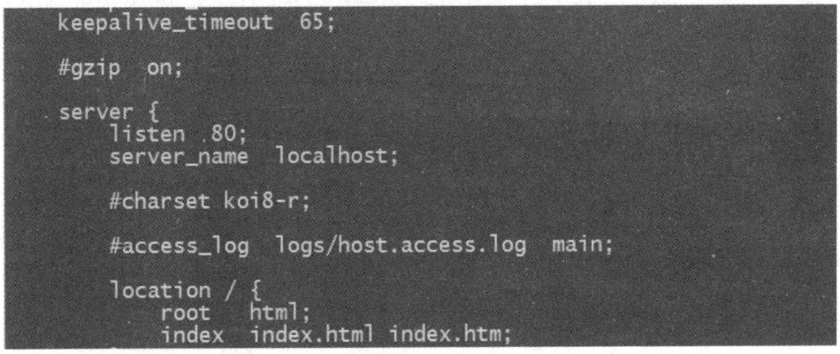

```
    keepalive_timeout  65;

    #gzip  on;

    server {
        listen  80;
        server_name  localhost;

        #charset koi8-r;

        #access_log  logs/host.access.log  main;

        location / {
            root   html;
            index  index.html index.htm;
```

（b）

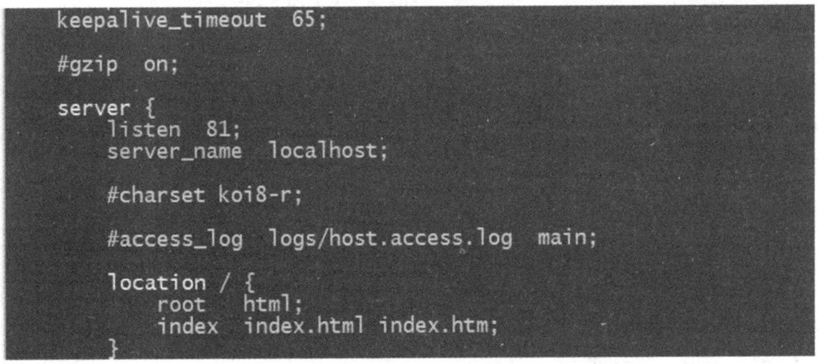

```
    keepalive_timeout  65;

    #gzip  on;

    server {
        listen  81;
        server_name  localhost;

        #charset koi8-r;

        #access_log  logs/host.access.log  main;

        location / {
            root   html;
            index  index.html index.htm;
        }
```

（c）

图 5-37　Ansible Playbook 执行剧本文件的执行结果

（a）Ansible Playbook 执行模板 YAML；（b）149.128 服务器 Nginx HTTP Port 80；（c）149.129 服务器 Nginx HTTP Port 81

5.16 Ansible 配置文件详解

Ansible 默认配置文件为/etc/ansible/ansible.cfg，配置文件中可以对 Ansible 进行各项参数的调整，包括并发线程、用户、模块路径、配置优化等。以下为 ansible.cfg 文件常用参数的详解。

```
[defaults]                                      #通用默认配置段
inventory=/etc/ansible/hosts                    #被控端 IP 地址列表或 DNS 列表
library=/usr/share/my_modules/                  #Ansible 默认搜寻模块的位置
remote_tmp=$HOME/.ansible/tmp                   #Ansible 远程执行临时文件
pattern=*                                       #对所有主机通信
forks=5                                         #并行进程数
poll_interval=15                                #回频率或轮训间隔时间
sudo_user=root                                  #sudo 远程执行用户名
ask_sudo_pass=True                              #使用 sudo,是否需要输入密码
ask_pass=True                                   #是否需要输入密码
transport=smart                                 #通信机制
remote_port=22                                  #远程 SSH 端口
module_lang=C                                   #模块和系统之间通信的语言
gathering=implicit                              #控制默认 facts 收集（远程系统变量）
roles_path=/etc/ansible/roles                   #Playbook 搜索 Ansible roles
host_key_checking=False                         #检查远程主机密钥
#sudo_exe=sudo                                  #sudo 远程执行命令
#sudo_flags=-H                                  #传递 sudo 之外的参数
timeout=10                                      #SSH 超时时间
remote_user=root                                #远程登录用户名
log_path=/var/log/ansible.log                   #日志文件的存放路径
module_name=command                             #Ansible 命令执行默认的模块
#executable=/bin/sh                             #执行的 Shell 环境,用户 Shell 模块
#hash_behaviour=replace                         #特定的优先级覆盖变量
#jinja2_extensions                              #允许开启 Jinja2 拓展模块
#private_key_file=/path/to/file                 #私钥文件的存储位置
#display_skipped_hosts=True                     #显示任何跳过任务的状态
#system_warnings=True                           #禁用系统运行 Ansible 潜在问题警告
#deprecation_warnings=True                      #Playbook 输出结果中禁用"不建议使用"警告
#command_warnings=False                         #command 模块 Ansible 默认发出警告
#nocolor=1                                      #输出带上颜色区别,开启/关闭：0/1
pipelining=False                                #开启 pipe SSH 通道优化
[accelerate]                                    #accelerate 缓存加速
accelerate_port=5099
```

```
accelerate_timeout=30
accelerate_connect_timeout=5.0
accelerate_daemon_timeout=30
accelerate_multi_key=yes
```

5.17　Ansible 性能调优

Ansible 企业实战环境中，如果管理的服务器越来越多，Ansibe 执行效率会变得越来越慢，可以通过优化 Ansible 提高工作效率。由于 Ansible 是基于 SSH 协议通信，SSH 连接慢会导致整个基于 Ansible 的执行变得缓慢，所以也需要对 OpenSSH 进行优化。具体优化方法如下。

（1）Ansible SSH 关闭密钥检测。

默认以 SSH 登录远程客户端，会检查远程主机的公钥（public key），并将该主机的公钥记录在~/.ssh/known_hosts 文件中。下次访问相同主机时，OpenSSH 会核对公钥，如果公钥不同，则 OpenSSH 会发出警告，如果公钥相同，则会提示输入密码。

SSH 对主机的 public key 的检查等级是根据 StrictHostKeyChecking 变量设定的，StrictHostKeyChecking 检查级别包括 no（不检查）、ask（询问）、yes（每次都检查）和 False（关闭检查）。

Ansible 配置文件中加入如下代码，即可关闭 StrictHostKeyChecking 检查。

```
host_key_checking=False
```

（2）OpenSSH 连接优化。

使用 OpenSSH 服务时，默认服务器配置文件中 UseDNS=YES 状态，该选项会导致服务器根据客户端的 IP 地址进行 DNS PTR 反向解析，得到客户端的主机名，然后根据获取到的主机名进行 DNS 正向 A 记录查询，并验证该 IP 地址是否与原始 IP 地址一致。关闭 DNS 解析的代码如下。

```
sed -i '/^GSSAPI/s/yes/no/g;/UseDNS/d;/Protocol/aUseDNS no' /etc/ssh/sshd_config
/etc/init.d/sshd restart
```

（3）SSH pipelining 加速 Ansible。

SSH pipelining 是一个加速 Ansible 执行速度的简单方法，SSH pipelining 默认是关闭的，关闭是为了兼容不同的 sudo 配置，主要是 requiretty 选项。

如果不使用 sudo 建议开启该选项，打开此选项可以减少 Ansible 执行没有文件传输时，SSH 在被控机器上执行任务的连接数。使用 sudo 操作的时候，必须在所有被管理的主机上将配置文件/etc/sudoers 中的 requiretty 选项禁用。

```
sed -i '/^pipelining/s/False/True/g' /etc/ansible/ansible.cfg
```

（4）Ansible Facts 缓存优化。

Ansible Playbook 在执行过程中，默认会执行 Gather facts，如果不需要获取客户端的 facts 数据，可以关闭获取 facts 数据功能，关闭之后可以加快 ansible-playbook 的执行效率。如需关闭获取 facts 功能，在 Playbook YAML 文件中加入以下代码即可。

```
gather_facts: nogather_facts: no
```

Ansible Facts 组件主要用于收集客户端设备的基础静态信息，以便在做配置管理的时候引用。facts 信息直接被当作 Ansible Playbook 变量信息引用，通过定制 facts 可以收集需要的信息，同时可以通过 Facter 和 Ohai 拓展 facts 信息，也可以将 facts 信息存入 Redis 缓存中。以下为 facts 使用 Redis 缓存的步骤。

① 部署 Redis 服务。

```
wget      http://download.redis.io/releases/redis-2.8.13.tar.gz
tar       zxf       redis-2.8.13.tar.gz
cd        redis-2.8.13
make      PREFIX=/usr/local/redis  install
cp        redis.conf    /usr/local/redis/
```

将/usr/local/redis/bin/目录加入环境变量配置文件/etc/profile 的末尾，然后 Shell 终端执行 source /etc/profile 让环境变量生效。

```
export PATH=/usr/local/redis/bin:$PATH
```

启动及停止 Redis 服务命令如下。

```
nohup /usr/local/redis/bin/redis-server  /usr/local/redis/redis.conf  &
```

② 安装 Python Redis 模块。

```
easy_install pip
pip install redis
```

③ Ansible 整合 Redis 配置。

在配置文件/etc/ansible/ansible.cfg 的 defaluts 段中加入代码，如果 Redis 密码为 admin，则开启 admin 密码行。

```
gathering=smart
fact_caching=redis
fact_caching_timeout=86400
fact_caching_connection=localhost:6379
#fact_caching_connection=localhost:6379:0:admin
```

④ 测试 Redis 缓存。

Ansible Playbook 执行 nginx_wget.yaml 剧本文件的代码如下，如图 5-38 所示。

```
ansible-playbook    nginx_wget.yaml
```

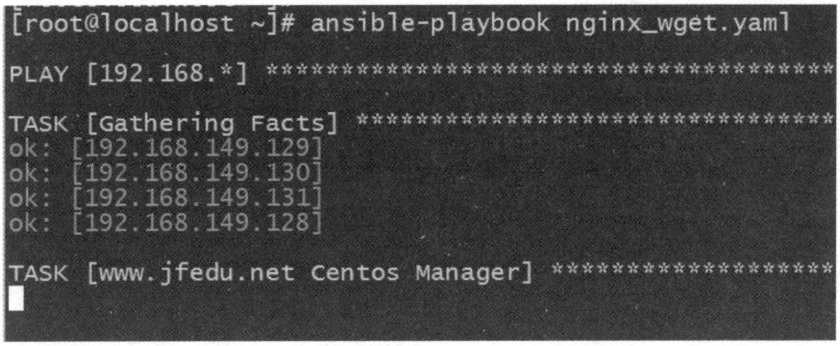

```
[root@localhost ~]# ansible-playbook nginx_wget.yaml

PLAY [192.168.*] ************************************

TASK [Gathering Facts] ******************************
ok: [192.168.149.129]
ok: [192.168.149.130]
ok: [192.168.149.131]
ok: [192.168.149.128]

TASK [www.jfedu.net Centos Manager] *****************
```

图 5-38 Ansible Playbook 执行 YAML

检查 Redis 服务器，facts key 已存入 Redis，如图 5-39 所示。

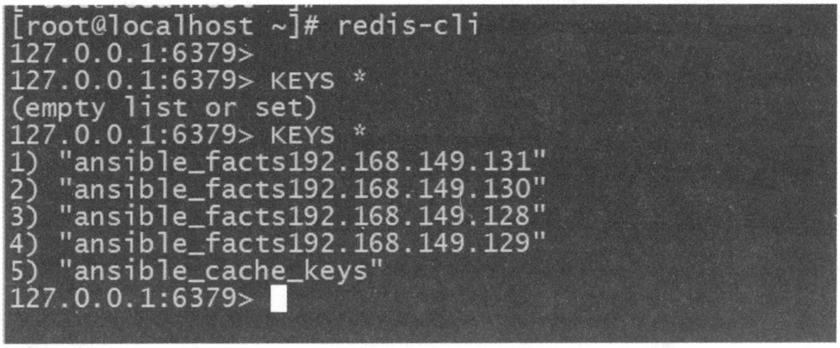

```
[root@localhost ~]# redis-cli
127.0.0.1:6379>
127.0.0.1:6379> KEYS *
(empty list or set)
127.0.0.1:6379> KEYS *
1) "ansible_facts192.168.149.131"
2) "ansible_facts192.168.149.130"
3) "ansible_facts192.168.149.128"
4) "ansible_facts192.168.149.129"
5) "ansible_cache_keys"
127.0.0.1:6379>
```

图 5-39 Redis 缓存服务器缓存 facts 主机信息

（5）ControlPersist SSH 优化。

ControlPersist 特性需要高版本的 SSH 支持，CentOS 6 操作系统默认是不支持的，如果要使用，则需要自行升级 OpenSSH。

ControlPersist，即持久化的 Socket，一次验证多次通信，只需要修改 SSH 客户端配置即可。

可使用 YUM 或源码编译升级 OpenSSH 服务，升级完毕后 ControlPersist 的设置方法如下，在其用户的主目录下创建 config 文件，如果 Ansible 以 root 用户登录客户端，则在客户端的/root/.ssh/config 文件中添加如下代码即可。

```
Host *
  Compression yes
  ServerAliveInterval 60
  ServerAliveCountMax 5
  ControlMaster auto
  ControlPath ~/.ssh/sockets/%r@%h-%p
  ControlPersist 4h
```

开启 ControlPersist 特性，SSH 在建立 sockets 后，节省了每次验证和创建的时间，对 Ansible 执行速度的提升是非常明显的。

第 6 章　SaltStack 自动运维企业实战

本章将介绍 SaltStack 工作原理、SaltStack 安装配置、生产环境模块讲解、SaltStack 企业场景案例、SaltStack SLS 案例实战等内容。

6.1　SaltStack 运维工具特点

SaltStack 与 Puppet 均属于 C/S 模式，需安装服务器与客户端。它是基于 Python 语言编写，并加入了 MQ 消息同步，这样可以使执行命令和执行结果高效返回，但其执行过程需等待客户端信息全部返回，如果客户端未及时返回信息或未响应，可能会导致部分机器没有执行结果。

6.2　SaltStack 运维工具简介

SaltStack 是一个服务器基础架构集中化管理平台，具备配置管理、远程执行和监控等功能，是基于 Python 语言实现，并结合轻量级消息队列（ZeroMQ）与 Python 第三方模块（PyZMQ、PyCrypto、Pyjinjia2、python-msgpack 和 PyYAML 等）构建的。

通过部署 SaltStack 平台，可以在成千上万台服务器上批量执行命令，根据不同业务进行配置集中化管理、分发文件、采集服务器数据、管理操作系统及软件包等。SaltStack 是运维人员提高工作效率、规范业务配置与操作的利器。

SaltStack 有一个配置管理系统，能够将远程节点维护在一个预定义的状态（如确保安装特定的软件包并运行特定的服务）。其底层采用动态的连接总线，使其可以用于编排、远程执行和配置管理等。

在大规模部署和小型系统之间提供多功能性似乎令人生畏，但无论项目规模如何，SaltStack

的设置和维护都非常简单。SaltStack 的体系结构旨在与任意数量的服务器协同工作，即从少数本地网络系统到跨不同数据中心的国际化部署。拓扑结构是一个简单的 C/S 模型，其中所需的功能都内置于一组守护进程中。虽然默认配置几乎不需要修改，但可以对 SaltStack 进行微调以满足特定需求。

SaltStack 利用了许多技术，网络层使用了优秀的 ZeroMQ 网络库构建，因此 SaltStack 守护程序包含了一个可行且透明的 AMQ 代理。SaltStack 使用公钥与 Master 守护程序进行身份验证，然后使用更快的 AES 加密算法对负载通信进行加密。

身份验证和加密都是 SaltStack 的组成部分。SaltStack 通过利用 Msgpack 数据编码格式进行通信，实现了更加快速轻便的网络流量。

6.3 SaltStack 运维工具原理

本书以 SaltStack 为案例，基于 SaltStack 构建企业自动化运维平台，实现大规模服务器的快速管理和部署。SaltStack 将平常复杂的配置工作变得简单、标准化且容易控制。

SaltStack 自动运维管理工具有以下优点。

（1）轻量级，只需要在操作机上更新。
（2）控端（Master）与被控端（Minion）基于证书认证。
（3）支持 API 及自定义 Python 模块，轻松实现功能扩展。
（4）尽可能使用最小和最快的网络负载。
（5）提供简单的编程接口。
（6）系统不仅可以通过主机名，还可以通过其系统属性进行定位。
（7）需要客户端安装 Agent，典型的 C/S 架构。
（8）使用 Python 编写，维护更简单。

SaltStack 客户端在启动时，会自动生成一套密钥，包含私钥和公钥。然后客户端将公钥发送给服务器，服务器验证并接收公钥，以此建立可靠且加密的通信连接。同时通过消息队列 ZeroMQ 在客户端与服务器之间建立消息发布连接。

Minion 是 SaltStack 需要管理的客户端安装组件，会主动连接 Master，并从 Master 得到资源状态信息，同步资源管理信息。Master 作为控制中心运行在主机服务器上，负责 SaltStack 命令运行和资源状态的管理。

ZeroMQ 是一款开源的消息队列软件，用于在 Minion 与 Master 之间建立系统通信桥梁。SaltStack 自动运维管理工具工作原理拓扑如图 6-1 所示。

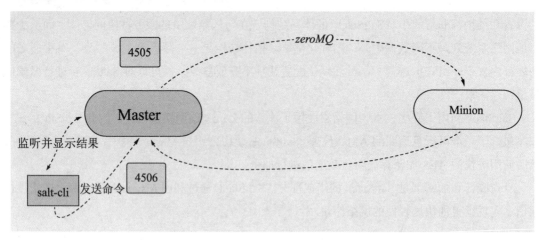

图 6-1 SaltStack 自动运维管理工具工作原理拓扑

6.4 SaltStack 平台配置实战

部署 SaltStack 自动化运维平台，至少需要准备两台服务器，一台 Master 节点和一台 Minion 节点，如下所示。

```
Salt Master 节点：192.168.1.145
Salt Minion 节点：192.168.1.146
```

6.5 SaltStack 节点 Hosts 及防火墙设置

对 Master 节点和 node1 节点进行如下配置。

```
#添加 hosts 解析
cat >/etc/hosts<<EOF
127.0.0.1 localhost localhost.localdomain
192.168.1.145 master
192.168.1.146 node1
EOF
#临时关闭 selinux 和防火墙
sed -i '/SELINUX/s/enforcing/disabled/g'  /etc/sysconfig/selinux
setenforce 0
systemctl    stop     firewalld.service
systemctl    disable  firewalld.service
#firewall-cmd --permanent --zone=public --add-port=4505-4506/tcp
#同步节点时间
yum install ntpdate -y
ntpdate pool.ntp.org
```

```
#修改对应节点主机名
hostname 'cat /etc/hosts|grep $(ifconfig|grep broadcast|awk '{print $2}'
|tail -1)|awk '{print $2}'';su
```

6.6 SaltStack 管理工具安装配置

SaltStack 可以工作在 Linux、BSD、macOS X 等操作系统上，对 Python 环境的版本要求为 Python 2.6 以上，如果操作系统的 Python 软件版本为 2.4，那么需要升级才能使用 SaltStack 工具。

（1）Red Hat 操作系统和 CentOS 操作系统可以直接基于 YUM 工具自动安装 SaltStack，CentOS 6.x 操作系统或 CentOS 7.x 操作系统在安装前，需先安装 SaltStack 源，其代码如下。

```
yum install -y https://repo.saltstack.com/py3/redhat/salt-py3-repo-latest.
el7.noarch.rpm
yum clean all
yum install -y salt-master salt-minion zeromq*
systemctl enable salt-master.service
systemctl enable salt-minion.service
systemctl restart salt-master.service
systemctl restart salt-minion.service
```

（2）SaltStack 工具默认配置文件：/etc/salt/master/和/etc/salt/minion，其中 master 为主配置文件，minion 为客户端配置文件，如图 6-2 所示。

```
[root@www-jfedu-net ~]#
[root@www-jfedu-net ~]# cd /etc/salt/
[root@www-jfedu-net salt]#
[root@www-jfedu-net salt]# ls
cloud            cloud.deploy.d   cloud.profiles.d    master
cloud.conf.d     cloud.maps.d     cloud.providers.d   master.d
[root@www-jfedu-net salt]#
[root@www-jfedu-net salt]#
[root@www-jfedu-net salt]#
```

图 6-2 SaltStack 主目录信息

（3）SaltStack 远程批量管理，首先要在所有的 Minion 配置文件中，指定 Master 的 IP 地址或者主机名，其代码如下，如图 6-3 所示。

```
sed -i -e '/master:/s/salt/master/g' -e '/master:/s/#//g' /etc/salt/minion
systemctl restart salt-minion.service
```

（4）启动 Minion 服务后，会产生一个密钥对，Minion 会根据配置的 Master 地址或者主机名连接 Master，并尝试将公钥发给 Master，Minion_id 表示 Minion 的身份。密钥认证后，Master 和 Minion 就可以通信了。此时可以通过 State 模块来管理 Minion。Minion 密钥存储位置为 /etc/salt/pki/minion/，Master 密钥存储位置为/etc/salt/pki/master/minions。

```
# as the main minion config file).
#default_include: minion.d/*.conf

# Set the location of the salt master server. If the master server
# resolved, then the minion will fail to start.
master: master

# Set http proxy information for the minion when doing requests
#proxy_host:
#proxy_port:
#proxy_username:
```

图 6-3 Minion 节点文件内容

（5）在 Master 执行指令 salt-key -A 并接收所有 Minion。salt-key 指令常见的参数和含义如图 6-4 所示。

```
[root@master ~]# salt-key -A
The following keys are going to be accepted:
Unaccepted Keys:
master
node1
Proceed? [n/Y] y
Key for minion master accepted.
Key for minion node1 accepted.
[root@master ~]#
[root@master ~]#
[root@master ~]#
```

图 6-4 salt-key 指令常见的参数和含义

6.7 SaltStack 工具参数详解

通过 SaltStack 自动运维工具管理客户端，由于 SaltStack 管理远程服务器基于 SSH，在登录远程服务器执行命令时需要远程服务器的用户名和密码，也可以加入-k 参数手动输入密码或者基于 ssh-keygen 生成免密钥。

SaltStack 自动化批量管理工具主要参数如下。

```
salt-key -L                              #列出所有 Minion 上的密钥
salt-key -a <证书名>                      #接收单个 Minion 证书
salt-key -d <证书名>                      #删除单个 Minion 证书
salt-key -D                              #删除所有 Minion 证书
salt-key -A                              #接收所有未验证的 Minion 证书
*                                        #指定 Minion(*代表所有 Minion)
salt '*' test.ping                       #test.ping 用来检测 Minion 是否连接正常
salt '*' disk.usage                      #disk.usage 用来查看磁盘使用情况
salt '*' network.interfaces              #列出 Minion 上的所有接口及其 IP 地址、子网掩码、
                                         #MAC 地址等
salt '*' cmd.run 'ls -l /etc'            #cmd.run       #查看/etc/文件和文件夹
salt '*' cmd.run 'yum install ntpdate -y' #安装 ntpdate 软件包
```

```
salt '*' pkg.version python              #显示软件包版本信息
salt '*' pkg.install vim                 #使用 YUM 安装包
salt 'node1' service.status mysql        #查看 MySQL 服务状态。也可以用 cmd.run,
                                         #效果是一样的
salt 'node[0-9]' cmd.run 'df -h'         #可以使用正则表达式
salt -L 'master,node1' cmd.run 'df -h'   #可以指定列表
salt -C 'G@os:Ubuntu and webser* or E@database.*' test.ping
                                         #在一个命令中混合使用多个选项
salt -G 'os:Ubuntu' test.ping            #可以使用 Grains 系统通过 Minion 的系统
                                         #信息进行过滤
salt-run manage.up                       #显示存活的 Minion
salt-run manage.down                     #查看死机的 Minion
salt-run manage.down removekeys=True     #查看死机的 Minion,并将其删除
salt-run manage.status                   #查看 Minion 的相关状态
salt-run manage.versions      #查看 SlatStack 的所有 Master 和 Minion 的版本信息
salt "*" cmd.script salt://shell.sh      #执行服务器的脚本;//注:默认 SaltStack
                                         #的脚本仓库目录在/srv/salt;
salt "*" cp.get_file salt://shell.sh /tmp/shell.sh
                      #复制文件到客户端,注:在复制文件时,如目标客户端目录不存在,可
                      #以在后面加上参数 makedirs=True,则会自动创建目录
salt "*" cp.get_dir salt://jfedu /tmp    #复制目录到客户端相应的目录
salt '*' file.copy /tmp/jfedu /tmp/jfedu #把 Master 端对应文件复制到
                                         #Minion 相应目录下
```

6.8 SaltStack ping 模块实战

SaltStack 最基础的模块为 ping 模块,主要用于判断远程客户端是否在线,用于 ping 本身服务器,返回值为 changed 和 ping。

SaltStack ping 服务器状态如图 6-5 所示。

```
salt '*' test.ping
salt '*' cmd.run 'ping -c1 www.baidu.com'
```

图 6-5 SaltStack ping 服务器状态

6.9 SaltStack cmd 模块实战

SaltStack cmd 模块为 SaltStack 的默认模块，主要用于执行 Linux 基础命令，可以执行远程服务器命令和任务等操作。SaltStack cmd 模块的企业常用案例如下。

（1）SaltStack cmd 模块远程执行 date 命令，执行结果如图 6-6 所示。

```
salt '*' cmd.run 'date'
```

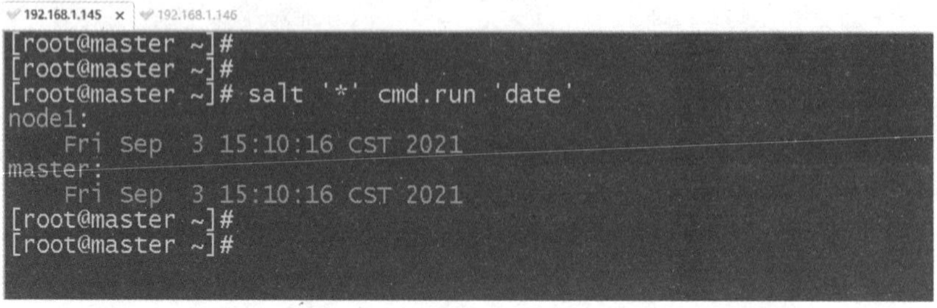

图 6-6 SaltStack cmd 模块远程执行 date 命令执行结果

（2）SaltStack cmd 模块远程执行 ping 命令，执行结果如图 6-7 所示。

```
salt '*' cmd.run 'ping -c1 www.jd.com'
```

图 6-7 SaltStack cmd 模块远程执行 ping 命令执行结果

（3）SaltStack Hosts 正则模式远程执行 df -h 命令，执行结果如图 6-8 所示。

```
salt '*' cmd.run 'df -h'
```

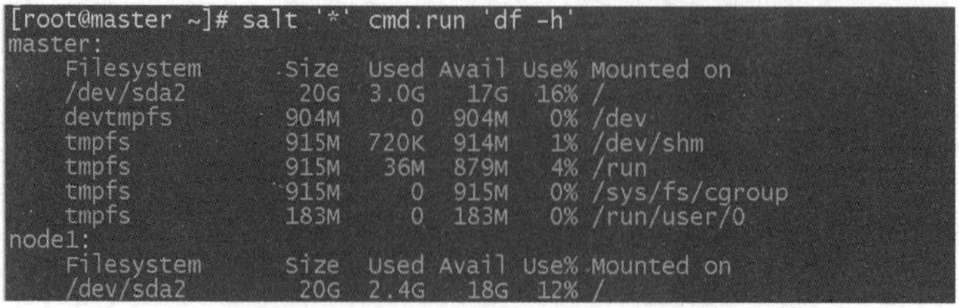

图 6-8 SaltStack cmd 正则模式远程执行 df -h 命令执行结果

6.10 SaltStack copy 模块实战

SaltStack copy 模块主要用于文件复制或目录复制，支持复制文件、目录、权限和用户。SaltStack copy 模块的企业常用案例如下。

（1）SaltStack copy 模块操作代码如下，执行结果如图 6-9 所示。

```
salt-cp '*' kube-flannel.yml /tmp/
```

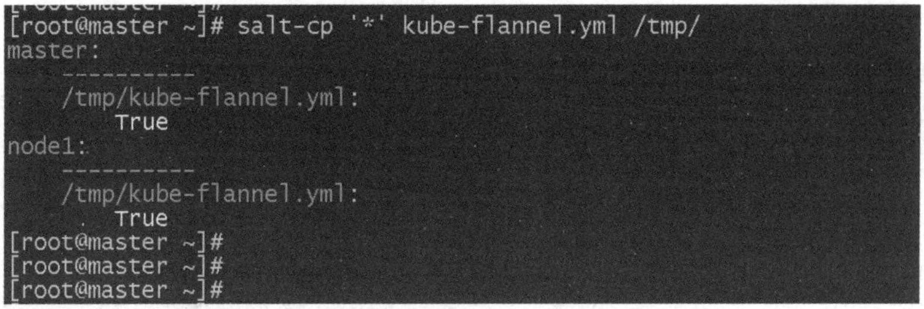

图 6-9　SaltStack copy 模块文件复制执行结果

（2）SaltStack copy 模块操作也支持目录复制，其代码如下，执行结果如图 6-10 所示。

```
salt-cp '*' /srv/salt/2021/ /tmp/ --chunked
salt "*" cp.get_file salt://kube-flannel.yml /tmp/
```

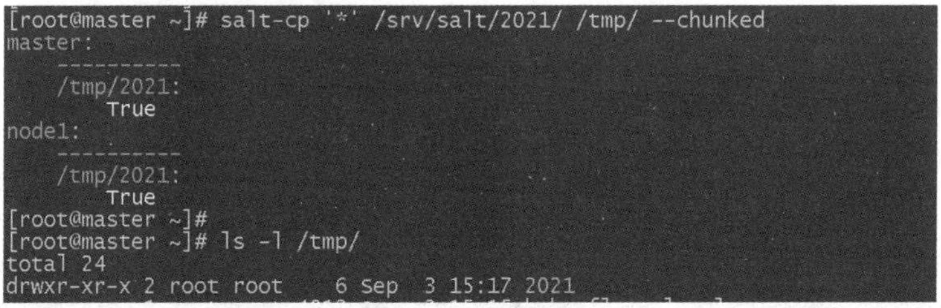

（a）

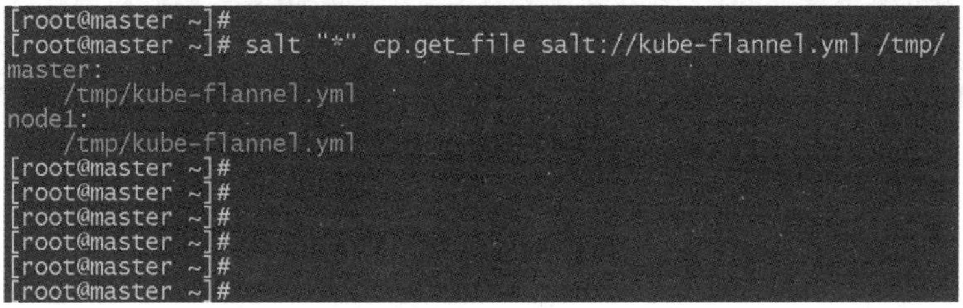

（b）

图 6-10　SaltStack copy 模块目录复制执行结果

（a）查看/tmp 目录内容；（b）将 kube-flannel.yml 文件复制到/tmp 目录

6.11 SaltStack pkg 模块实战

SaltStack pkg 模块主要用于软件的安装、升级和卸载,支持红帽软件,相当于 CentOS YUM 模块。企业常用案例如下。

(1) SaltStack pkg 模块操作,如远程安装 ntpdate 软件工具包的代码如下,执行结果如图 6-11 所示。

```
salt '*' pkg.install ntpdate
```

```
[root@master ~]# salt '*' pkg.install ntpdate
master:
    ----------
    ntpdate:
        ----------
        new:
            4.2.6p5-29.el7.centos.2
        old:
node1:
    ----------
    ntpdate:
        ----------
        new:
```

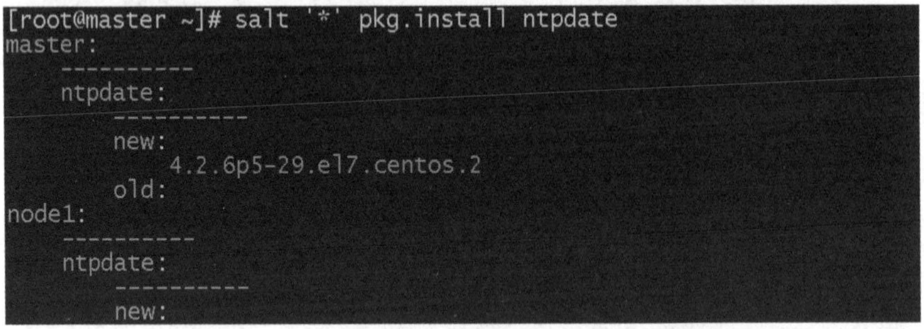

图 6-11　SaltStack YUM 安装软件包执行结果

(2) SaltStack pkg 模块操作,如远程卸载 ntpdate 软件工具包的代码如下,执行结果如图 6-12 所示。

```
salt '*' pkg.remove ntpdate
```

```
[root@master ~]# salt '*' pkg.remove ntpdate
master:
    ----------
    ntpdate:
        ----------
        new:
        old:
            4.2.6p5-29.el7.centos.2
node1:
    ----------
    ntpdate:
```

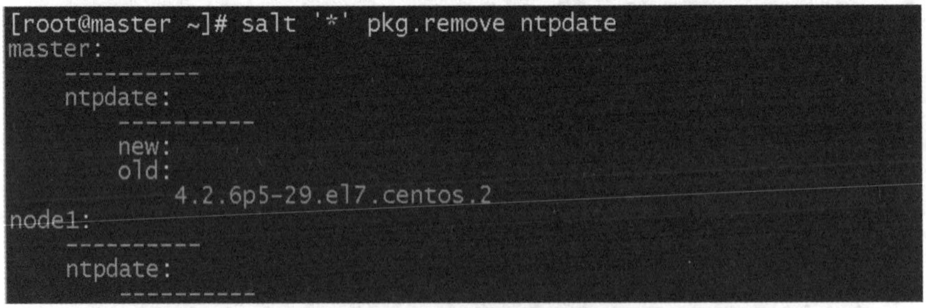

图 6-12　SaltStack YUM 卸载软件包执行结果

6.12 SaltStack service 模块实战

SaltStack service 模块主要用于对远程客户端进行各种服务管理,包括启动、停止、重启和重新加载等,SaltStack service 模块的企业常用案例如下。

(1) SaltStack service 模块操作,如远程重启 httpd 服务的代码如下,执行结果如图 6-13

所示。

```
salt 'node1' service.restart httpd
```

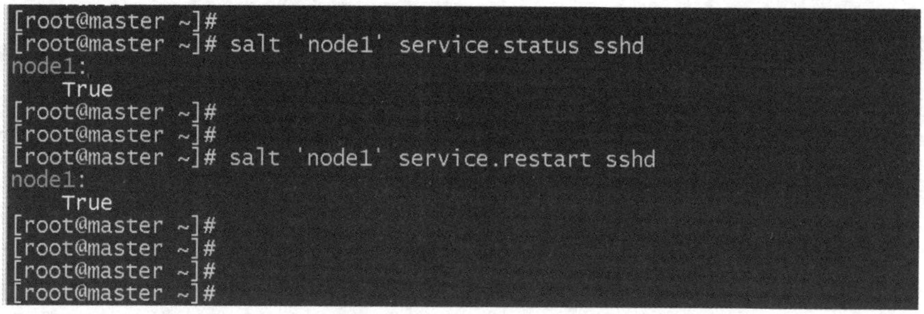

图 6-13　SaltStack service 重启 httpd 服务执行结果

（2）SaltStack service 模块操作，如远程停止 ntpd 服务，代码如下，执行结果如图 6-14 所示。

```
salt '*' service.start ntpd
salt '*' service.stop ntpd
```

图 6-14　SaltStack service 远程停止 ntpd 服务执行结果

6.13　SaltStack 配置文件详解

SaltStack 默认配置文件为/etc/SaltStack/Saltstack.cfg，配置文件中可以对 SaltStack 各项参数进行调整，包括并发线程、用户、模块路径和配置优化等。以下为 SaltStack.cfg 配置文件的常用参数详解。

```
interface: 192.168.1.145              #绑定到本地的某个网络地址
publish_port: 4505                    #默认端口 4505,设置 Master 与 Minion 通信端口
user: root                            #运行 SaltStack 进程的用户
max_open_files: 100000                #Master 可以打开的最大句柄数
worker_threads: 5                     #启动用来接收或应答 Minion 的线程数
ret_port: 4506                        #Master用来发送命令或接收Minions的命令执行返回
                                      #信息
```

```
    pidfile: /var/run/salt-master.pid    #指定Master的pid文件位置
    root_dir: /                          #该目录为SaltStack运行的根目录,改变它可以使
                                         #SaltStack在另外一个目录运行,好比chroot
    pki_dir: /etc/salt/pki/master        #存放pki认证密钥
    cachedir: /var/cache/salt            #存放缓存信息,SaltStack工作执行的命令信息
    verify_env: True                     #启动验证和设置权限配置目录
    keep_jobs: 24                        #保持工作信息的过期时间,单位为小时
    job_cache: True                      #设置Master维护的工作缓存。当Minions超过5000
                                         #台时,它将很好地承担这个大的架构
    timeout: 5                           #Master命令行可以接受的延迟时间
    output: nested                       #SaltStack命令的输出格式
    minion_data_cache: True              #关于Minion信息存储在Master上的参数,主要是
                                         #pilar数据和grains数据
    auto_accept: False                   #默认值为False。Master自动接收所有发送公钥的
                                         #Minion
    file_recv: False                     #允许Minion推送文件到Master上
    file_recv_max_size: 100              #默认值为100,设置一个hard-limit文档大小推送
                                         #到Master端
    state_top: top.sls                   #状态入口文件
    renderer: yaml_jinja                 #使用渲染器渲染Minions的状态数据
    failhard: False                      #当单个的状态执行失败后,将通知所有的状态停止运行
```

6.14 SaltStack State 自动化实战

SLS（SaltStack State 文件）是 SaltStack State 系统的核心。SLS 描述了系统的目标状态，由格式简单的数据构成。这经常被称作配置管理。

Top.sls 文件是配置管理的入口文件，一切都是从这里开始，在 Master 主机上，默认存放在 /srv/salt/ 目录下。Top.sls 文件默认从 base 标签开始解析执行，下一级是操作的目标，可以通过正则、grain 模块或分组名进行匹配，再下一级是要执行的 state 文件，不包括扩展名。

（1）创建/srv/salt/top.sls 文件，通过正则进行匹配的示例如下。

```
base:
  '*':
    - webserver
```

（2）通过分组名进行匹配的示例如下，必须要有 - match: nodegroup。

```
base:
  group1:
    - match: nodegroup
    - webserver
```

（3）通过 grain 模块匹配的示例如下，必须有- match: grain。

```
base:
  'os:Fedora':
    - match: grain
    - webserver
```

（4）准备好 top.sls 文件后，编写一个 state 文件，/srv/salt/webserver.sls 文件如下所示。

```
apache:                    #标签定义
  pkg:                     #state declaration
    - installed            #function declaration
```

注：第一行称为标签定义（ID declaration），在这里被定义为安装包的名。注意在不同发行版本中软件包命名不同。

第二行称为状态定义（state declaration），在这里被定义使用（pkg state module）。

第三行称为函数定义（function declaration），在这里被定义使用 installed 函数。

（5）最后可以在终端中执行如下命令，执行结果如图 6-15 所示。

```
salt '*' state.highstate
```

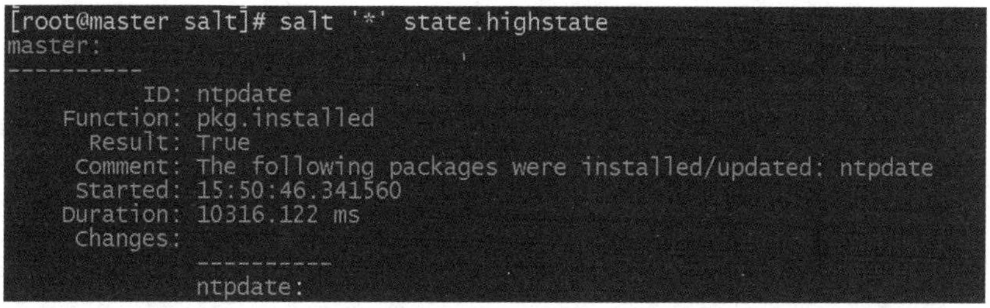

```
[root@master salt]# salt '*' state.highstate
master:
----------
          ID: ntpdate
    Function: pkg.installed
      Result: True
     Comment: The following packages were installed/updated: ntpdate
     Started: 15:50:46.341560
    Duration: 10316.122 ms
     Changes:
              ----------
              ntpdate:
```

（a）

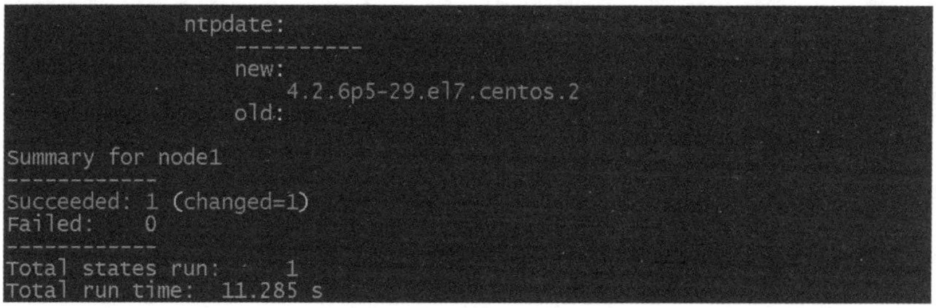

```
              ntpdate:
                  ----------
                  new:
                      4.2.6p5-29.el7.centos.2
                  old:

Summary for node1
------------
Succeeded: 1 (changed=1)
Failed:    0
------------
Total states run:     1
Total run time:  11.285 s
```

（b）

图 6-15 SaltStack SLS 案例实战

（a）SaltStack 执行 sls 内容；（b）查看 SaltStack 执行结果

采用 test=True 参数测试执行以下命令。

```
salt '*' state.highstate -v test=True
```

主控端对目标主机（targeted Minions）发出指令运行 state.highstatem 模块，目标主机首先会对 top.sls 文件下载并解析，然后按照 top.sls 文件匹配规则内定义的模块，将被下载、解析并执行的结果反馈给 Master。

（6）以上操作需要读取全局 top.sls 文件，有的情况需自定义读取某个 SLS 文件，可以在终端中执行如下命令，执行结果如图 6-16 所示。

```
salt '*' state.sls firewalld
```

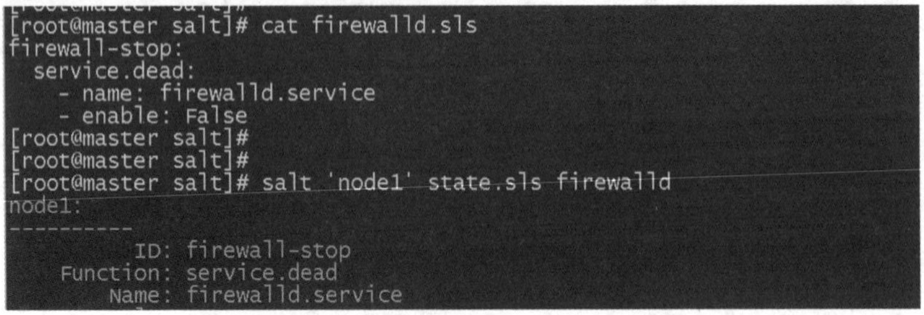

图 6-16　SaltStack SLS 文件案例实战

6.14.1　SLS 文件企业实战案例一

在企业生产环境中，Linux 操作系统安装完成之后，通常会将 DNS 设置为本地局域网的服务器 IP 地址，其操作方法和代码如下。

```
cat>dns.sls<<EOF
dns-config:
  file.managed:
    - name: /etc/resolv.conf
    - source: salt://init/files/resolv.conf
    - user: root
    - group: root
    - mode: 644
EOF
cp /etc/resolv.conf files/
```

6.14.2　SLS 文件企业实战案例二

在企业生产环境中，Linux 操作系统安装完成之后，通常会设置时间同步策略，安装 ntpd 服务，其操作方法和代码如下。

```
cat>ntp.sls<<EOF
ntp-install:
  pkg.installed:
    - name: ntpdate
```

```
cron-ntpdate:
  cron.present:
    - name: ntpdate poo.ntp.org
    - user: root
    - minute: 5
EOF
```

6.14.3 SLS 文件企业实战案例三

在企业生产环境中,Linux 操作系统安装完成之后,通常会设置 Selinux 策略,一般设置为关闭 Selinux 策略,其操作方法和代码如下。

```
mkdir -p /srv/salt/
mkdir -p init/files/
cd /srv/salt/init
cat>selinux.sls<<EOF
selinux-config:
  file.managed:
    - name: /etc/selinux/config
    - source: salt://init/files/selinux-config
    - user: root
    - group: root
    - mode: 0644
EOF
cp /etc/selinux/config files/selinux-config
```

6.14.4 SLS 文件企业实战案例四

在企业生产环境中,Linux 操作系统安装完成之后,通常会安装常见的软件包和工具,如 net-tools、gzip 等,其操作方法和代码如下。

```
cat>pkg-base.sls<<EOF
include:
  - init.yum-repo
base-install:
  pkg.installed:
    - pkgs:
      - screen
      - lrzsz
      - telnet
      - iftop
      - iotop
      - sysstat
      - wget
      - dos2unix
      - lsof
```

```
            - net-tools
            - unzip
            - zip
            - vim
        - require:
            - file: /etc/yum.repos.d/epel.repo
EOF
```

6.14.5　SLS 文件企业实战案例五

在企业生产环境中，Linux 操作系统安装完成之后，通常会进行 SSHD 服务优化，如关闭 DNS 反向查找等，其操作方法和代码如下。

```
cat>sshd.sls<<EOF
sshd-config:
  file.managed:
    - name: /etc/ssh/sshd_config
    - source: salt://init/files/sshd_config
    - user: root
    - gourp: root
    - mode: 0600
  service.running:
    - name: sshd
    - enable: True
    - reload: True
    - watch:
      - file: sshd-config
EOF
cp /etc/ssh/sshd_config files/
vim files/sshd_config
Port 6022
UseDNS no
PermitRootLogin no
PermitEmptyPasswords no
GSSAPIAuthentication no
```

6.14.6　SLS 文件企业实战案例六

在企业生产环境中，Linux 操作系统安装完成之后，通常会设置内核参数策略，如修改 Linux 系统最大文件数等，其操作方法和代码如下。

```
cat>limit.sls<<EOF
limit-config:
  file.managed:
    - name: /etc/security/limits.conf
    - source: salt://init/files/limits.conf
```

```
      - user: root
      - group: root
      - mode: 0644
EOF
cp /etc/security/limits.conf files/
echo "* - nofile 65535" >> files/limits.conf
```

6.14.7　SLS 文件企业实战案例七

在企业生产环境中，Linux 操作系统安装完成之后，通常会进行内核优化，如设置相关内核参数，其操作方法和代码如下。

```
cat>sysctl.sls<<EOF
net.ipv4.tcp_fin_timeout:
  sysctl.present:
    - value: 2
net.ipv4.tcp_tw_reuse:
  sysctl.present:
    - value: 1
net.ipv4.tcp_tw_recycle:
  sysctl.present:
    - value: 1
net.ipv4.tcp_syncookies:
  sysctl.present:
    - value: 1
net.ipv4.tcp_keepalive_time:
  sysctl.present:
- value: 600
EOF
```

6.14.8　SLS 文件企业实战案例八

在企业生产环境中，Linux 操作系统安装完成之后，通常会设置防火墙策略，如可以关闭 Firewalld，其操作方法和代码如下。

```
cat>firewalld.sls<<EOF
firewall-stop:
  service.dead:
    - name: firewalld.service
    - enable: False
EOF
```

第 7 章 企业邮件服务器实战

7.1 邮件服务器简介

邮件服务器是一种用来负责电子邮件收发管理的设备，自主构建的邮件服务器比网络上的免费邮箱更加安全和高效，因此邮件服务器几乎是每个公司必备的硬件之一。

邮件服务器构成了电子邮件系统的核心，每个收信人都有一个位于某个邮件服务器上的邮箱（mailbox），一个邮件消息的典型旅程是从发信人的用户代理开始，途经发信人的邮件服务器，中转到收信人的邮件服务器，然后投递到收信人的邮箱。

简单邮件传送协议（Simple Mail Transfer Protocol，SMTP）是因特网电子邮件系统首要的应用层协议。它使用由 TCP 提供的可靠的数据传输服务把邮件消息从发信人的邮件服务器传送到收信人的邮件服务器。

为了让用户的系统域名被正确解析为相应的服务器地址，邮件服务器需要在互联网上被识别和被查找到，这样邮件系统才能实现邮件的投递和接收。因此邮件服务器需要进行 DNS 设置，包括 MX 记录和 A 记录的设置。

整个邮件系统包括服务器和客户端，服务器基于 SMTP，客户端基于 POP3、IMAP 等协议。SMTP 监听端口为 TCP 25 端口，POP3 监听端口为 TCP 110 端口，IMAP 监听端口为 TCP 143 端口。

发送一封电子邮件信息时，信息会从一台服务器传递到另一台服务器，直到发送到收件人的电子邮件服务器。

更准确地说，信息是被发送到负责传输邮件的服务器，即邮件传输代理（Mail Transport Agent，MTA）经过若干 MTA 后，最终到达收件人的 MTA。在互联网上，MTA 之间使用 SMTP 进行通信，故称为 SMTP 服务器。

收件人的 MTA 会将电子邮件投递给邮件接收服务器，即邮件投递代理（Mail Delivery Agent，

MDA），MDA 会保存邮件并等待用户收取，如图 7-1 所示。MDA 主要有两种协议：邮局协议（Post Office Protocol，POP）和互联网应用协议（Internet Message Access Protocol，IMAP）。

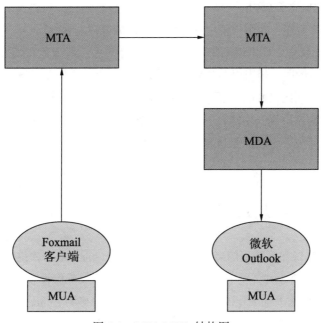

图 7-1　MTA-MUA 结构图

最简单的比喻就是：MTA 类似邮局，而 MDA 类似信箱，MDA 存储邮件并等待收件人检查信箱。发件人与收件人无须直接建立连接。为避免邮件被其他人看到，MDA 要验证用户名和密码才能访问。收取邮件的工作由一个叫作邮件用户代理（Mail User Agent，MUA）的程序完成。如果 MUA 是用户计算机或其他设备上的一个程序，则称它为邮件客户端（如 Mozilla Thunderbird、网易邮箱大师、Foxmail、微软的 Outlook）。

POP3 允许电子邮件客户端下载服务器上的邮件，但是客户端的操作（如移动邮件、标记已读等）不会反馈到服务器上。比如通过客户端收取了邮箱中的 3 封邮件并移动到其他文件夹，邮箱服务器上的这些邮件是没有同时被移动的，所以有很多用户反馈使用 Foxmail 客户端配置 POP3 收取邮件的时候，有时候非常快，而使用 IMAP 收取邮件则非常慢。

IMAP 提供 WebMail 与电子邮件客户端之间的双向通信，客户端的操作都会反馈到服务器上，服务器上的邮件也会做相应的动作。

目前主流免费邮箱服务商有网易邮箱、QQ 邮箱、新浪邮箱、搜狐邮箱等。个人和中小型公司可以直接使用网易邮箱或者 QQ 企业邮箱，因为维护一个邮件服务器也是一项非常庞大的工程。

Linux 平台开源免费的邮件服务器包括 Sendmail、Postfix、Q-mail，而 Windows 平台主要

为 Exchange 服务器（正版需要收费），后续章节主要是基于 Linux 平台构建企业独立的邮件服务器系统。

邮件服务器相关概念汇总如下。

MUA：邮件用户代理，即用户使用的写信、收信的客户端软件。

MTA：邮件传输代理，即常说的邮件服务器，用于转发、收取用户邮件。

MDA：邮件投递代理，相当于 MUA 和 MTA 的中间人，可用于过滤垃圾邮件。

POP：邮局协议，用于 MUA 连接服务器收取用户邮件，通信端口为 110。

IMAP：互联网应用协议，功能较 POP 更多，通信端口为 143。

SMTP：简单邮件传送协议，MUA 连接 MTA（或 MTA 连接 MTA）发送邮件时使用此协议，通信端口为 25。

7.2 Sendmail 安装配置

作为 Linux/UNIX 操作系统下的老牌邮件服务器，Sendmail 是一款免费的邮件服务器软件，已被广泛应用于各种服务器中，它在稳定性、可移植性，以及确保没有 Bug 等方面具有一定的特色，还可以在网络中搜索到大量的使用资料，是一款最经典的 Linux 系统下的邮件服务器。

（1）Sendmail 环境版本。

```
系统版本：CentOS 6.8 64 位；
Sendmail 版本：Sendmail-8.14；
Open WebMail 版本：openwebmail-2.53-6、openwebmail-data-2.53-6。
```

（2）Sendmail 安装的代码如下，如图 7-2 所示。

```
yum install sendmail* -y
```

（3）Sendmail 服务配置。

配置 sendmail.cf 服务，通过 local-host-names 来设置邮件服务器提供邮件服务的域名为 jfteach.com(jingfengjiaoyu.com)。

```
cp   /etc/mail/sendmail.mc /etc/mail/sendmail.mc.back
cp   /etc/mail/sendmail.cf /etc/mail/sendmail.cf.back
echo "jfteach.com" >>/etc/mail/local-host-names
```

配置 Sendmail 监听服务器网卡地址为 0.0.0.0。

```
sed -i 's/Addr = 127.0.0.1/Addr=0.0.0.0/g' /etc/mail/sendmail.mc
```

修改 vi /etc/mail/sendmail.mc 文件中的如下两行，开启 SMTP 的所有用户必须认证。

```
Dnl TRUST_AUTH_MECH('EXTERNAL DIGEST-MD5 CRAM-MD5 LOGIN PLAIN')dnl
Dnl define('confAUTH_MECHANISMS', 'EXTERNAL GSSAPI DIGEST-MD5 CRAM-MD5 LOGIN PLAIN')dnl
```

```
[root@localhost ~]# yum install sendmail*

Loaded plugins: fastestmirror
Setting up Install Process
Determining fastest mirrors
base
extras
updates
Resolving Dependencies
--> Running transaction check
---> Package sendmail.x86_64 0:8.14.4-9.el6_8.1 will be insta
--> Processing Dependency: procmail for package: sendmail-8.14
--> Processing Dependency: libhesiod.so.0()(64bit) for package
---> Package sendmail-cf.noarch 0:8.14.4-9.el6_8.1 will be ins
---> Package sendmail-devel.x86_64 0:8.14.4-9.el6_8.1 will be
---> Package sendmail-doc.noarch 0:8.14.4-9.el6_8.1 will be in
---> Package sendmail-milter.x86_64 0:8.14.4-9.el6_8.1 will be
```

（a）

```
  Verifying  : sendmail-devel-8.14.4-9.el6_8.1.x86_64
  Verifying  : sendmail-8.14.4-9.el6_8.1.x86_64
  Verifying  : procmail-3.22-25.1.el6_5.1.x86_64
  Verifying  : sendmail-milter-8.14.4-9.el6_8.1.x86_64
  Verifying  : sendmail-doc-8.14.4-9.el6_8.1.noarch
  Verifying  : sendmail-cf-8.14.4-9.el6_8.1.noarch

Installed:
  sendmail.x86_64 0:8.14.4-9.el6_8.1                          sen
  sendmail-devel.x86_64 0:8.14.4-9.el6_8.1                    sen
  sendmail-milter.x86_64 0:8.14.4-9.el6_8.1

Dependency Installed:
  hesiod.x86_64 0:3.1.0-19.el6                        procmail

Complete!
```

（b）

```
[root@localhost yum.repos.d]# yum install openwebmail
Loaded plugins: fastestmirror
Setting up Install Process
Loading mirror speeds from cached hostfile
openwebmail
openwebmail/primary_db
Resolving Dependencies
--> Running transaction check
---> Package openwebmail.noarch 0:2.53-6.el6 will be installe
--> Processing Dependency: openwebmail-data = 2.53-6.el6 for
--> Processing Dependency: perl-suidperl for package: openweb
--> Processing Dependency: perl-Text-Iconv for package: openw
--> Processing Dependency: perl-Compress-Zlib for package: op
--> Processing Dependency: perl-CGI for package: openwebmail-
--> Running transaction check
---> Package openwebmail-data.noarch 0:2.53-6.el6 will be ins
```

（c）

图 7-2　Sendmail 安装过程

（a）安装 Sendmail；（b）查看 Sendmail 是否安装成功；（c）安装 Open WebMail

将 vi/ etc/mail/sendmail.mc 文件修改为：

```
TRUST_AUTH_MECH('EXTERNAL DIGEST-MD5 CRAM-MD5 LOGIN PLAIN')dnl
define('confAUTH_MECHANISMS', 'EXTERNAL GSSAPI DIGEST-MD5 CRAM-MD5 LOGIN PLAIN')dnl
```

即去掉首行的 Dnl。Sendmail 配置完毕。

Sendmail.mc 文件修改完毕后，使用 m4 命令生成 sendmail.cf 主配置文件。

```
m4 sendmail.mc >sendmail.cf
/etc/init.d/sendmail restart
```

（4）配置 SMTP 认证。

saslauthd 服务作用：提供 SMTP 用户验证，检查用户名和密码是否正确，并基于系统 shadow 文件验证配置。

```
service saslauthd restart
```

7.3 Dovecot 服务配置

Dovecot 是一个开源的 IMAP 和 POP3 邮件服务器，支持 Linux/UNIX 系统。作为 IMAP 和 POP3 服务器，Dovecot 为 MUA 提供了一种访问服务器上存储邮件的方法。但是 Dovecot 并不负责从其他邮件服务器接收邮件（类似 MDA）。

Dovecot 只是将已存储在邮件服务器上的邮件通过 MUA 显示出来，IMAP 和 POP3 是用于连接 MUA 与邮件存储服务器的两种常见的协议。

Dovecot 安装命令如下。

```
yum install dovecot* -y
```

去掉/etc/dovecot.conf 文件中如下行命令前面的#号。

```
#protocols = imap pop3 lmtp
```

（1）配置 Dovecot 禁止 SSL，并设置邮箱。

```
vim /etc/dovecot/conf.d/10-ssl.conf
ssl = no
vim /etc/dovecot/conf.d/10-auth.conf
disable_plaintext_auth = no
vim /etc/dovecot/conf.d/10-mail.conf
mail_location = mbox:~/mail:INBOX=/var/mail/%u
```

（2）Sendmail 配置完毕。

Sendmail 建立邮箱用户后，可以用客户端收发邮件即表示正常。

```
groupadd mailgroup
useradd -g mailgroup -s /sbin/nologin jfedu
```

```
echo 123456|passwd --stdin jfedu
mkdir -p /home/jfedu/mail/.imap/INBOX
chown -R jfedu.jfedu /home/jfedu/
service sendmail restart
service dovecot restart
service saslauthd restart
```

7.4 Sendmail 别名配置

Sendmail 服务器中可以使用 aliases 机制实现邮件别名和邮件群发功能，也可以创建用户组，将用户加入某个组中，实现组邮件的群发。

在/etc/目录下存在 aliases 文件和 aliases.db 文件两个文件，aliases 文件是文本文件，其内容是可阅读和可编辑的，aliases.db 是数据库文件，由 aliases 文件生成。在/etc/aliases 文件中添加如下代码。

```
jfedu:   jf1,jf2
```

给 jfedu@jfteach.com 邮箱发送邮件，将群发到 jf1@jfteach.com 邮箱和 jf2@jfteach.com 邮箱。设置完成之后，通过 newaliases 命令生成新的 aliases.db 文件。

7.5 测试邮件收发

通过 Mail 命令发送邮件和收取邮件的前提是能 ping 通 jfteach.com 域名，可以添加 hosts 或者映射 DNS。

```
echo "This first test Mail"|mail -s "Test Mail Postfix" jfedu1@jfteach.com
```

测试邮件收发，如图 7-3 所示。

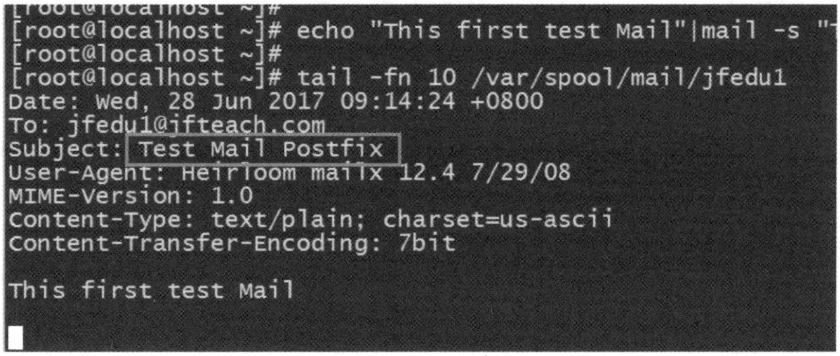

图 7-3 测试邮件收发

基于 Foxmail 的邮件收发，如图 7-4 所示。

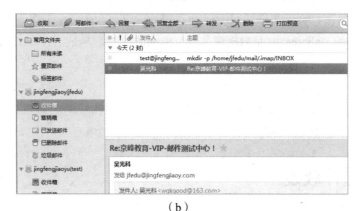

图 7-4　基于 Foxmail 的邮件收发

（a）设置邮箱账号和邮件服务器；（b）查看是否接收到邮件

7.6　配置 Open WebMail

配置完成 Sendmail 邮件服务器，即可通过 Outlook 和 Foxmail 收发邮件。如果需要通过 Web 页面收发邮件，那么可以基于 Open WebMail 实现。例如，访问邮件域名地址 http://mail.jfteach.com/，以用户名和密码登录；进行邮件的发送与收取，则需配置 Open WebMail 软件。

（1）Open WebMail 服务安装。

```
cd /etc/yum.repos.d
wget -q http://openwebmail.org/openwebmail/download/redhat/rpm/release/openwebmail.repo
yum install openwebmail -y
```

（2）Open WebMail 软件配置。

```
yum install httpd httpd-devel -y
/var/www/cgi-bin/openwebmail/openwebmail-tool.pl --init
```

```
vi  /var/www/cgi-bin/openwebmail/etc/openwebmail.conf
domainnames                 jfteach.com
default_language            zh_CN.GB2312
default_iconset             Cool3D.Chinese.Simplified
/var/www/cgi-bin/openwebmail/openwebmail-tool.pl --init
```

通过浏览器访问 Open WebMail 的界面，如图 7-5 所示。

（a）

（b）

（c）

图 7-5　通过浏览器访问 Open WebMail 的界面

（a）登录 Open WebMail；（b）发送测试邮件；（c）查看是否发送成功

Sendmail 外网域名配置方法如图 7-6 所示。

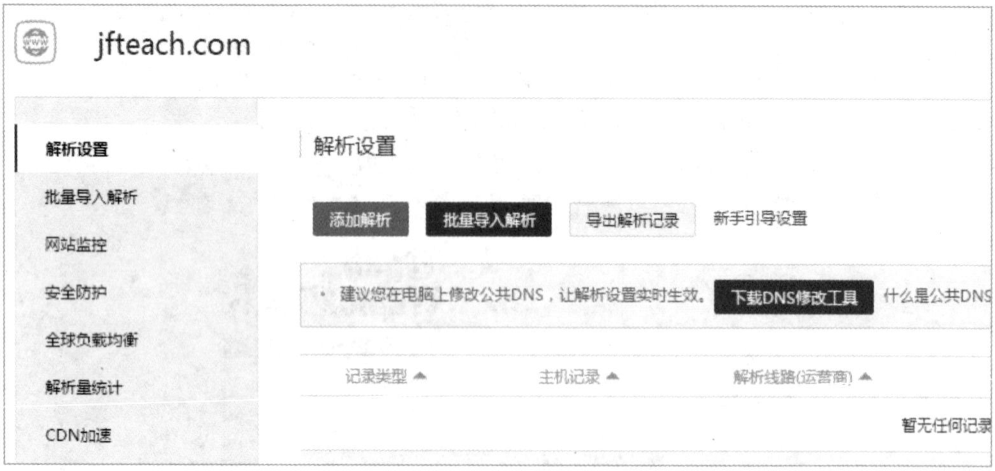

（a）

（b）

图 7-6 Sendmail 外网域名配置方法

（a）外网域名解析；（b）添加 MX 邮件记录

同时配置 Apache rewrite 规则，开启访问邮件域名地址 mail.jfteach.com 时直接访问 Open WebMail 页面的规则如下。

```
RewriteEngine On
ProxyPreserveHost On
RewriteRule ^/$ http://mail.jfteach.com/cgi-bin/openwebmail/openwebmail.pl [P,L,NC]
```

通过浏览器再次访问邮件域名地址，如图 7-7 所示。

图 7-7　再次访问邮件域名地址

7.7　Postfix 入门简介

Postfix 是 Wietse Venema 在 IBM 的 GPL 协议之下开发的 MTA 软件。Postfix 是 Wietse Venema 为使用最广泛的 Sendmail 提供替代品的一个尝试。

在互联网世界中，大部分的电子邮件都是通过 Sendmail 投递的，大约有 100 万用户使用 Sendmail，每天投递上亿封邮件。这是一个让人吃惊的数字。Postfix 试图更快、更容易管理、更安全，同时与 Sendmail 保持足够的兼容性。Postfix 的拓扑结构如图 7-8 所示。

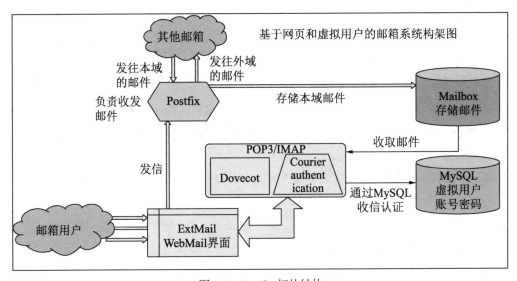

图 7-8　Postfix 拓扑结构

7.8 Postfix 服务安装

Postfix 服务安装代码如下。

```
yum -y install postfix*
```

7.9 Postfix 服务器配置

编辑 Postfix 主配置文件，命令为 vi /etc/postfix/main.cf。

以下配置案例域名为 jfteach.com，邮件服务器主机名为 mail.jfteach.com。

（1）修改 myhostname 参数。

```
myhostname = mail.jfteach.com
```

myhostname 参数即是邮件服务器的主机名称。

（2）修改 mydomain 参数。

```
mydomain = jfteach.com
```

mydomain 参数设定网域名称（domain name），主机名称通常建立在网域名称之内，例如，www.google.com 是网页服务，mail.google.com 则是邮件主机服务。域名通常是主机名称（hostname）去掉第一个点及其前面的文字部分，如 www.google.com 的域名是 google.com。

（3）修改 myorigin 参数。

```
myorigin = $mydomain
```

myorigin 参数是邮件地址中"@"后面的文字内容，如对于 wugk@163.com，163.com 即是 myorigin 参数。

（4）修改 SMTP 监听端口。

```
inet_interfaces = all
```

inet_interfaces 参数指定了 Postfix 系统监听的网络接口。Postfix 预设只会监听来自本机端所传出的封包，必须使用上述设定，才可以监听所有来自网络端的封包。

（5）修改 inet_protocols 参数。

修改 Postfix 的通信协定。目前网络的主流协定有 IPv4 与 IPv6，大部分情况都是使用 IPv4，如果 Mail Server 不需要使用 IPv6，可以做以下修改。

```
inet_protocols = ipv4
```

（6）修改 mydestination 参数。

```
mydestination = $myhostname, localhost.$mydomain, localhost, $mydomain
```

mydestination 参数设定了能够接收信件的主机名称，Postfix 预设只能收到设定的主机名称、网域名称以及本机端的信件，此步骤是增加能收信件的网络名称。

（7）设定信任用户端。

```
mynetworks = 127.0.0.0/8, 192.168.1.0/24, hash:/etc/postfix/access
```

mynetworks 参数设定了信任的用户端，寄信时会参考此值，若非信任的用户，则不会将信件转给其他 MTA 主机。

（8）设定 relay_domain（转发邮件域名）。

规范可以转发的 MTA 主机位址（收取邮件的程序为 MUA），通常直接设为 mydestination。

```
relay_domains = $mydestination
```

（9）设定邮件别名的路径。

检查 alias_maps 参数是否为以下设定。

```
alias_maps = hash:/etc/aliases
```

（10）设定指定邮件别名表资料库路径。

检查 alias_database 参数是否为以下设定。

```
alias_database = hash:/etc/aliases
```

（11）设定邮件主机使用权限、过滤机制及邮件别名。

执行以下命令。

```
#postmap   hash:/etc/postfix/access
#postalias hash:/etc/aliases
```

（12）重启 Postfix 服务。

Postfix 配置完毕，同样需要配置 Dovecot 和启动 saslauthd 服务后，才能进行邮件收发。

```
/etc/init.d/postfix    restart
/etc/init.d/saslauthd  restart
/etc/init.d/dovecot    restart
```

以下为 Postfix 邮件服务器 main.cf 配置文件的完整代码。

```
queue_directory = /var/spool/postfix
command_directory = /usr/sbin
daemon_directory = /usr/libexec/postfix
data_directory = /var/lib/postfix
mail_owner = postfix
myhostname = mail.jfteach.com
mydomain = jfteach.com
myorigin = $mydomain
inet_interfaces = all
inet_protocols = ipv4
mydestination = $myhostname, localhost.$mydomain, localhost, $mydomain
```

```
unknown_local_recipient_reject_code = 550
mynetworks = 0.0.0.0/0
relay_domains = $mydestination
alias_maps = hash:/etc/aliases
alias_database = hash:/etc/aliases
debug_peer_level = 2
debugger_command =
    PATH = /bin:/usr/bin:/usr/local/bin:/usr/X11R6/bin
    ddd $daemon_directory/$process_name $process_id & sleep 5
sendmail_path = /usr/sbin/sendmail.postfix
newaliases_path = /usr/bin/newaliases.postfix
mailq_path = /usr/bin/mailq.postfix
setgid_group = postdrop
html_directory = no
manpage_directory = /usr/share/man
sample_directory = /usr/share/doc/postfix-2.6.6/samples
readme_directory = /usr/share/doc/postfix-2.6.6/README_FILES
```

测试 Postfix 邮件收发如图 7-9 所示。

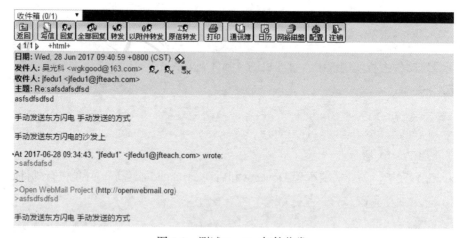

图 7-9　测试 Postfix 邮件收发

7.10　Foxmail 本地邮箱配置

（1）Linux 系统创建用户 jfedu1、用户 jfedu2，同时设置密码。

```
useradd  jfedu1 ;echo 123456|passwd --stdin jfedu1
useradd  jfedu2 ;echo 123456|passwd --stdin jfedu2
mkdir -p /home/jfedu1/mail/.imap/INBOX/
mkdir -p /home/jfedu2/mail/.imap/INBOX/
chown -R jfedu1.jfedu1 /home/jfedu1/
chown -R jfedu2.jfedu2 /home/jfedu2/
```

（2）Dovecot Smtp 验证设置。

```
vim /etc/dovecot/conf.d/10-auth.conf
disable_plaintext_auth = no
vim /etc/dovecot/conf.d/10-mail.conf
mail_location = mbox:~/mail:INBOX=/var/mail/%u
vim /etc/dovecot/conf.d/10-ssl.conf
ssl = no
```

（3）重启 Postfix 和 Dovecot 服务。

```
/etc/init.d/postfix   restart
/etc/init.d/dovecot   restart
```

在 Foxmail 客户端中设置 POP 服务器及 SMTP 服务器为 mail.jingfengjiaoyu.com，如图 7-10 所示。

（a） （b）

图 7-10　Foxmail 客户端设置

（a）设置邮件服务器地址；（b）邮箱账号添加完成

设置完毕后，发送测试邮件给邮箱 wgkgood@163.com，如图 7-11 所示。

图 7-11　Foxmail 邮件测试 1

登录邮箱 wgkgood@163.com 收取邮件，如图 7-12 所示。

图 7-12　Foxmail 邮件测试 2

从邮箱 wgkgood@163.com 发送邮件给邮箱 jfedu1@jingfengjiaoyu.com，收件界面如图 7-13 所示。

图 7-13　Foxmail 邮件测试 3

7.11　PostfixAdmin 配置

　　Postfix 一般管理均是基于命令行管理，对于运维人员来说比较麻烦。有没有 Postfix 图形界面管理工具呢？PostfixAdmin 就是为 Postfix 邮件服务器提供的图形界面管理工具，用它可以很方便地管理 Postfix 服务器。

以下为 PostfixAdmin 的安装配置方法。由于 Postfixadmin 是基于 PHP 语言编写的，所以需要安装 LAMP（LNMP、LEMP）环境，同时最新版本的 PostfixAdmin 需要 PHP 5.4 以上才能支持运行。

（1）安装 LAMP 及 PostfixAdmin 软件。

```
rpm -Uvh http://repo.webtatic.com/yum/el6/latest.rpm
yum remove php*
yum install php56w.x86_64 php56w-cli.x86_64 php56w-common.x86_64 php56w-gd.x86_64 php56w-ldap.x86_64 php56w-mbstring.x86_64 php56w-mcrypt.x86_64 php56w-mysql.x86_64 php56w-pdo.x86_64 -y
yum install httpd httpd-devel httpd-tools mysql mysql-devel mysql-server -y
wget https://jaist.dl.sourceforge.net/project/postfixadmin/postfixadmin/postfixadmin-3.1/postfixadmin-3.1.tar.gz
tar xzf postfixadmin-3.1.tar.gz -C /var/www/html/
cd /var/www/html/
mv postfixadmin-3.1/ postfixadmin
mkdir -p /var/www/html/postfixadmin/templates_c
chmod -R 777 /var/www/html/postfixadmin/templates_c
chown -R root.root /var/www/html/
```

（2）修改 PostfixAdmin 配置。

将 /var/www/html/postfixadmin/config.inc.php 文件中的 $CONF['configured'] 选项修改如下。

```
$CONF['configured'] = true;
```

（3）创建 Postfix 数据库。

```
create database postfix charset = utf8;
grant all on *.* to postfix@'localhost' identified by "postfixadmin";
flush privileges;
```

（4）创建 PostfixAdmin 管理员。

访问 PostfixAdmin 网页地址 http://113.209.20.234/postfixadmin/setup.php，如图 7-14 所示。

图 7-14　访问 PostfixAdmin 网页

修改 config.inc.php 配置文件并添加密码，其代码如下。PostfixAdmin 的界面操作如图 7-15 所示。

```
$CONF['setup_password'] = 'c11c1f22771ecc03d4507f366aeead44:3300e64c1a1e1
1a446ca2e964dadd4468ec163d3';
```

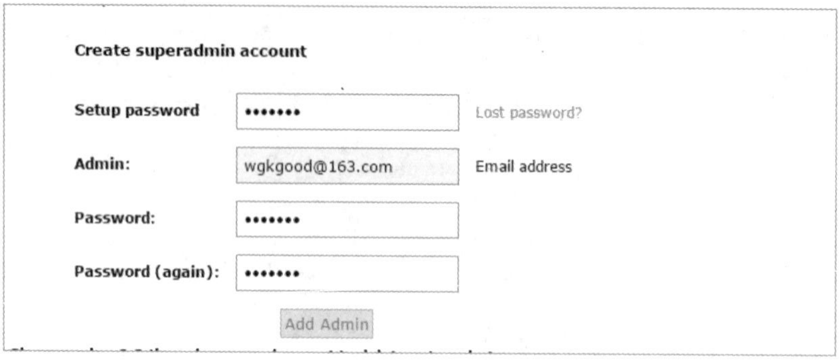

（a）

（b）

（c）

图 7-15　PostfixAdmin 的界面操作

（a）创建 PostfixAdmin 账号；（b）设置管理员账号；（c）新增管理员信息

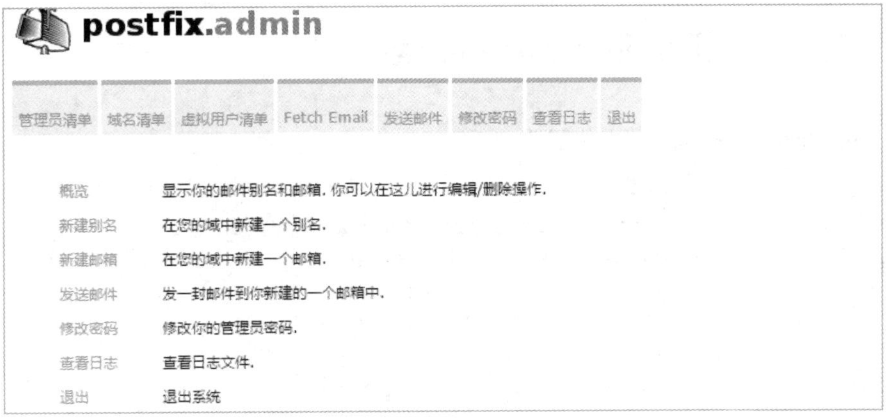

（d）

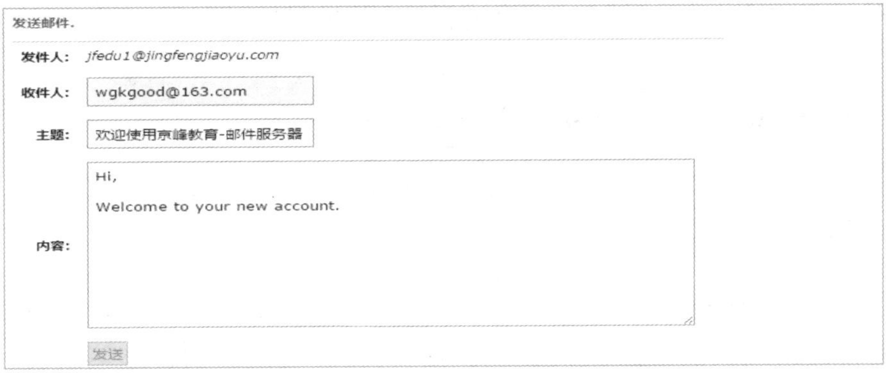

（e）

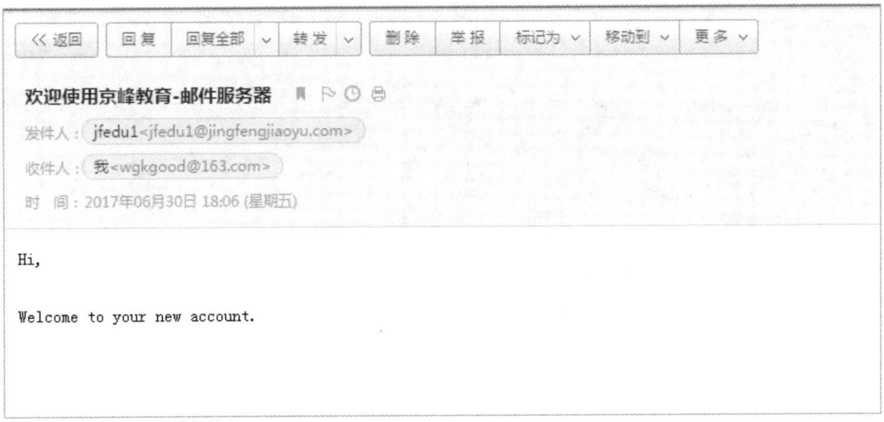

（f）

图 7-15　PostfixAdmin 的界面操作（续）

（d）postfixadm 清单；（e）发送测试邮件；（f）查看邮件

7.12 Roundcube GUI Web 配置

（1）Roundcube WebMail 的安装配置如下。

```
wget
https://jaist.dl.sourceforge.net/project/roundcubemail/roundcubemail/1.1.4/
roundcubemail-1.1.4-complete.tar.gz
tar -xzf roundcubemail-1.1.4-complete.tar.gz
mv roundcubemail-1.1.4 /var/www/html/webmail/
yum install php56w-dom epel-release libmcrypt* php56w-intl php56w-mbstring
php56w-ldap php56w-mcrypt -y
sed -i 's#;date.timezone =#date.timezone = PRC#g' /etc/php.ini
/etc/init.d/httpd restart
```

（2）配置 WebMail，访问 URL 地址 http://113.209.20.234/webmail/installer/index.php，如图 7-16 所示。

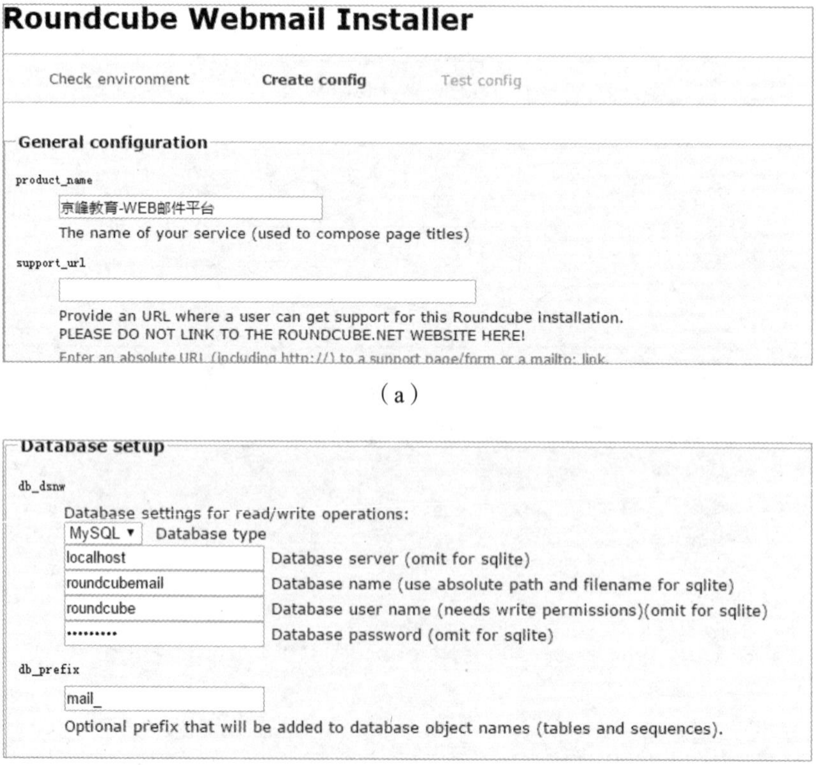

（a）

（b）

图 7-16 Roundcube 界面

（a）设置 Roundcube 名称；（b）设置 MySQL 数据库

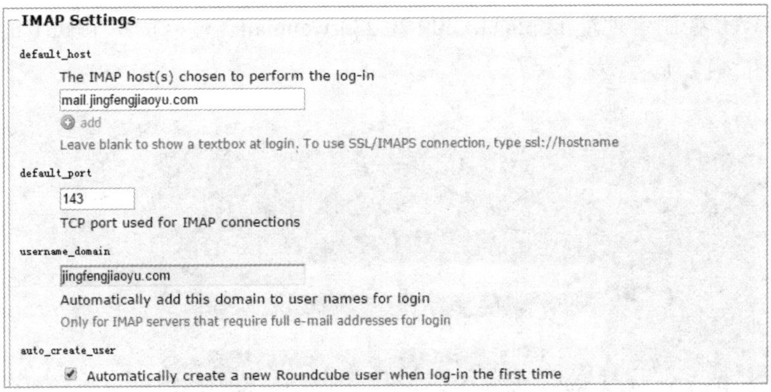

图 7-16　Roundcube 界面（续）

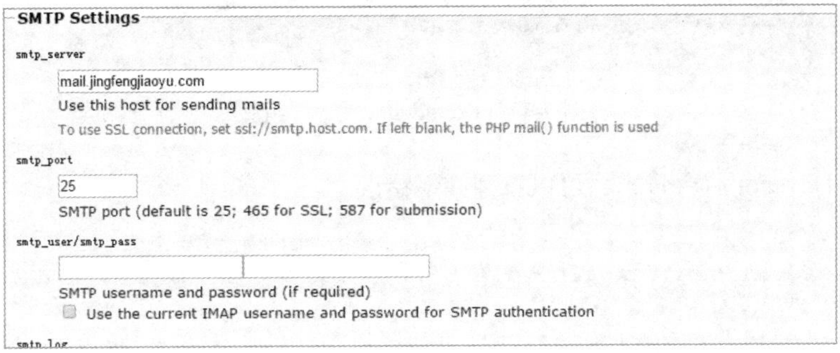

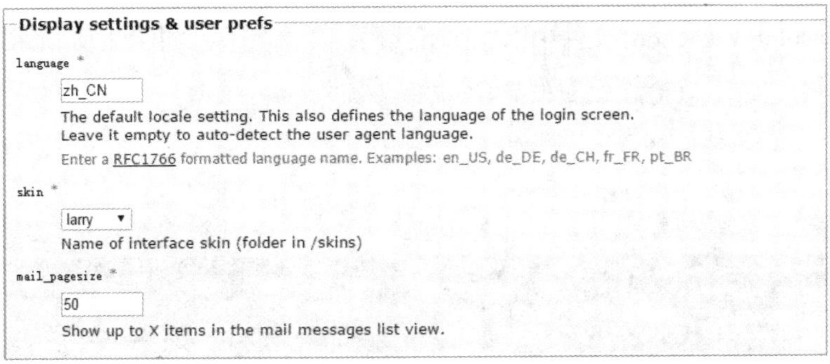

（c）填写邮件服务器地址；（d）填写 SMTP 地址和端口；（e）设置字符编码

（3）创建 WebMail 数据库信息的命令如下。

```
create database roundcubemail charset = utf8;
grant all on *.* to roundcube@'localhost' identified by "roundcube";
flush privileges;
chmod 777 -R temp/ logs/
```

（4）登录 Web 控制台地址 http://113.209.20.234/webmail，输入用户 jfedu1 的用户名和密码登录失败，如图 7-17 所示。

图 7-17　Roundcube 登录失败

（5）查看 Dovecot 后台日志，可以看到如下错误。

```
Jun 30 18:34:58 imap-login: Info: Disconnected (auth failed, 1 attempts):
user = <jfedu1@jingfengjiaoyu.com>, method = PLAIN, rip = 113.209.20.234,
lip = 192.168.0.3
Jun 30 18:35:17 imap-login: Info: Disconnected (auth failed, 1 attempts):
user = <jfedu2@jingfengjiaoyu.com>, method = PLAIN, rip = 113.209.20.234,
lip = 192.168.0.3
```

在/etc/dovecot/dovecot.conf 文件中添加如下代码，再次登录，如图 7-18 所示。

```
#表示去掉用户名后面其他信息,只保留用户名称,例如 jfedu1
auth_username_format = %n
```

图 7-18　Roundcube 登录成功

（6）创建测试邮件，如图 7-19 所示，然后发送邮件。

图 7-19　Roundcube 创建邮件

（7）登录 163 邮箱收取邮件，如图 7-20 所示。

图 7-20　Roundcube 邮件接收

（8）设置 Roundcube WebMail 界面风格，如图 7-21 所示。

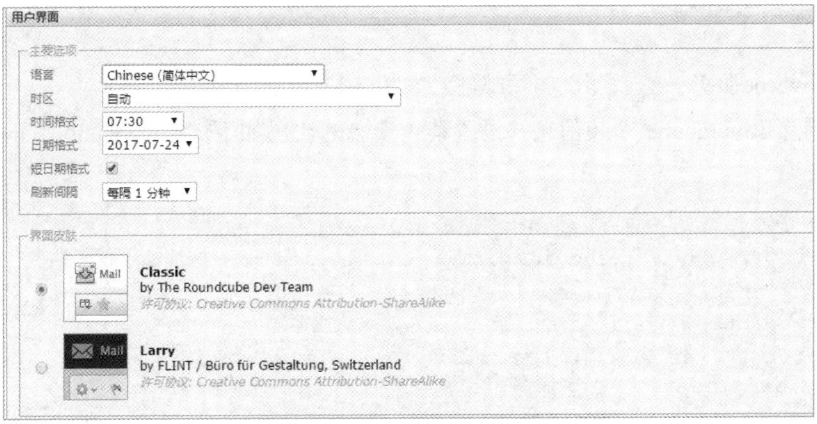

图 7-21　Roundcube 界面设置

(9) Roundcube WebMail 添加通讯录,如图 7-22 所示。

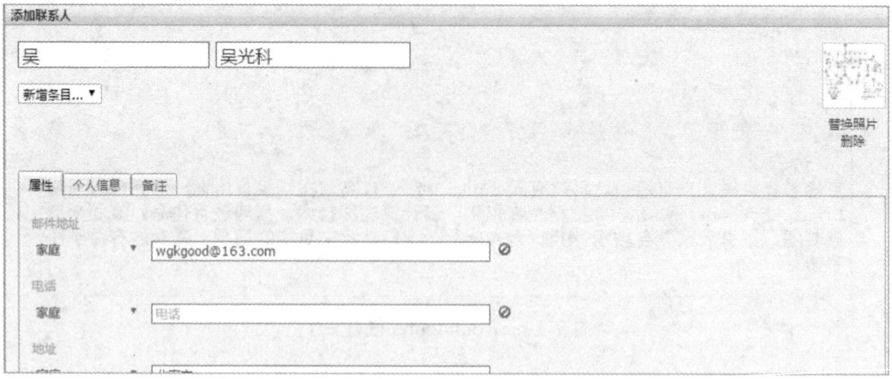

图 7-22 Roundcube 通讯录设置

(10) 创建"垃圾邮件"及"已删除邮件"邮件夹。

通过 Roundcube 删除邮件时,默认会报错,报错信息如下。

```
UID COPY: Mailbox doesn't exist: Trash
```

基于 Dovecot 自动创建"垃圾邮件"及"已删除邮件"邮件夹,代码如下。

```
vim /etc/dovecot/conf.d/20-imap.conf
mail_plugins = $mail_plugins autocreate
plugin {
  autocreate = Trash
  autocreate2 = Junk
  autocreate3 = Drafts
  autocreate4 = Sent
  autosubscribe = Trash
  autosubscribe2 = Junk
  autosubscribe3 = Drafts
  autosubscribe4 = Sent
}
```

重启 Dovecot 服务,然后测试删除邮件,如图 7-23 所示。

(11) 删除 Roundcube 安装信息并设置禁止安装的代码如下。Roundcube 禁止重新安装如图 7-24 所示。

```
rm -rf /var/www/html/webmail/installer/
#vim /var/www/html/webmail/config/config.inc.php 添加如下代码
$config['enable_installer'] = false;
#vim /etc/httpd/conf/httpd.conf 加入如下代码
<Directory "/var/www/html/webmail/installer/">
    Options Indexes FollowSymLinks
    AllowOverride None
    Order allow,deny
```

```
    Deny from all
</Directory>
```

(a)

(b)

图 7-23　Roundcube 邮件删除

(a) 收件箱 Roundcube；(b) Roundcub 已删除邮件

图 7-24　Roundcube 禁止重新安装

（12）实现域名访问邮件服务器。

在 Apache 配置文件 httpd.conf 的末尾加入如下代码。然后重启 Apache，访问 WebMail，如图 7-25 所示。

```
<IfModule alias_module>
    Alias  /  /var/www/html/webmail/
</IfModule>
```

（a）

（b）

图 7-25　Roundcube 域名访问

（a）登录 Roundcub；（b）访问 Roundcub

7.13　Postfix 虚拟用户配置

通过以上配置，用户可以登录 Postfix 邮件服务器，并进行邮件的发送和接收。为了管理的方便和系统的安全，一般通过 Postfix 虚拟用户管理邮件用户。

Postfix 虚拟用户的原理是基于系统创建一个映射用户,该用户不能登录系统,然后将其他虚拟用户全部映射到该系统用户所属目录。

对于操作系统来说,所有的操作均是通过该系统用户进行的,但是对于 Postfix 的邮件用户来说又是各自独立的。Postfix 虚拟用户与 Vsftpd 虚拟用户相似,所有的虚拟用户或系统用户均可以通过 PostfixAdmin 进行管理。

(1)配置 Postfix SMTP 认证基于 MySQL 数据库,读取 MySQL 虚拟用户配置,修改 Postfix 真实用户的 UID 和 GID,其命令如下。

```
usermod -u 1000 postfix
groupmod -g 1000 postfix
chown -R postfix.postfix /var/spool/mail/
```

(2)创建 Postfix 主配置文件 mail.cf,文件内容如下。

```
queue_directory = /var/spool/postfix
command_directory = /usr/sbin
daemon_directory = /usr/libexec/postfix
data_directory = /var/lib/postfix
mail_owner = postfix
myhostname = mail.jingfengjiaoyu.com
mydomain = jingfengjiaoyu.com
myorigin = $mydomain
inet_interfaces = all
inet_protocols = ipv4
#mydestination = localhost
#mydestination = $myhostname,localhost.$mydomain,localhost,$mydomain
unknown_local_recipient_reject_code = 550
#mynetworks = 192.168.0.0/24,127.0.0.0/8
mynetworks = 0.0.0.0/0
relay_domains = $mydestination
alias_maps = hash:/etc/aliases
alias_database = hash:/etc/aliases
debug_peer_level = 2
debugger_command =
    PATH = /bin:/usr/bin:/usr/local/bin:/usr/X11R6/bin
    ddd $daemon_directory/$process_name $process_id & sleep 5
sendmail_path = /usr/sbin/sendmail.postfix
newaliases_path = /usr/bin/newaliases.postfix
mailq_path = /usr/bin/mailq.postfix
setgid_group = postdrop
html_directory = no
manpage_directory = /usr/share/man
sample_directory = /usr/share/doc/postfix-2.6.6/samples
readme_directory = /usr/share/doc/postfix-2.6.6/README_FILES
#Config Virtual Mailbox Settings 2021
```

```
virtual_minimum_uid = 100
virtual_mailbox_base = /var/spool/mail
virtual_mailbox_maps = mysql:/etc/postfix/mysql_virtual_mailbox_maps.cf
virtual_mailbox_domains = mysql:/etc/postfix/mysql_virtual_domains_maps.cf
virtual_alias_domains = $virtual_alias_maps
virtual_alias_maps = mysql:/etc/postfix/mysql_virtual_alias_maps.cf
virtual_uid_maps = static:1000
virtual_gid_maps = static:1000
virtual_transport = virtual
maildrop_destination_recipient_limit = 1
maildrop_destination_concurrency_limit = 1
#Config QUOTA 2021
message_size_limit = 52428800
mailbox_size_limit = 209715200
virtual_mailbox_limit = 209715200
virtual_create_maildirsize = yes
virtual_mailbox_extended = yes
virtual_mailbox_limit_maps = mysql:/etc/postfix/mysql_virtual_mailbox_
limit_maps.cf
virtual_mailbox_limit_override = yes
virtual_maildir_limit_message = Sorry, the user's maildir has overdrawn his
diskspace quota, please try again later.
virtual_overquota_bounce = yes
#Config SASL 2021
broken_sasl_auth_clients = yes
smtpd_recipient_restrictions = permit_mynetworks,permit_sasl_authenticated,
reject_invalid_hostname,reject_non_fqdn_hostname,reject_unknown_sender_
domain,reject_non_fqdn_sender,reject_non_fqdn_recipient,reject_unknown_
recipient_domain,reject_unauth_pipelining,reject_unauth_destination,permit
smtpd_sasl_auth_enable = yes
smtpd_sasl_type = dovecot
smtpd_sasl_path = /var/run/dovecot/auth-client
smtpd_sasl_local_domain = $myhostname
smtpd_sasl_security_options = noanonymous
smtpd_sasl_application_name = smtpd
smtpd_banner=$myhostname SMTP "Version not Available"
```

（3）创建 Postfix 读取 MySQL 配置信息，内容如下。

```
cat>/etc/postfix/mysql_virtual_alias_maps.cf<<EOF
user = postfix
password = postfix
hosts = localhost
dbname = postfix
table = alias
select_field = goto
where_field = address
```

```
EOF
cat>/etc/postfix/mysql_virtual_domains_maps.cf<<EOF
user = postfix
password = postfix
hosts = localhost
dbname = postfix
table = domain
select_field = description
where_field = domain
EOF
cat>/etc/postfix/mysql_virtual_mailbox_limit_maps.cf<<EOF
user = postfix
password = postfix
hosts = localhost
dbname = postfix
table = mailbox
select_field = quota
where_field = username
EOF
cat>/etc/postfix/mysql_virtual_mailbox_maps.cf<<EOF
user = postfix
password = postfix
hosts = localhost
dbname = postfix
table = mailbox
select_field = maildir
where_field = username
EOF
```

相关配置文件的内容如下。

① mysql_virtual_mailbox_maps.cf 配置文件：服务器邮箱文件的存储路径。

② mysql_virtual_domains_maps.cf 配置文件：邮件服务器上所有的虚拟域。

③ mysql_virtual_alias_maps.cf 配置文件：邮件服务器上虚拟别名和实际邮件地址间的对应关系。

④ mysql_virtual_mailbox_limit_maps.cf 配置文件：服务器上邮箱的一些限制参数。

以上配置文件可以供 Postfix 读取数据库表。默认是基于 hash:/etc/aliases 查询表，哈希文件的路径是/etc/aliases。Postfix 在需要的时候读取这些配置信息，然后根据这些配置信息的指示，到另外的文件或者数据库中读取实际的数据。

（4）创建数据库并授权。

```
create database postfix charset = utf8;
grant all privileges on postfix.* to postfix@'localhost' identified by
"postfix";
flush privileges;
```

（5）修改 PostfixAdmin 配置文件的数据库连接信息，如图 7-26 所示。

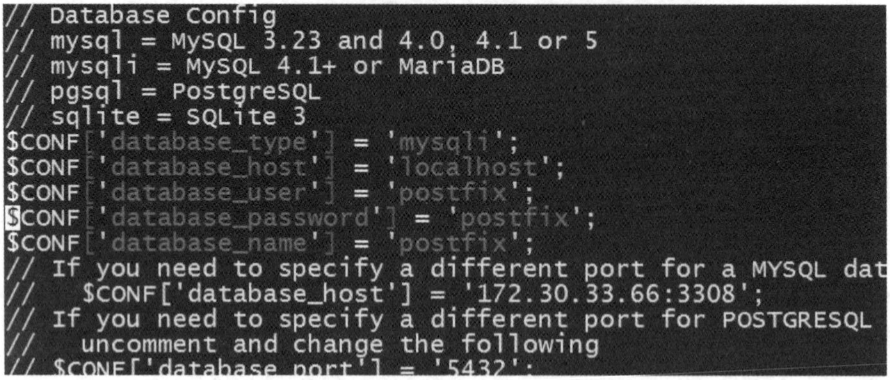

```
// Database Config
// mysql = MySQL 3.23 and 4.0, 4.1 or 5
// mysqli = MySQL 4.1+ or MariaDB
// pgsql = PostgreSQL
// sqlite = SQLite 3
$CONF['database_type'] = 'mysqli';
$CONF['database_host'] = 'localhost';
$CONF['database_user'] = 'postfix';
$CONF['database_password'] = 'postfix';
$CONF['database_name'] = 'postfix';
// If you need to specify a different port for a MYSQL dat
//    $CONF['database_host'] = '172.30.33.66:3308';
// If you need to specify a different port for POSTGRESQL
//    uncomment and change the following
// $CONF['database_port'] = '5432';
```

图 7-26 修改 PostfixAdmin 配置文件的数据库连接信息

（6）通过 PostfixAdmin 创建 4 个虚拟用户，如图 7-27 所示。

图 7-27 通过 PostfixAdmin 创建 4 个虚拟用户

（7）配置 Dovecot 服务，修改 dovecot.conf 配置文件的代码如下。

```
#/etc/dovecot/dovecot.conf
#CentOS: Linux 2.6.32
auth_mechanisms = PLAIN LOGIN
#auth_mechanisms = PLAIN LOGIN CRAM-MD5 DIGEST-MD5
disable_plaintext_auth = no
first_valid_uid = 1000
listen = *
mail_location = maildir:/var/spool/mail/%d/%n
managesieve_notify_capability = mailto
managesieve_sieve_capability = fileinto reject envelope encoded-character
vacation subaddress comparator-i;ascii-numeric relational regex imap4flags
copy include variables body enotify environment mailbox date
```

```
passdb {
  args = /etc/dovecot/mysql.conf
  driver = sql
}
protocols = imap pop3
service auth {
  unix_listener auth-client {
    group = postfix
    mode = 0660
    user = postfix
  }
}
ssl = no
userdb {
  args = /etc/dovecot/mysql.conf
  driver = sql
}
```

（8）创建 Dovecot 连接 MySQL 认证的配置文件/etc/dovecot/mysql.conf 的代码如下。

```
driver = mysql
connect = host = /var/lib/mysql/mysql.sock dbname = postfix user = postfix
password = postfix
default_pass_scheme = MD5
password_query = SELECT password FROM mailbox WHERE username = '%u'
user_query = SELECT maildir, 1000 AS uid, 1000 AS gid FROM mailbox WHERE
username = '%u'
```

（9）使用虚拟用户登录 Roundcube，如图 7-28 所示。

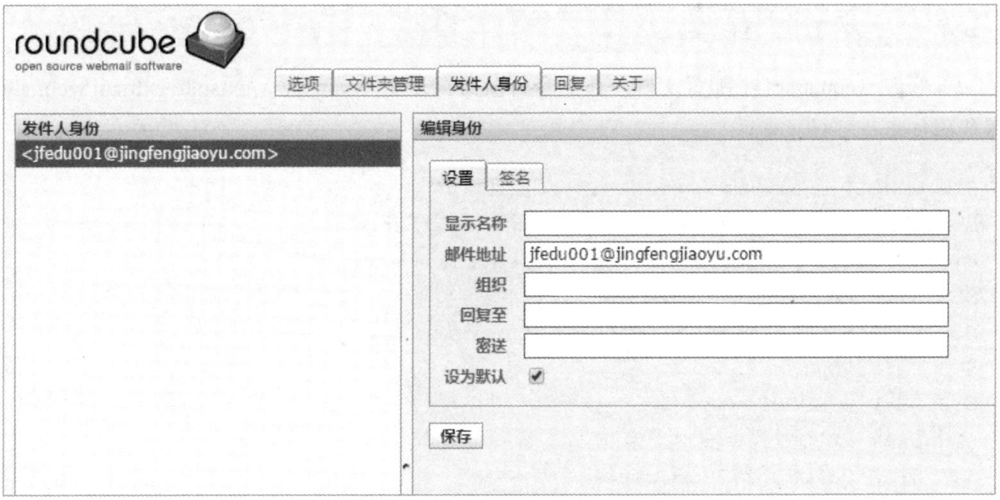

图 7-28　使用虚拟用户登录 Roundcube

7.14 Postfix+ExtMail 配置实战

通过本章的部署,已经建成了一个基本的邮件服务器系统,它能够发送、接收邮件,还能够对用户进行身份验证等。用户可以使用 Outlook、Foxmail、Roundcube 等工具发送和接收邮件。

Roundcube WebMail 页面功能相对比较少,可以使用 ExtMail 实现 Web 端邮件的收取和发送。

ExtMail 是一个用 Perl 语言编写、面向大容量/ISP 级应用、免费的、高性能 WebMail 软件。ExtMail 套件用于提供从浏览器中登录、使用邮件系统的 Web 操作界面,它以 GPL 版权释出,设计初衷是设计一个能适应当前高速发展的 IT 应用环境,能满足用户多变的需求,能快速进行开发、改进和升级且适应能力强的 WebMail 系统。

对于国内的电子邮件系统来说,无论是从系统功能、易用性还是中文化等方面,ExtMail 平台都是一个相当不错的选择。ExtMail 套件可以供普通邮件用户使用,而 ExtMan 套件可以供邮件系统的管理员使用。

ExtMail 的配置方法如下。

(1) 从官网下载文件 extmail-1.2.tar.gz,并解压安装。

```
tar -zxf extmail-1.2.tar.gz
mkdir -p /var/www/extsuite
mv  extmail-1.2 /var/www/extsuite/extmail
cd  /var/www/extsuite/extmail/
cp  webmail.cf.default   webmail.cf
yum install perl-devel perl perl-Unix-Syslog -y
```

(2) 修改 webmail.cf 主配置文件。在命令行窗口打开文件 var/www/extsuite/extmail/webmail.cf,配置代码如下。

```
SYS_CONFIG = /var/www/extsuite/extmail/
SYS_LANGDIR = /var/www/extsuite/extmail/lang
SYS_TEMPLDIR = /var/www/extsuite/extmail/html
SYS_HTTP_CACHE = 0
SYS_SMTP_HOST = mail.jingfengjiaoyu.com
SYS_SMTP_PORT = 25
SYS_SMTP_TIMEOUT = 5
SYS_SPAM_REPORT_ON = 0
SYS_SPAM_REPORT_TYPE = dspam
SYS_SHOW_WARN = 0
SYS_IP_SECURITY_ON = 1
```

```
SYS_PERMIT_NOQUOTA = 1
SYS_SESS_DIR = /tmp
SYS_UPLOAD_TMPDIR = /tmp
SYS_LOG_ON = 1
SYS_LOG_TYPE = syslog
SYS_LOG_FILE = /var/log/extmail.log
SYS_SESS_TIMEOUT = 0
SYS_SESS_COOKIE_ONLY = 1
SYS_USER_PSIZE = 10
SYS_USER_SCREEN = auto
SYS_USER_LANG = en_US
SYS_APP_TYPE = Webmail
SYS_USER_TEMPLATE = default
SYS_USER_CHARSET = utf-8
SYS_USER_TRYLOCAL = 1
SYS_USER_TIMEZONE = +0800
SYS_USER_CCSENT = 1
SYS_USER_SHOW_HTML = 1
SYS_USER_COMPOSE_HTML = 1
SYS_USER_CONV_LINK =1
SYS_USER_ADDR2ABOOK = 1
SYS_MESSAGE_SIZE_LIMIT = 5242880
SYS_MIN_PASS_LEN = 2
SYS_MFILTER_ON = 1
SYS_NETDISK_ON = 1
SYS_SHOW_SIGNUP = 1
SYS_DEBUG_ON = 1
SYS_AUTH_TYPE = mysql
SYS_MAILDIR_BASE = /var/spool/mail/
SYS_AUTH_SCHEMA = virtual
SYS_CRYPT_TYPE = md5crypt
SYS_MYSQL_USER = postfix
SYS_MYSQL_PASS = postfix
SYS_MYSQL_DB = postfix
SYS_MYSQL_HOST = localhost
SYS_MYSQL_SOCKET = /var/lib/mysql/mysql.sock
SYS_MYSQL_TABLE = mailbox
SYS_MYSQL_ATTR_USERNAME = username
SYS_MYSQL_ATTR_DOMAIN = domain
SYS_MYSQL_ATTR_PASSWD = password
```

```
SYS_MYSQL_ATTR_CLEARPW = clearpwd
SYS_MYSQL_ATTR_QUOTA = quota
SYS_MYSQL_ATTR_NDQUOTA = netdiskquota
SYS_MYSQL_ATTR_HOME = homedir
SYS_MYSQL_ATTR_MAILDIR = maildir
SYS_MYSQL_ATTR_DISABLEWEBMAIL = disablewebmail
SYS_MYSQL_ATTR_DISABLENETDISK = disablenetdisk
SYS_MYSQL_ATTR_DISABLEPWDCHANGE = disablepwdchange
SYS_MYSQL_ATTR_ACTIVE = active
SYS_MYSQL_ATTR_PWD_QUESTION = question
SYS_MYSQL_ATTR_PWD_ANSWER = answer
SYS_LDAP_BASE = o = extmailAccount,dc = example.com
SYS_LDAP_RDN = cn = Manager,dc = example.com
SYS_LDAP_PASS = secret
SYS_LDAP_HOST = localhost
SYS_LDAP_ATTR_USERNAME = mail
SYS_LDAP_ATTR_DOMAIN = virtualDomain
SYS_LDAP_ATTR_PASSWD = userPassword
SYS_LDAP_ATTR_CLEARPW = clearPassword
SYS_LDAP_ATTR_QUOTA = mailQuota
SYS_LDAP_ATTR_NDQUOTA = netdiskQuota
SYS_LDAP_ATTR_HOME = homeDirectory
SYS_LDAP_ATTR_MAILDIR = mailMessageStore
SYS_LDAP_ATTR_DISABLEWEBMAIL = disablewebmail
SYS_LDAP_ATTR_DISABLENETDISK = disablenetdisk
SYS_LDAP_ATTR_DISABLEPWDCHANGE = disablePasswdChange
SYS_LDAP_ATTR_ACTIVE = active
SYS_LDAP_ATTR_PWD_QUESTION = question
SYS_LDAP_ATTR_PWD_ANSWER = answer
SYS_AUTHLIB_SOCKET = /var/spool/authdaemon/socket
SYS_G_ABOOK_TYPE = file
SYS_G_ABOOK_LDAP_HOST = localhost
SYS_G_ABOOK_LDAP_BASE = ou = AddressBook,dc = example.com
SYS_G_ABOOK_LDAP_ROOTDN = cn = Manager,dc = example.com
SYS_G_ABOOK_LDAP_ROOTPW = secret
SYS_G_ABOOK_LDAP_FILTER = objectClass = OfficePerson
SYS_G_ABOOK_FILE_PATH = /var/www/extsuite/extmail/globabook.cf
SYS_G_ABOOK_FILE_LOCK = 1
SYS_G_ABOOK_FILE_CONVERT = 0
SYS_G_ABOOK_FILE_CHARSET = utf-8
```

常用配置文件参数的详解如下。

```
SYS_MESSAGE_SIZE_LIMIT = 5242880
#用户可以发送的最大邮件
SYS_USER_LANG = en_US
#语言选项,可改作 SYS_USER_LANG = zh_CN
SYS_MAILDIR_BASE = /home/domains
#此处即为7.13节所设置的用户邮件的存放目录,可改作 SYS_MAILDIR_BASE = /var/spool/mail/
SYS_MYSQL_USER = postfix
SYS_MYSQL_PASS = db_pass
#以上语句用来设置连接数据库服务器所使用的用户名、密码和邮件服务器所用到的数据库,这里修
#改为
SYS_MYSQL_USER = postfix
SYS_MYSQL_PASS = postfix
SYS_MYSQL_HOST = localhost
#指明数据库服务器主机名,这里默认即可
SYS_MYSQL_TABLE = mailbox
SYS_MYSQL_ATTR_USERNAME = username
SYS_MYSQL_ATTR_DOMAIN = domain
SYS_MYSQL_ATTR_PASSWD = password
#以上语句用来指定验证用户登录时所用到的表,以及用户名、域名和用户密码所对应的表中列的名
#称。这里默认即可
```

Apache 配置文件 httpd.conf 的设置方式一:

```
User postfix
Group postfix
<VirtualHost *:80>
ServerName     mail.jingfengjiaoyu.com
DocumentRoot   /var/www/extsuite/extmail/html/
ScriptAlias    /extmail/cgi /var/www/extsuite/extmail/cgi
Alias          /extmail /var/www/extsuite/extmail/html
</VirtualHost>
```

Apache 配置文件 httpd.conf 的设置方式二:由于 ExtMail 要进行本地邮件的投递操作,必须将运行 Apache 服务器用户的身份修改为 MDA 的用户,本案例为 Postfix;开启 Apache 服务器的 suexec 功能,使用以下方法实现虚拟主机运行身份的指定。

```
<VirtualHost *:80>
ServerName mail.jingfengjiaoyu.com
DocumentRoot /var/www/extsuite/extmail/html/
ScriptAlias    /extmail/cgi /var/www/extsuite/extmail/cgi
Alias          /extmail  /var/www/extsuite/extmail/html
SuexecUserGroup postfix postfix
</VirtualHost>
```

浏览器访问 ExtMail,如图 7-29 所示。

（a）

（b）

（c）

图 7-29　浏览器访问 ExtMail

（a）ExtMail 登录界面；（b）ExtMail 后台界面；（c）ExtMail 发送邮件

7.15　Postfix+ExtMan 配置实战

ExtMan 是 ExtMail 项目组在 ExtMail WebMail 之后推出的一个用来管理 ExtMail 虚拟账号的管理软件。

ExtMan 是用 Perl 语言编写的，目前支持 MySQL 和 OpenLDAP 作为账号信息存储源，新的存储源正在开发之中。

ExtMan 可以让用户通过浏览器轻松地管理 ExtMail 系统中的虚拟域和账号信息，同时还加入了图形化日志监控工具，使邮件的管理更加方便。

```
tar xzf extman-1.1.tar.gz -C /var/www/extsuite/
cd  /var/www/extsuite/
mv  extman-1.1 extman
chown -R postfix.postfix /var/www/extsuite/extman/cgi/
cd  extman/
cp  webman.cf.default  webman.cf
mkdir  -p  /tmp/extman
chown postfix.postfix /tmp/extman
yum  install  perl-rrdtool*  -y
```

创建 ExtMan 所需数据库。

```
cd  /var/www/extsuite/extman/docs
mysql  -uroot -p <extmail.sql
mysql  -uroot -p <init.sql
#进入 MySQL 命令行执行:
grant all on extmail.* to webman@'localhost' identified by "webman";
flush  privileges;
```

打开/var/www/extsuite/extman/webmail.cf 文件，其配置代码如下。

```
SYS_CONFIG = /var/www/extsuite/extman/
SYS_LANGDIR = /var/www/extsuite/extman/lang
SYS_TEMPLDIR = /var/www/extsuite/extman/html
SYS_MAILDIR_BASE = /var/spool/mail
SYS_SHOW_WARN = 0
SYS_SESS_DIR = /tmp/extman/
SYS_CAPTCHA_ON = 1
SYS_CAPTCHA_KEY = r3s9b6a7
SYS_CAPTCHA_LEN = 6
SYS_PURGE_DATA = 0
SYS_PSIZE = 20
SYS_APP_TYPE = ExtMan
SYS_TEMPLATE_NAME = default
SYS_DEFAULT_EXPIRE = 1y
```

```
SYS_GROUPMAIL_SENDER = postmaster@extmail.org
SYS_DEFAULT_SERVICES = webmail,smtpd,smtp,pop3,netdisk
SYS_ISP_MODE = no
SYS_DOMAIN_HASHDIR = yes
SYS_DOMAIN_HASHDIR_DEPTH = 2x2
SYS_USER_HASHDIR = yes
SYS_USER_HASHDIR_DEPTH = 2x2
SYS_MIN_UID = 500
SYS_MIN_GID = 100
SYS_DEFAULT_UID = 1000
SYS_DEFAULT_GID = 1000
SYS_QUOTA_MULTIPLIER = 1048576
SYS_QUOTA_TYPE = courier
SYS_DEFAULT_MAXQUOTA = 500
SYS_DEFAULT_MAXALIAS = 100
SYS_DEFAULT_MAXUSERS = 100
SYS_DEFAULT_MAXNDQUOTA = 500
SYS_USER_DEFAULT_QUOTA = 5
SYS_USER_DEFAULT_NDQUOTA = 5
SYS_USER_DEFAULT_EXPIRE = 1y
SYS_BACKEND_TYPE = mysql
SYS_CRYPT_TYPE = md5crypt
SYS_MYSQL_USER = webman
SYS_MYSQL_PASS = webman
SYS_MYSQL_DB = extmail
SYS_MYSQL_HOST = localhost
SYS_MYSQL_SOCKET = /var/lib/mysql/mysql.sock
SYS_MYSQL_TABLE = manager
SYS_MYSQL_ATTR_USERNAME = username
SYS_MYSQL_ATTR_PASSWD = password
SYS_LDAP_BASE = dc=extmail.org
SYS_LDAP_RDN = cn=Manager,dc=extmail.org
SYS_LDAP_PASS = secret
SYS_LDAP_HOST = localhost
SYS_LDAP_ATTR_USERNAME = mail
SYS_LDAP_ATTR_PASSWD = userPassword
SYS_RRD_DATADIR = /var/lib
SYS_RRD_TMPDIR = /tmp/viewlog
SYS_RRD_QUEUE_ON = yes
SYS_CMDSERVER_SOCK = /tmp/cmdserver.sock
SYS_CMDSERVER_MAXCONN = 5
SYS_CMDSERVER_PID = /var/run/cmdserver.pid
SYS_CMDSERVER_LOG = /var/log/cmdserver.log
SYS_CMDSERVER_AUTHCODE = your_auth_code_here
SYS_IGNORE_SERVER_LIST = web
```

修改 httpd.conf 配置文件的代码如下。

```
User postfix
Group postfix
<VirtualHost *:80>
ServerName      mail.jingfengjiaoyu.com
DocumentRoot    /var/www/extsuite/extmail/html/
ScriptAlias     /extmail/cgi /var/www/extsuite/extmail/cgi
Alias           /extmail /var/www/extsuite/extmail/html
ScriptAlias     /extman/cgi /var/www/extsuite/extman/cgi
Alias           /extman /var/www/extsuite/extman/html
</VirtualHost>
```

使用默认的用户名和密码登录。

ExtMan 后台管理员的初始用户名为 root@extmail.org。

ExtMan 后台管理员的初始密码为 extmail*123*。

Postmaster 用户的初始密码为 extmail。

通过浏览器访问地址 http://mail.jingfengjiaoyu.com/extman/，如图 7-30 所示。

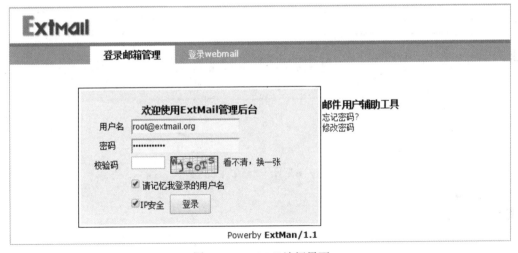

图 7-30　ExtMail 访问界面

7.16 MailGraph_ext 安装配置

在安装 MailGraph_ext 之前，必须要先安装其所依赖的一些软件包，否则 MailGraph_ext 就不能正常工作，如图 7-31 所示。

（1）RRDtool 及 RRDtool 的 Perl 包的地址如下。

http://people.ee.ethz.ch/~oetiker/webtools/rrdtool/

（2）启动 ExtMan 服务。

```
cp -r /var/www/extsuite/extman/addon/mailgraph_ext/ /usr/local/mailgraph_ext/
/usr/local/mailgraph_ext/mailgraph-init start
/var/www/extsuite/extman/daemon/cmdserver --daemon
```

将 MailGraph_ext 及 cmdserver 加入系统自启动。

```
echo "/usr/local/mailgraph_ext/mailgraph-init start" >> /etc/rc.d/rc.local
echo "/var/www/extsuite/extman/daemon/cmdserver -v -d" >> /etc/rc.d/rc.local
```

>> [投递成功信件分析图][失败投递信件分析图][邮件投递流量分析图]
>> [**IMAP/POP3** 登录日志图][邮件队列分析图][**WebMail** 登录日志图]

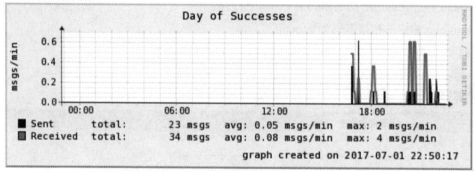

（a）

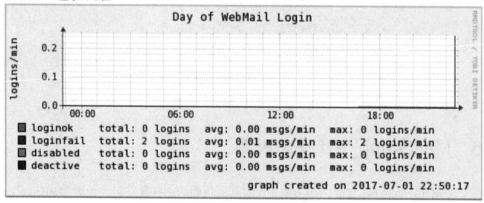

（b）

图 7-31　WebMail 日志界面

（a）投递信息分析；（b）WebMail 登录日志

7.17　Postfix+ExtMan 虚拟用户注册

经过 7.15 节和 7.16 节的操作步骤和配置，Postfix+ExtMan 可以实现虚拟用户的注册，如图 7-32 所示。

（a）

（b）

图 7-32　ExtMan 虚拟用户注册

（a）ExtMan 开启注册；（b）ExtMan 注册用户

7.18 基于 ExtMan 自动注册并登录

(1) 通过 ExtMan 自动注册用户并设置密码后,发现用户无法登录 ExtMail Web 平台,提示用户名或密码错误,如图 7-33 所示。

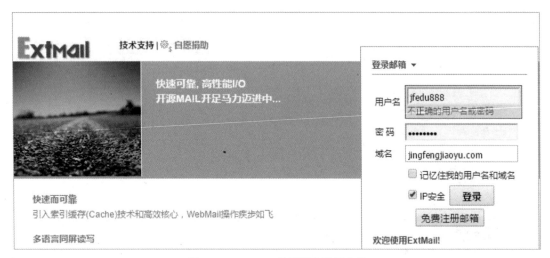

图 7-33 ExtMan 虚拟用户登录失败

问题分析如下。

由于 ExtMan 数据库连接的是 ExtMail 数据库,而虚拟用户读取的是 Postfix 数据库,因此导致数据不同步。可以同步数据,或者合并数据库后登录。

```
mysqldump -uroot -p postfix >postfix.sql
mysqldump -uroot -p extmail >extmail.sql
#进入 MySQL 命令行
drop database extmail;
use postfix;
source /tmp/extmail.sql;
```

修改 webman.cf 配置文件中的数据库连接信息如下。

```
SYS_MYSQL_USER = postfix
SYS_MYSQL_PASS = postfix
SYS_MYSQL_DB = postfix
```

重新登录 ExtMail,使用用户名 jfedu888 登录,如图 7-34 所示。

重新注册虚拟用户,并验证登录。注册用户名为 jfedu999,密码为 jfedu999,如图 7-35 所示。

图 7-34　ExtMan 虚拟用户登录

（a）

（b）

图 7-35　ExtMan 虚拟用户登录（更改用户名后）

（a）查看 ExtMan 虚拟用户；（b）登录 jfedu999 用户邮箱

（2）PostfixAdmin 无法登录的报错信息如下，如图 7-36 所示。

```
PHP Warning: Unknown: Failed to write session data (files). Please verify
that the current setting of session.save_path is correct (/var/lib/php/
session) in Unknown on line 0, referer: http://mail.jingfengjiaoyu.com/
postfixadmin/login.php
[Sat Jul 01 23:24:42 2021] [error] [client 36.102.227.109] PHP Warning:
session_destroy(): Session object destruction failed in /var/www/html/
postfixadmin/common.php on line 30, referer: http://mail.jingfengjiaoyu.
com/postfixadmin/main.php
```

图 7-36　PostfixAdmin 访问报错

在 Apache 配置文件中添加如下代码。

```
User postfix
Group postfix
<VirtualHost *:80>
    ServerName          mail.jingfengjiaoyu.com
    DocumentRoot        /var/www/extsuite/extmail/html/
    ScriptAlias         /extmail/cgi /var/www/extsuite/extmail/cgi
    Alias               /extmail /var/www/extsuite/extmail/html
    ScriptAlias         /extman/cgi /var/www/extsuite/extman/cgi
    Alias               /extman /var/www/extsuite/extman/html
    Alias               /postfixadmin /var/www/html/postfixadmin/
</VirtualHost>
```

重启 Apache 服务，访问 PostfixAdmin，如图 7-37 所示。

无法登录 PostfixAdmin 时，查看 maillog 日志，其代码如下，结果如图 7-38 所示。

```
tail -fn 100 /var/log/httpd/error_log
```

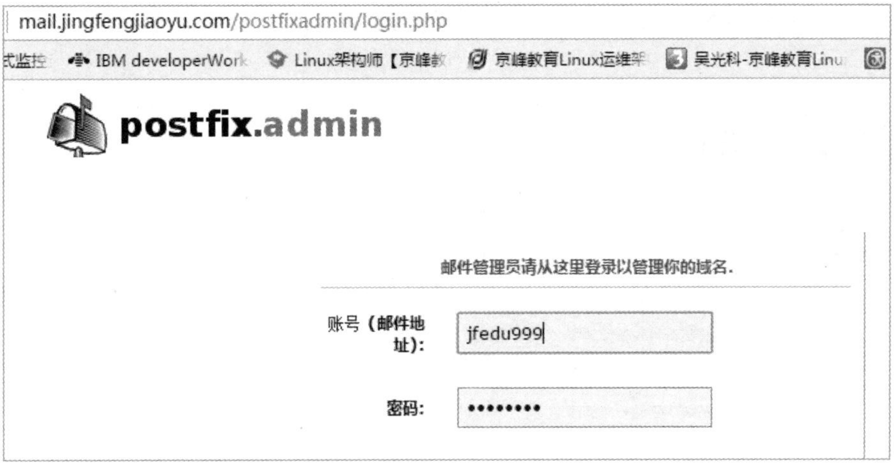

（a）

（b）

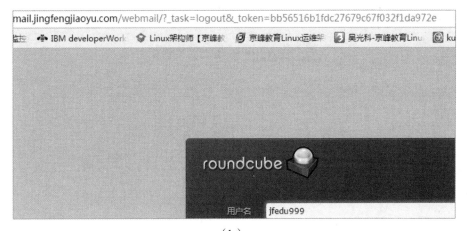

图 7-37　PostfixAdmin 登录

（a）登录 PostfixAdmin；（b）登录 Roundcube

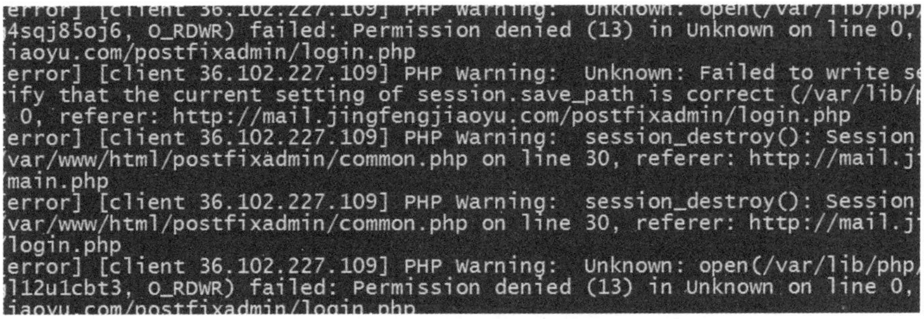

图 7-38　PostfixAdmin 报错信息

解决方法为执行命令 chmod 757 -R /var/lib/php/session/，然后再次登录。结果如图 7-39 所示。

图 7-39 PostfixAdmin 登录成功

第 8 章 Jenkins 持续集成企业实战

构建企业自动化部署平台，可以大大提升企业网站的部署效率。企业生产环境每天需要更新各种系统，传统更新网站的方法是使用 Shell+Rsync 实现网站代码备份和更新，更新完成后，运维人员手动发送邮件给测试人员、开发人员以及相关的业务人员。使用传统方法更新网站会耗费大量的人力，同时由于误操作偶尔会出现细节问题，因此构建自动化部署平台迫在眉睫。

本章将介绍传统部署网站方法、企业主流部署方法、Jenkins 持续集成简介、Jenkins 持续集成平台构建、Jenkins 插件部署、Jenkins 自动化部署网站、Jenkins 多实例及 Ansible+Jenkins 批量自动部署等内容。

8.1 传统部署网站的流程

服务器部署网站是运维工程师的主要工作之一。传统运维部署网站主要靠手动部署，手动部署网站的流程大致为：需求分析→原型设计→开发代码→提交测试→内网部署→确认上线→备份数据→外网更新→外网测试→发布完成等。如果发现外网部署的代码有异常，需要及时回滚，如图 8-1 所示。

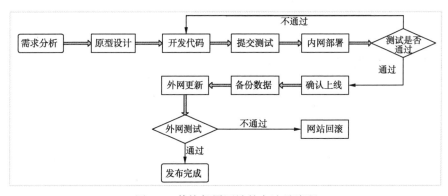

图 8-1 传统部署网站的方法及流程

服务器部署是基于 YUM 安装 LAMP 架构,并部署 Discuz。最终效果如图 8-2 所示。

图 8-2　YUM 部署 LAMP+Discuz 网站

通过 SecureCRT 登录网站服务器,并将 logo.png 文件上传至网站目录,然后手动备份网站,并更新网站的 Logo,如图 8-3 所示。

图 8-3　手动更新 LAMP 网站 Logo 文件

8.2　目前主流部署网站的流程

传统部署网站的方法对于单台或者几台服务器的更新很容易,如果服务器规模超过百台甚至千台,或者更新网站代码很频繁,那么手动更新就非常消耗时间。

基于主流的 Hudson/Jenkins 工具平台可实现全自动网站部署、网站测试和网站回滚等操作,这大大减轻了部署网站的工作量。Jenkins 的前身为 Hudson,Hudson 主要用于商业版,而 Jenkins 为开源免费版。

Jenkins 是一个可扩展的持续集成引擎、框架,是一个开源软件项目,旨在提供一个开放易用的软件平台,使软件的持续集成变成可能。Jenkins 平台的安装和配置非常容易,其使用也非常简单。构建 Jenkins 平台可以解放以下人员的双手。

(1)开发人员。对于开发人员来说,只需要负责网站代码的编写,而不需要手动对源码进行编译、打包、单元测试等工作,直接将写好的代码分支存放在 SVN 仓库或 Git 仓库即可。

(2)运维人员。对于运维人员来说,使用 Jenkins 自动部署,可以降低人工干预的错误率,同时省去繁杂的上传代码、手动备份、手动更新等工作。

(3)测试人员。对于测试人员来说,可以通过 Jenkins 进行代码测试、网站的功能或者性能测试。

基于 Jenkins 自动部署网站的流程大致为:需求分析→原型设计→开发代码→提交测试→Jenkins 内网部署→确认上线→Jenkins 备份数据→Jenkins 外网部署→外网测试→发布完成等。如果发现外网部署的代码有异常,可以通过 Jenkins 及时回滚,如图 8-4 所示。

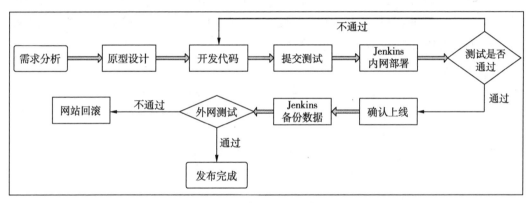

图 8-4　Jenkins 部署网站的方法及流程

8.3　Jenkins 持续集成简介

持续集成(Continuous Integration,CI)是一种软件开发实践,为提高软件开发效率并保障软件开发质量提供了理论基础。

(1)持续集成中的任何一个环节都是自动完成的,无须太多的人工干预,有利于减少重复过程,节省时间、费用和工作量。

(2)持续集成保障了在每个时间点团队成员提交的代码都是能成功集成的。换言之,在任何情况下都能第一时间发现软件的集成问题,它使在任意时间发布可部署的软件成为可能。

(3)持续集成符合软件本身的发展趋势,这点在需求不明确或是频繁性变更的情景中尤其重要。持续集成的质量能帮助团队进行有效决策,同时还能建立团队对开发产品的信心。

8.4　Jenkins 持续集成组件

（1）自动构建过程作业（job）。job 的功能主要是获取 SVN/Git 源码、自动编译、自动打包、部署分发和自动测试等。

（2）源代码存储库。开发编写的代码需上传至 SVN 代码库或 Git 代码库中，以供 Jenkins 获取。

（3）Jenkins 持续集成服务器，用于部署 Jenkins UI、存放 job 工程、各种插件、编译打包的数据等。

8.5　Jenkins 平台实战部署

从官网获取 Jenkins 软件。官网地址为 http://mirrors.jenkins-ci.org/，从中下载稳定的 Jenkins 软件版本。由于 Jenkins 是基于 Java 语言开发的一种持续集成工具，所以 Jenkins 服务器需安装 Java JDK 开发软件。Jenkins 平台搭建步骤如下。

（1）Jenkins 稳定版下载的代码如下。

```
https://mirrors.tuna.tsinghua.edu.cn/jenkins/war/2.260/jenkins.war
```

（2）官网下载 Java JDK，并解压安装。执行命令 vi /etc/profile，在文件末尾添加如下语句。

```
export JAVA_HOME = /usr/java/jdk1.8.0_131
export CLASSPATH = $CLASSPATH:$JAVA_HOME/lib:$JAVA_HOME/jre/lib
export PATH = $JAVA_HOME/bin:$JAVA_HOME/jre/bin:$PATH
```

（3）配置 Java 环境变量，在/etc/profile 配置文件末尾加入如下语句。

```
export JAVA_HOME = /usr/java/jdk1.8.0_131
export CLASSPATH = $CLASSPATH:$JAVA_HOME/lib:$JAVA_HOME/jre/lib
export PATH = $JAVA_HOME/bin:$JAVA_HOME/jre/bin:$PATH:$HOME/bin
```

配置并查看环境变量的命令如下。

```
source /etc/profile
java --version
```

（4）Tomcat Java 容器配置如下。

```
wget https://dlcdn.apache.org/tomcat/tomcat-8/v8.5.72/bin/apache-tomcat-8.5.72.tar.gz
tar xzf apache-tomcat-8.5.72.tar.gz
mv apache-tomcat-8.5.72  /usr/local/tomcat
```

（5）Tomcat 发布 Jenkins，将 Jenkins.war 文件复制到 Tomcat 默认发布目录下，并使用 jar 工具解压，然后启动 Tomcat 服务，其代码如下。

```
rm    -rf  /usr/local/tomcat/webapps/*
mkdir -p /usr/local/tomcat/webapps/ROOT/
mv    jenkins.war  /usr/local/tomcat/webapps/ROOT/
cd    /usr/local/tomcat/webapps/ROOT/
jar   -xvf  jenkins.war;rm -rf  Jenkins.war
sh    /usr/local/tomcat/bin/startup.sh
```

（6）如果安装新版，提示 Jenkins 已经离线，可以用以下方法解决该问题。

① 在命令行窗口打开文件 root/.jenkins/updates/default.json 并修改。

Jenkins 在安装插件时需要检查网络，默认是访问地址 www.google.com，但国内服务器连接比较慢，可以改用国内的地址作为测试地址，此处改为地址 www.baidu.com。

```
{"connectionCheckUrl":"http://www.baidu.com/"
```

② 在命令行窗口打开文件 root/.jenkins/hudson.model.UpdateCenter.xml 并修改。

默认该文件为 Jenkins 下载插件的源地址，其中默认地址为 https://updates.jenkins.io/update-center.json，因为 https 连接慢的问题，所以此处将其改为 http。

```
<url>http://updates.jenkins.io/update-center.json</url>
```

③ 重启 Jenkins 所在的 Tomcat 服务的命令如下。

```
/usr/local/tomcat/bin/shutdown.sh
/usr/local/tomcat/bin/startup.sh
```

（7）根据提示完成安装，最终通过客户端浏览器访问 Jenkins 服务器的 IP 地址，如图 8-5 所示。

图 8-5　Jenkins 自动部署平台

8.6　Jenkins 相关概念

要熟练掌握 Jenkins 持续集成的配置、使用和管理，需要了解相关的概念，如代码开发、编译、打包、构建等。常见的代码相关概念包括 Make、Ant、Maven、Jenkins、Eclipse 等。

（1）Make 编译工具。

Make 编译工具是 Linux 操作系统和 Windows 操作系统最原始的编译工具，在 Linux 操作系统下编译程序常用 Make，在 Windows 操作系统下对应的工具为 nmake。读取本地 makefile 文件，该文件决定了源文件之间的依赖关系，Make 负责根据 makefile 文件组织构建软件，负责指挥编译器如何编译、连接器如何连接，以及最后如何生成可用的二进制代码。

（2）Ant 编译工具。

Make 工具在编译比较复杂的工程时使用起来不方便，语法也很难理解，故延伸出了 Ant 工具。Ant 工具属于 Apache 基金会软件成员之一，是一个将软件编译、测试、部署等步骤联系在一起加以自动化的工具，大多用于 Java 环境中的软件开发。

Ant 构建文件是 XML 文件。每个构建文件定义一个唯一的项目（Project）元素。每个项目下可以定义很多目标元素，这些目标元素之间可以有依赖关系。

构建一个新的项目时，首先应该编写 Ant 构建文件。因为构建文件定义了构建过程，并能为团队开发中每个人所使用。

Ant 构建文件的默认命名为 build.xml，也可以取其他名字，只不过在运行的时候需要把这个命名当作参数传给 Ant。构建文件可以放在任意位置，一般做法是放在项目顶层目录，即根目录，这样可以保持项目的简洁和清晰。

（3）Maven 编译工具。

Maven 工具是对 Ant 工具的进一步改进。在 Make 工具中，如果要编译某些源文件，首先要安装编译器等工具。在 Java 环境中不同版本的编译器需要各种不同的数据包的支持，如果把每个数据包都下载下来，在 makefile 中进行配置制定，那么当需要的数据包非常多时，将很难管理。

Maven 与 Ant 类似，都是构建（build）工具。它如何调用各种不同的编译器和连接器呢？使用 Maven Plugin（Maven 插件），Maven 项目对象模型（Project Object Model，POM），可以通过一小段描述信息来管理项目的构建、报告和文档的软件项目管理工具。Maven 除了以程序构建能力为特色外，还提供了高级项目管理工具。

POM 是 Maven 项目中的文件，用 XML 表示，名称为 pom.xml。在 Maven 中构建的 Project 不仅仅是一堆包含代码的文件，还包含了 pom.xml 配置文件，该文件包括 Project 与开发者有关的缺陷跟踪系统、组织与许可、项目的 URL、项目依赖，以及其他配置。

在基于 Maven 构建编译时，Project 可以什么都没有，甚至没有代码，但是必须包含 pom.xml 文件。由于 Maven 的默认构建规则有较高的可重用性，因此常常只用两三行 Maven 构建脚本就可以构建简单的项目。

由于 Maven 是面向项目的方法，许多 Apache Jakarta 项目发文时会使用 Maven，且公司项目采用 Maven 的比例也在持续增长。

（4）Jenkins 框架工具。

Maven 可以对软件代码进行编译、打包和测试，Maven 可以控制编译、控制连接，可以生成各种报告，可以进行代码测试，但是不能控制完整的流程，也没有顺序定义——是先编译还是先连接？是先进行代码测试，还是先生成报告？所以可以使用 Jenkins 对 Maven 进行控制，将这些流程关联起来。

（5）Eclipse 工具。

Eclipse 是一个开放源代码的、基于 Java 语言的可扩展开发平台。就其本身而言，它只是一个框架和一组服务，用于通过插件组件构建开发环境。Eclipse 附带了一个标准的插件集，包括 Java 开发工具（Java Development Kit，JDK），主要用于开发网站代码。

8.7 Jenkins 平台设置

Jenkins 持续集成平台部署完毕后，需要进行简单配置，例如，配置 Java 路径，安装 Maven，指定 SVN 仓库、Git 仓库地址等。以下为 Java 路径和 Maven 的设置步骤。

（1）Jenkins 服务器安装 Maven。

```
wget
http://mirrors.tuna.tsinghua.edu.cn/apache/maven/maven-3/3.3.9/binaries/
apache-maven-3.3.9-bin.tar.gz
tar -xzf apache-maven-3.3.9-bin.tar.gz
mv apache-maven-3.3.9  /usr/maven/
```

（2）Jenkins 系统设置环境变量，如图 8-6 所示。

(a)

图 8-6　Jenkins 系统设置环境变量

(a) 设置 Java 路径

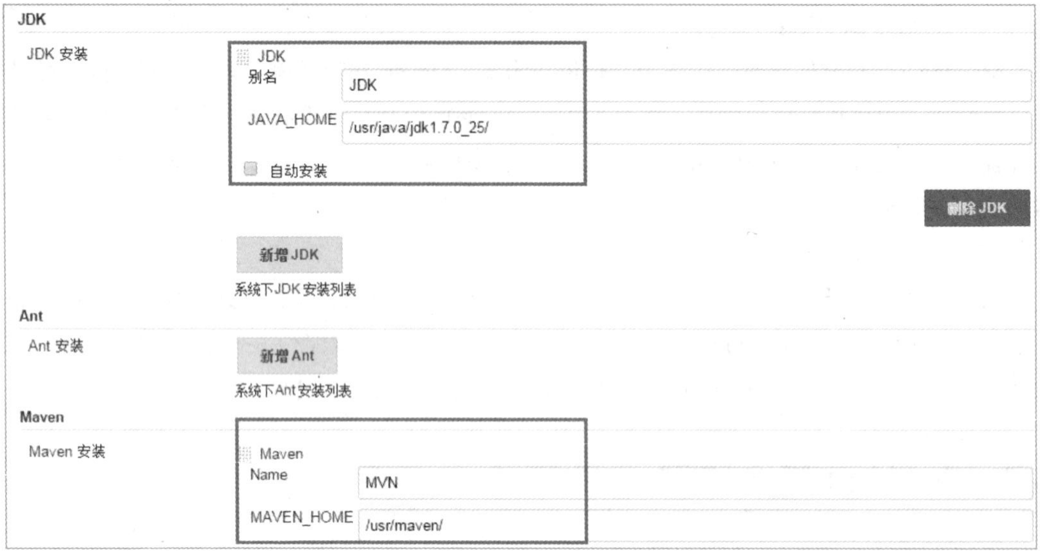

图 8-6 Jenkins 系统设置环境变量（续）

（b）设置 Maven 路径

（3）Jenkins 系统设置完毕后，需要创建 job 工程。

首先在 Jenkins 平台首页单击"创建一个新任务"按钮，然后在弹出对话框的"Item 名称"文本框中填入 Item 的名称，接下来选中"构建一个 maven 项目"单选按钮，最后单击 OK 按钮，如图 8-7 所示。

图 8-7 Jenkins 创建 job 工程

（4）创建 job 工程后，需要对其进行配置，如图 8-8 所示。

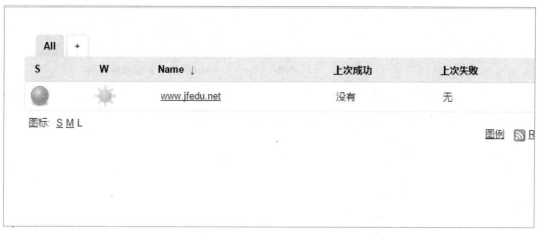

图 8-8　Jenkins 配置 job 工程

（5）单击工程名 www.jfedu.net 选择配置，进入 job 工程详细配置，打开"源码管理"对话框，选择 Subversion 配置 SVN 仓库地址，如果报错，需要输入 SVN 仓库的用户名和密码，如图 8-9 所示。

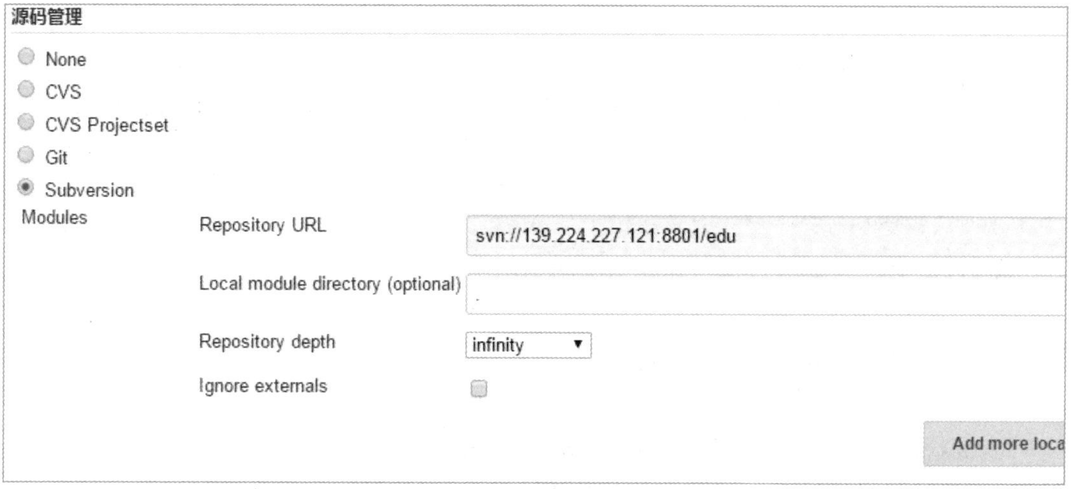

图 8-9　Jenkins 配置 SVN 仓库地址

SVN 代码迁出参数详解如下。

```
Respository url                    #配置 SVN 仓库地址
Local module directory             #存储 SVN 源码的路径
Ignore externals option            #忽略额外参数
```

```
Check-out Strategy              #代码检出策略
Repository browser              #仓库浏览器,默认为 Auto
add more locations              #源码管理,允许下载多个地址的代码
Repository depth                #获取 SVN 源码的目录深度,默认为 infinity
empty                           #不检出项目的任何文件,files:所有文件;immediates:
                                #目录第一级;infinity:整个目录所有文件
```

（6）配置 Maven 编译参数，依次选择 Build→Goals and options，输入 clean install -Dmaven.test.skip=true，此处为 Maven 自动编译、打包并跳过单元测试选项，如图 8-10 所示。

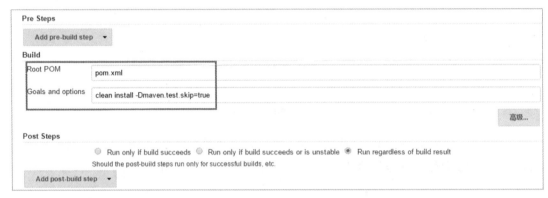

图 8-10　Jenkins 配置 Maven 编译参数

Maven 工具常用命令如下：

```
mvn clean                                       #打包清理（删除 target 目录内容）
mvn compile                                     #编译项目
mvn package                                     #打包发布
clean install -Dmaven.test.skip=true            #打包时跳过测试
```

通过以上步骤的配置，完成了 job 工程的创建。

8.8　Jenkins 构建 job 工程

Jenkins job 工程创建完毕后，直接运行构建，Jenkins 将从 SVN 仓库获取 SVN 代码，然后通过 Maven 进行编译、打包，并生成可以使用的压缩包。其操作步骤如下：

（1）单击工程名 www.jfedu.net，进入 job 工程详细配置界面，单击"立即构建"按钮，如图 8-11 所示。

（2）查看 Build History，单击最新一次的百分比进度条任务，如图 8-12 所示。

（3）进入 job 工程编译界面，单击 Console Output 按钮，如图 8-13 所示。

图 8-11　Jenkins job 工程配置界面

图 8-12　Jenkins job 工程构建界面

图 8-13　Jenkins job 工程编译界面

（4）查看 Jenkins 构建实时日志，如图 8-14 所示。

```
Started by user anonymous
Building on master in workspace /root/.jenkins/workspace/www.jfedu.net
Updating svn://139.224.227.121:8801/edu at revision '2017-06-04T15:17:11.704 +0800'
At revision 200
no change for svn://139.224.227.121:8801/edu since the previous build
No emails were triggered.
Parsing POMs
[www.jfedu.net] $ /usr/java/jdk1.8.0_131//bin/java -cp /root/.jenkins/plugins/maven-plugin/WEB-INF/lib/maven31-ag
1.5.jar:/data/maven/boot/plexus-classworlds-2.5.2.jar:/data/maven/conf/logging jenkins.maven3.agent.Maven31Main /
usr/local/tomcat_jenkins/webapps/ROOT/WEB-INF/lib/remoting-2.57.jar /root/.jenkins/plugins/maven-plugin/WEB-INF/
/root/.jenkins/plugins/maven-plugin/WEB-INF/lib/maven3-interceptor-commons-1.5.jar 27365
<===[JENKINS REMOTING CAPACITY]===>channel started
Executing Maven:  -B -f /root/.jenkins/workspace/www.jfedu.net/pom.xml clean install -Dmaven.test.skip=true
[INFO] Scanning for projects...
[INFO]
[INFO] ------------------------------------------------------------------------
[INFO] Building edu Maven Webapp 0.0.1-SNAPSHOT
[INFO] ------------------------------------------------------------------------
[INFO]
[INFO] --- maven-clean-plugin:2.5:clean (default-clean) @ edu ---
[INFO] Deleting /root/.jenkins/workspace/www.jfedu.net/target
```

（a）

```
[INFO]
[INFO] --- maven-install-plugin:2.4:install (default-install) @ edu ---
[INFO] Installing /root/.jenkins/workspace/www.jfedu.net/target/edu.war to /root/.m2/repository/com/shareku/edu/0
SNAPSHOT.war
[INFO] Installing /root/.jenkins/workspace/www.jfedu.net/pom.xml to /root/.m2/repository/com/shareku/edu/0.0.1-SN
[INFO] ------------------------------------------------------------------------
[INFO] BUILD SUCCESS
[INFO] ------------------------------------------------------------------------
[INFO] Total time: 13.733 s
[INFO] Finished at: 2017-06-04T15:17:35+08:00
[INFO] Final Memory: 25M/171M
[INFO] ------------------------------------------------------------------------
[JENKINS] Archiving /root/.jenkins/workspace/www.jfedu.net/pom.xml to com.shareku/edu/0.0.1-SNAPSHOT/edu-0.0.1-SN
[JENKINS] Archiving /root/.jenkins/workspace/www.jfedu.net/target/edu.war to com.shareku/edu/0.0.1-SNAPSHOT/edu-0
channel stopped
[www.jfedu.net] $ /bin/sh -xe /usr/local/tomcat_jenkins/temp/hudson6629705887694115145.sh
Archiving artifacts
```

（b）

图 8-14 Jenkins 构建实时日志

（a）Jenkins 控制台日志获取 SVN 代码；（b）编译、安装和打包 edu 程序包

控制台日志打印"Finished: SUCCESS"，则表示 Jenkins 持续集成构建完成，会在 Jenkins 服务器目录的 www.jfedu.net 工程名目录下生成网站可用的压缩文件，将该压缩文件部署至其他服务器，压缩文件路径为 /root/.jenkins/workspace/www.jfedu.net/target/edu.war。

至此，Jenkins 持续集成平台完成了自动构建软件。但该步骤只是生成了文件，并没有实现自动将该文件部署至其他服务器，如果要实现自动部署，需要基于 Jenkins 插件或者基于 Shell、Python 等自动部署脚本。

8.9 Jenkins 自动部署

8.8 节的手动构建 Jenkins job 工程，自动编译、打包生成压缩文件，并不能实现自动部署，如需实现自动部署可以基于自动部署插件或者 Shell 脚本、Python 脚本等实施。

以下用 Shell 脚本实现 Jenkins 自动部署压缩文件至其他多台服务器，并自动启动 Tomcat，实现最终 Web 浏览器访问。Jenkins 自动部署的完整操作步骤如下。

（1）单击工程名 www.jfedu.net，依次选择"配置"→"构建后操作"→Add post-build step→Archive the artifacts 命令，在"用于存档的文件"文本框中输入"**/target/*.war"，该选项主要用于 Jenkins 编译后将压缩文件存档一份到 target 目录下，该文件可以通过 Jenkins Tomcat 的 HTTP 端口访问，如图 8-15 所示。

（a）

（b）

图 8-15　Jenkins job 工程编译控制台

（a）增加文件归档；（b）设置文件归档路径

（2）Jenkins 构建完毕，访问 Jenkins war 存档的文件，URL 地址如下。

```
http://139.224.227.121:7001/job/www.jfedu.net/lastSuccessfulBuild/artifact/target/edu.war
```

（3）选择 Add post-build step→Execute shell 命令，在 Command 文本框中输入以下代码，实现 Jenkins edu.war 压缩文件自动部署。以下为 139.199.228.59 客户端单台服务器部署 edu.war 压缩文件，多台服务器可以使用 ip.txt 文件里的列表，将 IP 地址加入 ip.txt 文件，通过 for 循环实现批量部署，如图 8-16 所示。

```
cp /root/.jenkins/workspace/www.jfedu.net/target/edu.war /root/.jenkins/jobs/www.jfedu.net/builds/lastSuccessfulBuild/archive/target/
ssh root@ 139.199.228.59 'bash -x -s' < /data/sh/auto_deploy.sh
#for I in 'cat ip.txt';do ssh root@${I} 'bash -x -s' < /data/sh/auto_deploy.sh ;done
```

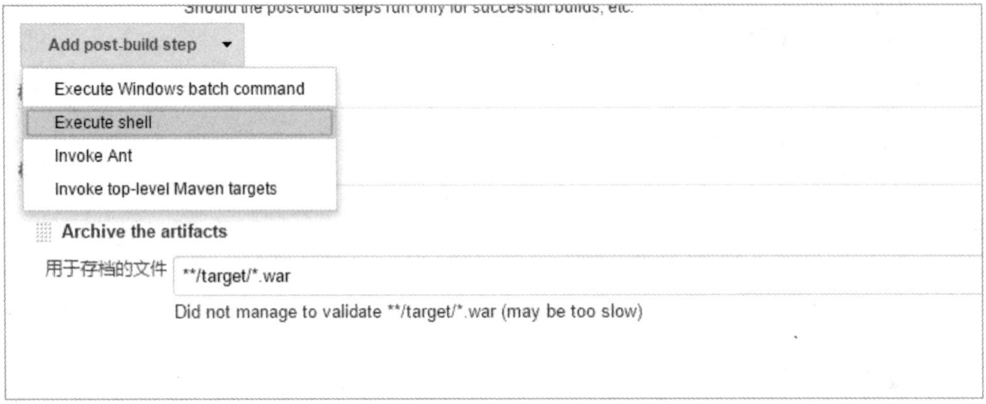

（a）

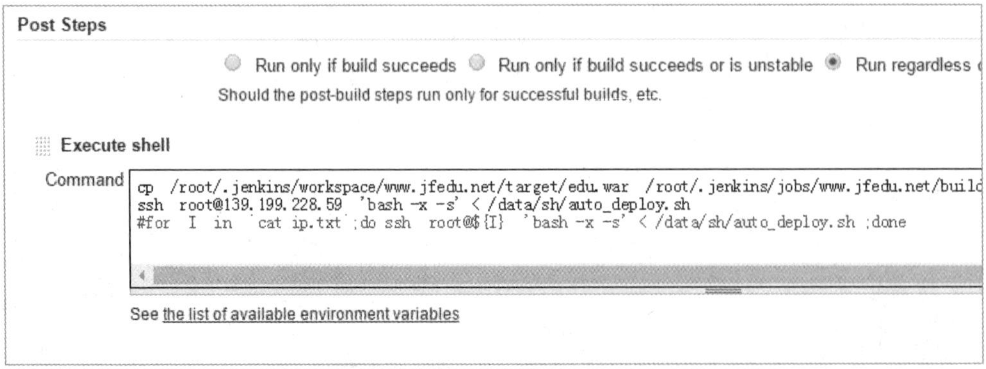

（b）

图 8-16　Jenkins job 构建完毕执行 Shell

（a）执行 Shell 指令；（b）编写 Shell 代码

（4）基于 Jenkins 将 edu.war 压缩文件自动部署至 139.199.228.59 服务器 Tomcat 发布目录，需提前配置登录远程客户端免密钥。免密钥配置是首先在 Jenkins 服务器执行 ssh-keygen 命令，然后按 Enter 键生成公钥和私钥，最后将公钥 id_rsa.pub 复制到客户端/root/.ssh/目录下，并重命名为 authorized_keys，操作命令如下。

```
ssh-keygen -t rsa -P '' -f /root/.ssh/id_rsa
ssh-copy-id -i /root/.ssh/id_rsa.pub 139.199.228.59
```

（5）Shell 脚本需放在 Jenkins 服务器/data/sh/目录下，而无须放在客户端，Shell 脚本内容如下。

```
#!/bin/bash
#Auto deploy Tomcat for jenkins
#By author jfedu.net 2021
export JAVA_HOME=/usr/java/jdk1.6.0_25
TOMCAT_PID='/usr/sbin/lsof -n -P -t -i :8081'
TOMCAT_DIR="/usr/local/tomcat/"
FILES="edu.war"
DES_DIR="/usr/local/tomcat/webapps/ROOT/"
DES_URL="http://139.224.227.121:7001/job/www.jfedu.net/lastSuccessfulBuild/artifact/target/"
BAK_DIR="/export/backup/'date +%Y%m%d-%H%M'"
[ -n "$TOMCAT_PID" ] && kill -9 $TOMCAT_PID
cd $DES_DIR
rm -rf $FILES
mkdir -p $BAK_DIR;\cp -a $DES_DIR/* $BAK_DIR/
rm -rf $DES_DIR/*
wget $DES_URL/$FILES
/usr/java/jdk1.6.0_25/bin/jar -xvf $FILES
#####################
cd $TOMCAT_DIR;rm -rf work
/bin/sh $TOMCAT_DIR/bin/start.sh
sleep 10
tail -n 50 $TOMCAT_DIR/logs/catalina.out
```

通过 Shell+for 循环可以实现网站简单的异步部署。如果需要将 Jenkins edu.war 压缩文件批量快速部署至 100 台甚至 500 台服务器，该如何实现呢？8.13 节会讲解此问题。

8.10 Jenkins 插件安装

Jenkins 最大的功能莫过于插件丰富，基于各种插件可以满足各项需求。Jenkins 本身只是一个框架，真正发挥作用的是各种插件。Jenkins 默认自带很多插件，如果需添加新插件，可以在 Jenkins 平台的主页面进行操作，操作步骤如下。

如图 8-17 所示，单击"管理插件"按钮，在可选插件中搜索 Email-ext plugin，选择并安装插件；如果没有该插件，则需要单击"高级"按钮，手动上传插件并安装。

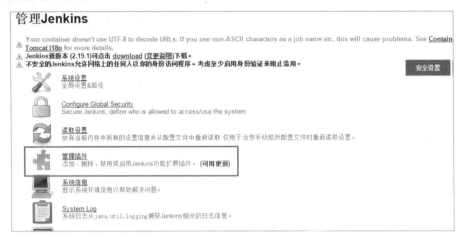

图 8-17　Jenkins 添加新插件

访问 Jenkins 官网，手动下载插件，将下载的插件上传到服务器 Jenkins 根目录/root 下的 plugins 目录，即/root/.jenkins/plugins 目录，然后重启 Jenkins。Jenkins 插件下载地址为 https://wiki.jenkins-ci.org/display/JENKINS/Plugins。

（1）下载 Email-ext 插件、Token-Macro 插件和 Email-ext template 插件，可以输入插件名称搜索某个插件，如图 8-18 所示。

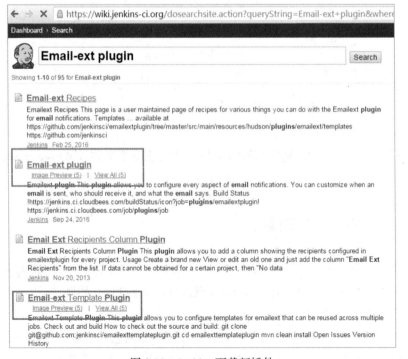

图 8-18　Jenkins 下载新插件

（2）Email-ext 插件、Token-Macro 插件和 Email-ext template 插件的下载地址如下。

```
https://wiki.jenkins-ci.org/display/JENKINS/Email-ext+plugin
https://wiki.jenkins-ci.org/display/JENKINS/Token+Macro+Plugin
https://wiki.jenkins-ci.org/display/JENKINS/Email-ext+Template+Plugin
```

（3）安装 Token-Macro 插件，如图 8-19 所示。

（a）

（b）

图 8-19　安装 Token-Macro 插件

（a）上传插件；（b）安装插件

（4）安装 Email-ext 插件，如图 8-20 所示。

图 8-20　安装 Email-ext 插件

（5）Email-ext 插件、Token-Macro 插件和 Email-ext template 插件安装完毕，如图 8-21 所示。

图 8-21　插件安装完毕

（6）Email-ext 插件安装完毕，在 Jenkins 主界面依次选择"系统管理"→"系统设置"命令，弹出 Extended E-mail Notification 界面，如图 8-22 所示。

如需安装 Git 插件、Publish Over 插件或者 Jenkins 其他任意插件，其安装方法与 Email-ext 插件的安装方法一致。

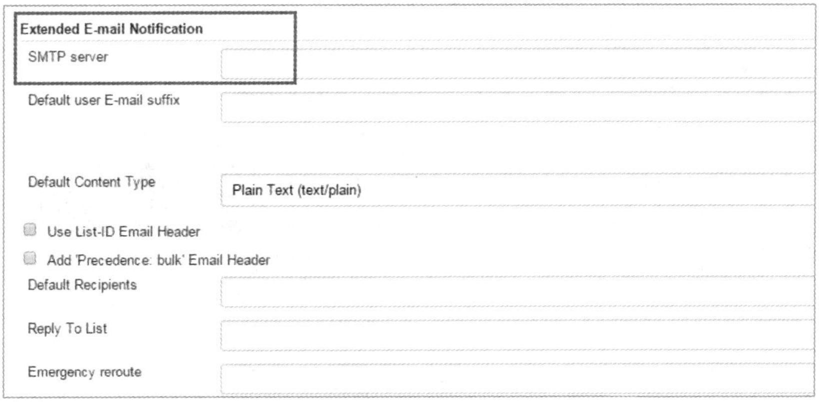

图 8-22 Extended E-mail Notification 界面

8.11　Jenkins 邮件配置

Jenkins 持续集成配置完毕后，可以进行网站代码的自动更新、部署、升级及回滚操作，通过控制台信息可以查看每个 job 工程构建的状态。

如果网站项目很多，人工查看工程构建状态就变得不可取，可以借助 Jenkins Email-ext 插件，实现网站构建之后自动发送邮件给相应的开发人员、运维人员或者测试人员。Jenkins 发送邮件需安装 Email 邮件插件 Email-ext、Token-Macro 和 Email-ext template。Jenkins Email 邮件配置常见参数如下。

```
SMTP server                    #邮件服务器地址
Default Content Type           #内容展现的格式,一般选择 HTML
Default Recipients             #默认收件人
Use SMTP Authentication        #使用 SMTP 身份验证
User Name                      #邮件发送账户的用户名
Password                       #邮件发送账户的密码
SMTP port                      #SMTP 服务器口
```

Jenkins Email 邮件配置方法如下。

（1）设置 Jenkins 邮件发送者，在 Jenkins 平台首页单击"系统管理"按钮，然后单击"系统设置"按钮，在弹出的"Maven 项目配置"对话框的 Jenkins Location 选项组中的 Jenkins URL 和"系统管理员邮件地址"文本框中填写相关的内容，如图 8-23 所示。

（2）设置发送邮件的 SMTP 服务器、邮箱后缀、发送类型、接收者或抄送者。在 Jenkins 平台首页单击"系统管理"按钮，然后单击"系统设置"按钮，在 Extended E-mail Notification 选项组中填写如图 8-24 所示的内容，包括 SMTP server、Default user E-mail Suffix、Use SMTP Authentication、Default Recipients 等信息。

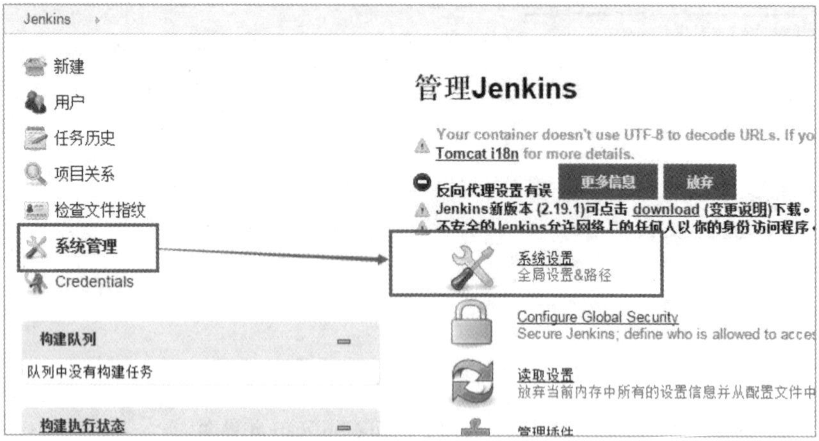

（a）

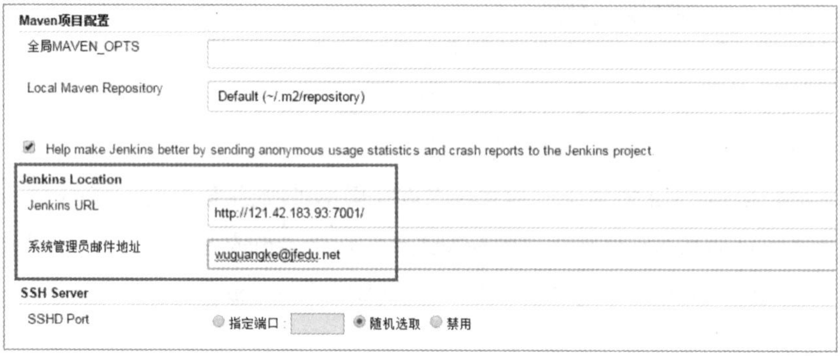

（b）

图 8-23　Jenkins Email 邮件配置 1

（a）Jenkins 系统管理；（b）添加邮件服务器地址

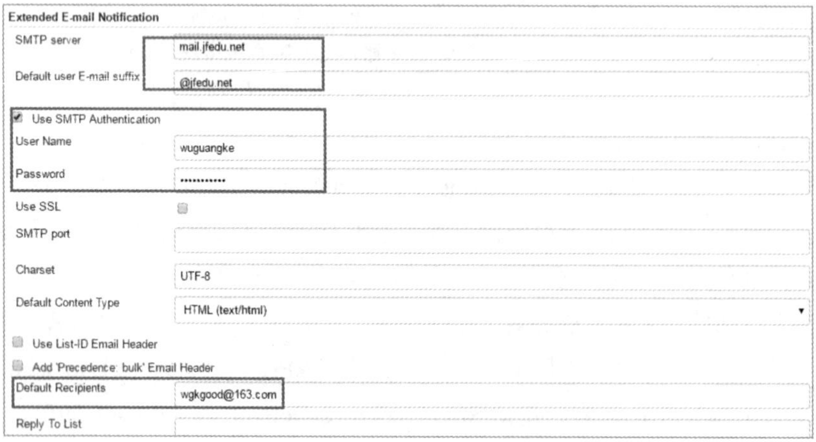

图 8-24　Jenkins Email 邮件配置 2

（3）设置邮件的默认标题（default subject）如下。

```
#构建通知
$PROJECT_NAME - Build # $BUILD_NUMBER - $BUILD_STATUS
```

（4）设置发送邮件的默认内容（default content）如下。

```
<hr/>
<h3>(本邮件是程序自动下发的,请勿回复！)</h3><hr/>
#项目名称
$PROJECT_NAME<br/><hr/>
#构建编号
$BUILD_NUMBER<br/><hr/>
#构建状态
$BUILD_STATUS<br/><hr/>
#触发原因
${CAUSE}<br/><hr/>
构建日志地址：<a href="${BUILD_URL}console">${BUILD_URL}console</a><br/><hr/>
#构建地址
<a href="$BUILD_URL">$BUILD_URL</a><br/><hr/>
#变更集
${JELLY_SCRIPT,template="html"}<br/>
<hr/>
```

（5）设置每个 job 工程邮件时，单击 www.jfedu.net job 名称，然后依次选择"配置"→"构建后操作"命令，在 Editable Email Notification 选项组中的信息保持默认，如图 8-25 所示。

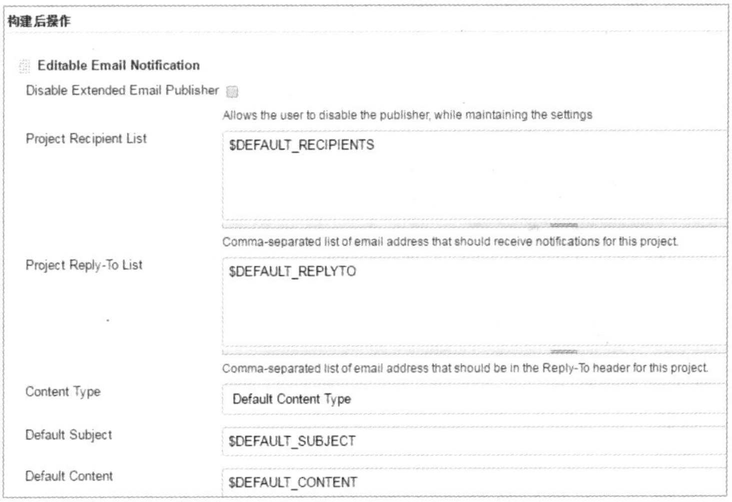

图 8-25 Jenkins Email job 邮件模板配置

（6）选择 Advanced Settings，设置 Trigger 阈值，选择发送邮件的触发器，默认触发器包括第一次构建、构建失败、总是发送邮件、构建成功等，一般选择 Always（总是发送邮件），发送给 Developers 组，如图 8-26 所示。

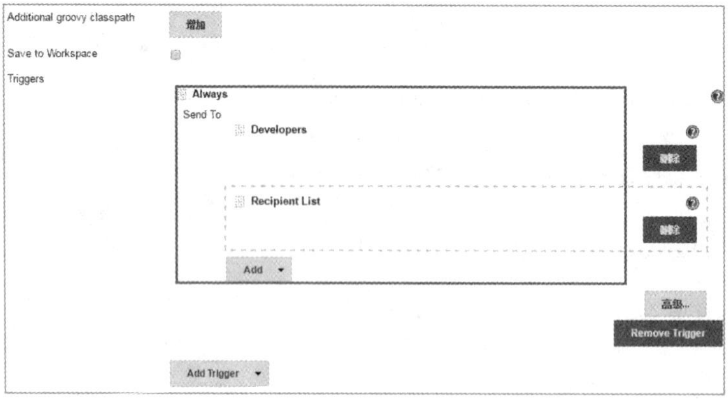

图 8-26　Jenkins Email 触发器设置

（7）Jenkins 构建邮件验证，如图 8-27 所示。

（a）

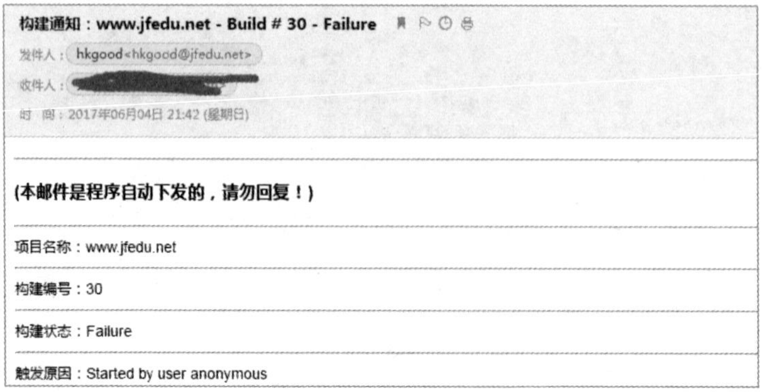

（b）

图 8-27　Jenkins 构建邮件验证

（a）Jenkins 构建报错触发邮件；（b）Jenkins Email 邮件信息 1

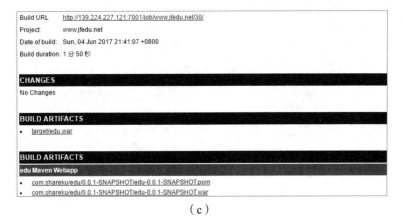

（c）

图 8-27　Jenkins 构建邮件验证（续）

（c）Jenkins Email 邮件信息 2

8.12　Jenkins 多实例配置

单台 Jenkins 服务器可以满足企业测试环境及生产环境使用 Jenkins 自动部署+测试平台，如果每天更新发布多个 Web 网站，Jenkins 需要同时处理很多任务。

基于 Jenkins 分布式（多 Slave 方式）可以缓解 Jenkins 服务器的压力。Jenkins 多 Slave 架构如图 8-28 所示，可以在 Windows、Linux、macOS 等操作系统上执行 Slave。

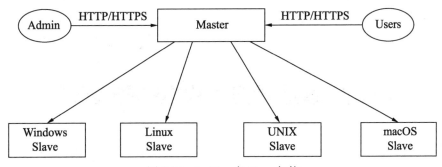

图 8-28　Jenkins 多 Slave 架构

Jenkins 多 Slave 原理是将原本在 Jenkins Master 端的构建项目分配给 Slave 端去执行，Jenkins Master 端分配任务时，Jenkins Master 端通过 SSH 远程控制 Slave，在 Slave 端启动 slave.jar 程序，通过 slave.jar 程序实现对网站工程的构建编译以及自动部署。所以，在 Slave 端服务器必须安装 Java JDK 环境以执行 Master 端分配的构建任务。配置多 Slave 服务器的方法和步骤如下。

（1）在 Slave 服务器上创建远程执行 Jenkins 任务的用户，其名称为 Jenkins。Jenkins 的工作目录为/home/Jenkins，Jenkins Master 免密钥登录 Slave 服务器或者通过用户名和密码登录 Slave。

（2）Slave 服务器安装 Java JDK 环境，并将其软件路径加入系统环境变量。

（3）Jenkins Master 端平台添加管理节点，在"系统管理"对话框中单击"管理节点"按钮，然后单击"新建节点"按钮，输入节点名称，如图 8-29 所示。

（a）

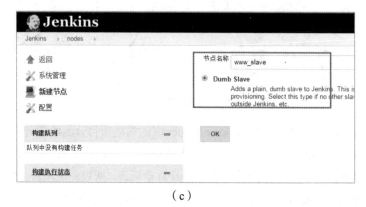

（b）

（c）

图 8-29　Jenkins Slave 配置

（a）Jenkins 管理节点；（b）Jenkins 新增节点；（c）输入 Jenkins 节点名称

（4）配置 www_slave 节点，指定其 Jenkins 编译工作目录，设置 IP 地址，添加登录 Slave 用户名和密码，如图 8-30 所示。

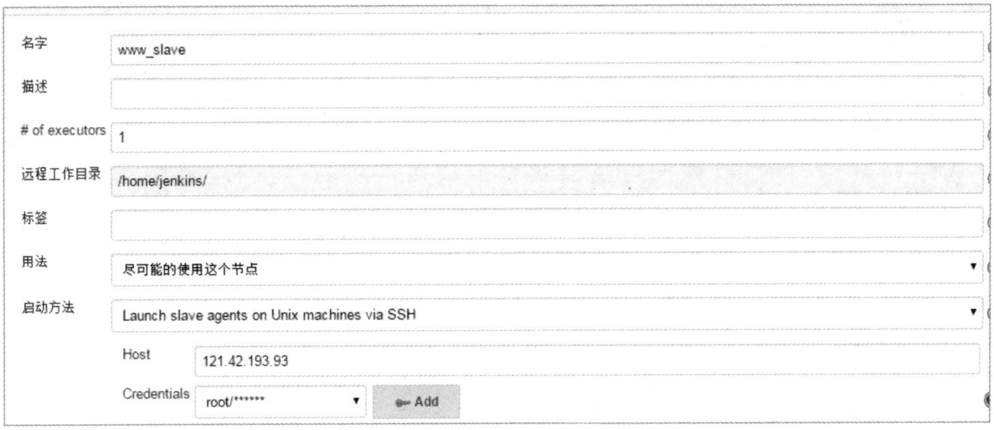

（a）

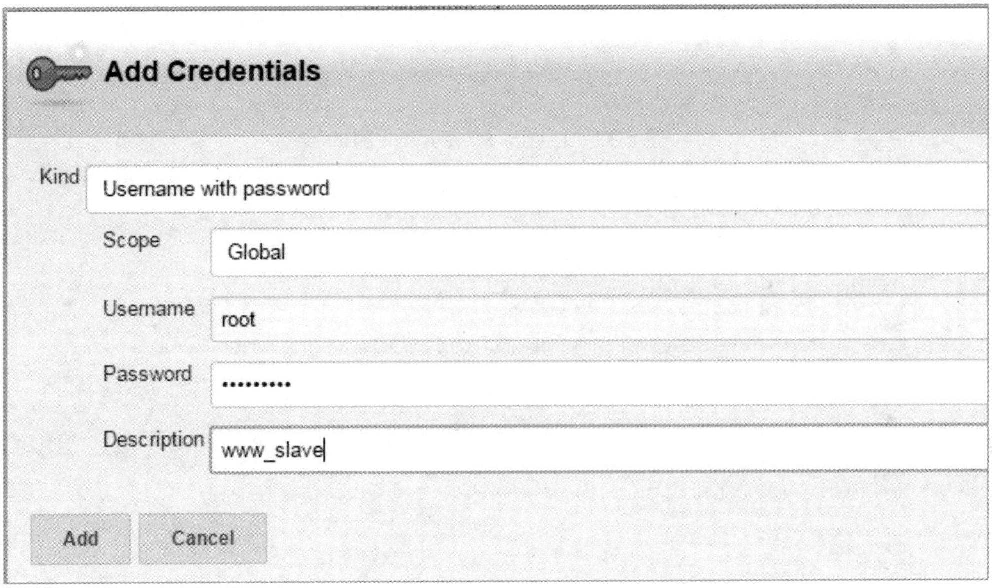

（b）

图 8-30　配置 www_slave 节点

（a）设置 Jenkins 节点工作目录；（b）添加节点登录密码信息

（5）Jenkins Slave 配置完毕，查看 Slave 状态信息，如图 8-31 所示。

（6）单击 www_slave 节点，然后单击 Launch slave agent 按钮，测试 Slave Agent 是否正常工作，如图 8-32 所示。

S	名称 ↓	Architecture	Clock Difference	Free Disk Space	Free Swap Space	Free Temp Space	R
	master	Linux (amd64)	In sync	14.24 GB	0 B	14.24 GB	
	www_slave		N/A	N/A	N/A	N/A	
	获取到的数据	15 分	15 分	15 分	15 分	15 分	

图 8-31　Jenkins Slave 状态信息

图 8-32　Jenkins Slave Agent 测试

（7）出现如图 8-33 所示的提示，证明 Slave 添加成功。

```
[01/08/17 16:34:48] [SSH] Opening SSH connection to 121.42.183.93:22.
[01/08/17 16:34:48] [SSH] Authentication successful.
[01/08/17 16:34:48] [SSH] The remote users environment is:
BASH=/bin/bash
BASHOPTS=cmdhist:extquote:force_fignore:hostcomplete:interactive_comments:progcomp:promptvars:sourcepath
BASH_ALIASES=()
BASH_ARGC=()
BASH_ARGV=()
BASH_CMDS=()
BASH_EXECUTION_STRING=set
BASH_LINENO=()
BASH_SOURCE=()
BASH_VERSINFO=([0]="4" [1]="1" [2]="2" [3]="1" [4]="release" [5]="x86_64-redhat-linux-gnu")
BASH_VERSION='4.1.2(1)-release'
CVS_RSH=ssh
DIRSTACK=()
EUID=0
GROUPS=()
G_BROKEN_FILENAMES=1
HOME=/root
HOSTNAME=BeiJing-JFEDU-NET-WEB-001.COM
HOSTTYPE=x86_64
ID=0
IFS=$' \t\n'
LANG=en_US.UTF-8
```

（a）

图 8-33　Jenkins Slave 测试

（a）Jenkins 节点启动

```
PIPESTATUS=([0]="0")
PPID=5325
PS4='+ '
PWD=/root
SHELL=/bin/bash
SHELLOPTS=braceexpand:hashall:interactive-comments
SHLVL=1
SSH_CLIENT='139.224.227.121 21604 22'
SSH_CONNECTION='139.224.227.121 21604 121.42.183.93 22'
TERM=dumb
UID=0
USER=root
_=/etc/bashrc
[01/08/17 16:34:48] [SSH] Starting sftp client.
[01/08/17 16:34:48] [SSH] Remote file system root /home/jenkins does not exist. Will try to create it...
[01/08/17 16:34:48] [SSH] Copying latest slave.jar...
[01/08/17 16:34:51] [SSH] Copied 522,364 bytes.
Expanded the channel window size to 4MB
[01/08/17 16:34:51] [SSH] Starting slave process: cd "/home/jenkins" && /usr/java/jdk1.7.0_25/bin/java -jar slave.jar
<===[JENKINS REMOTING CAPACITY]===>channel started
Slave.jar version: 2.57
This is a Unix slave
```

（b）

图 8-33 Jenkins Slave 测试（续）

（b）Jenkins 节点测试

（8）配置完毕，Jenkins Master 端通过 SSH 方式启动 Slave 的 slave.jar 文件命令为 java -jar slave.jar，Slave 等待 Master 端的任务分配，单击 www.jfedu.net 按钮，如图 8-34 所示，然后构建任务。

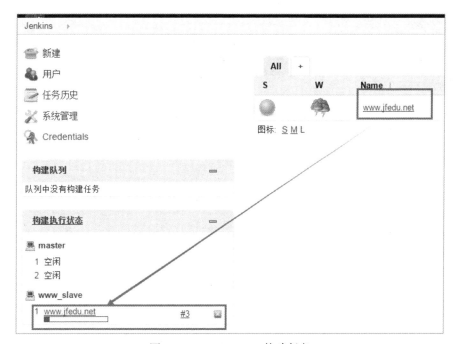

图 8-34 Jenkins Slave 构建任务

（9）Jenkins+Slave 配置完毕后，如果同时运行多个任务，将只运行一个任务，而其他任务在等待。如果想多个任务同时运行，需要怎么调整呢？答案是需要在配置 job 工程时，勾选"在必要的时候并发构建"复选框，如图 8-35 所示。

（a）

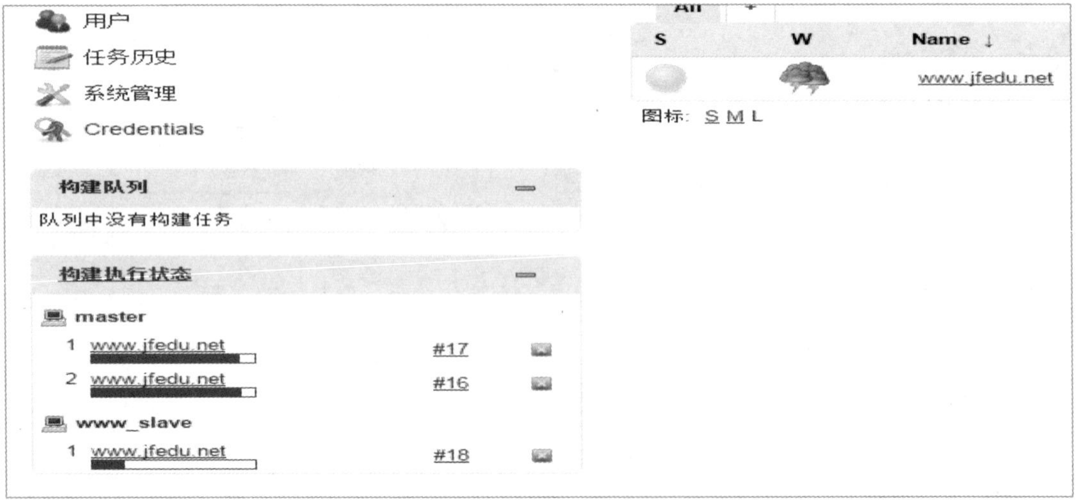

（b）

图 8-35　Jenkins Slave 构建多任务

（a）选择节点并发执行；（b）查看节点运行任务状态

8.13　Jenkins+Ansible 高并发构建

　　Jenkins 基于 Shell+for 循环的方式自动部署 10 台以下的 Java 客户端服务器，其效率是可以接受的，但如果是大规模服务器需要部署或者更新网站，那么通过 for 循环串行执行的效率会大打折扣，所以需要考虑并行机制。

　　Ansible 是一款极为灵活的开源工具套件，能够大大简化 UNIX 管理员的自动化配置管理与流程控制方式。它利用推送方式对客户系统加以配置，这样所有工作都可在主服务器完成。使用 Ansible+Jenkins 架构方式实现网站自动部署，可满足上百台、甚至上千台服务器的网站部署和更新。

　　Ansible 服务需要部署在 Jenkins 服务器上，客户端无须安装 Ansible。Ansible 是基于 SSH 工作，所以需要提前设置免密钥或者通过 sudo 用户远程更新网站。此处省略 Ansible 安装方法，Ansible 相关知识请参考本书 5.3 节。

　　Ansible 自动部署网站有两种方式，一种是基于 Ansible 远程执行 Shell 脚本，另外一种是 Ansible 编写 Playbook 剧本，实现网站自动部署。以下为 Ansible+Shell 脚本方式自动部署网站的方法。

　　（1）Jenkins 服务器安装 Ansible 软件，Red Hat 操作系统和 CentOS 操作系统可以直接基于 YUM 工具自动安装 Ansible，CentOS 6.x 操作系统或者 CentOS 7.x 操作系统安装前，需要先安装 EPEL 扩展源，安装代码如下。

```
rpm -Uvh http://mirrors.ustc.edu.cn/fedora/epel/6/x86_64/epel-release-6-8.noarch.rpm
yum    install    epel-release    -y
yum    install    ansible         -y
```

　　（2）添加客户端，在/etc/ansible/hosts 文件中添加需要部署的客户端 IP 地址列表的代码如下。

```
[www_jfedu]
139.199.228.59
139.199.228.60
139.199.228.61
139.199.228.62
```

　　（3）在 Jenkins 平台首页单击 www.jfedu.net 项目，选择"配置"命令，再单击 Post Steps 按钮，选择 Execute shell 命令，在 Command 文本框中输入以下代码，www_jfedu 为 Ansible Hosts 组模块名称。

```
cp /root/.jenkins/workspace/www.jfedu.net/target/edu.war /root/.jenkins/jobs/www.jfedu.net/builds/lastSuccessfulBuild/archive/target/
ansible www_jfedu -m copy -a "src=/data/sh/auto_deploy.sh dest=/tmp/"
ansible www_jfedu -m shell -a "cd /tmp ;/bin/bash auto_deploy.sh"
```

（4）Jenkins 服务器/data/sh/auto_deploy.sh Shell 文件的脚本内容如下。

```bash
#!/bin/bash
#Auto deploy Tomcat for jenkins
#By author jfedu.net 2021
export JAVA_HOME=/usr/java/jdk1.6.0_25
TOMCAT_PID=`/usr/sbin/lsof -n -P -t -i :8081`
TOMCAT_DIR="/usr/local/tomcat/"
FILES="edu.war"
DES_DIR="/usr/local/tomcat/webapps/ROOT/"
DES_URL="http://139.224.227.121:7001/job/www.jfedu.net/lastSuccessfulBuild/artifact/target/"
BAK_DIR="/export/backup/`date +%Y%m%d-%H%M`"
[ -n "$TOMCAT_PID" ] && kill -9 $TOMCAT_PID
cd $DES_DIR
rm -rf $FILES
mkdir -p $BAK_DIR;\cp -a $DES_DIR/* $BAK_DIR/
rm -rf $DES_DIR/*
wget $DES_URL/$FILES
/usr/java/jdk1.6.0_25/bin/jar -xvf $FILES
####################
cd $TOMCAT_DIR;rm -rf work
/bin/sh $TOMCAT_DIR/bin/start.sh
sleep 10
tail -n 50 $TOMCAT_DIR/logs/catalina.out
```

（5）单击 www.jfedu.net 构建任务，并查看控制台信息，如图 8-36 所示。

```
[www.jfedu.net] $ /bin/sh -xe /usr/local/tomcat_jenkins/temp/hudson2957878034347171274.sh
+ cp /root/.jenkins/workspace/www.jfedu.net/target/edu.war /root/.jenkins/jobs/www.jfedu.net/builds/lastSuccessfu
+ ansible www_jfedu -m copy -a 'src=/data/sh/auto_deploy.sh dest=/tmp/'
/usr/lib/python2.6/site-packages/cryptography-1.7.1-py2.6-linux-x86_64.egg/cryptography/__init__.py:26: Deprecati
supported by the Python core team, please upgrade your Python. A future version of cryptography will drop support
 DeprecationWarning
139.199.228.59 | SUCCESS => {
    "changed": true,
    "checksum": "2ee998fb7ef482a498f1a312db913cce3e0d821b",
    "dest": "/tmp/auto_deploy.sh",
    "gid": 0,
    "group": "root",
    "md5sum": "5271b4d091746545353f79a96207a16b",
    "mode": "0644",
    "owner": "root",
    "size": 715,
    "src": "/root/.ansible/tmp/ansible-tmp-1496587744.11-232708497021134/source",
    "state": "file",
    "uid": 0
}
+ ansible www_jfedu -m shell -a 'cd /tmp ;/bin/bash auto_deploy.sh'
/usr/lib/python2.6/site-packages/cryptography-1.7.1-py2.6-linux-x86_64.egg/cryptography/__init__.py:26: Deprecati
supported by the Python core team, please upgrade your Python. A future version of cryptography will drop support
 DeprecationWarning
```

（a）

图 8-36　Jenkins Ansible 自动部署

（a）查看 Jenkins 任务构建状态

```
24700K .........                                          97%  246K 2s
24750K .........                                          98%  297K 2s
24800K .........                                          98%  291K 2s
24850K .........                                          98%  248K 1s
24900K .........                                          98%  297K 1s
24950K .........                                          98%  251K 1s
25000K .........                                          99%  362K 1s
25050K .........                                          99%  251K 1s
25100K .........                                          99%  247K 1s
25150K .........                                          99%  297K 0s
25200K .........                                          99%  296K 0s
25250K .........                                         100%  259K=94s

2017-06-04 22:50:42 (270 KB/s) - 已保存 "edu.war" [25900686/25900686])
Archiving artifacts
Email was triggered for: Always
Sending email for trigger: Always
Sending email to: wgkgood@163.com
Finished: SUCCESS
```

(b)

(c)

图 8-36　Jenkins Ansible 自动部署（续）

（b）Jenkins 任务构建过程；（c）访问构建后的网站

第 9 章 SVN 版本管理实战

9.1 SVN 服务器简介

Subversion（SVN）是一个自由、开源的版本控制系统。

在实际使用分布式控制系统的时候，其实很少在两台计算机之间推送版本库的修改，可能因为这两台计算机不在一个局域网内而无法互相访问，也可能因为你的同事今天病了，他的计算机压根没有开机。因此，分布式版本控制系统通常会有一台充当"中央服务器"的计算机，但这个服务器的作用仅仅是方便"交换"大家的修改，没有它大家也一样工作，只是不方便交换修改而已。

在 SVN 管理下，文件和目录可以超越时空。SVN 允许数据恢复到早期版本，或者是检查数据修改的历史。正因如此，许多人将版本控制系统当作一种神奇的"时间机器"。

SVN 的版本库可以通过网络访问，从而让用户可以在不同的计算机上进行操作。从某种程度上来说，允许用户在各自的空间里修改和管理同一组数据，能够有效促进团队协作。因为修改不再是单线进行，所以开发速度会更快。

此外，由于所有的工作都已经版本化，也就不必担心由于错误的更改而影响软件质量的问题，即使出现不正确的更改，只要撤销那一次更改操作（回滚）即可。

9.2 SVN 的功能特性

SVN 版本库有以下特点。

（1）版本化的目录。CVS 只能跟踪单个文件的变更历史，但是 SVN 实现的"虚拟"版本化文件系统可以跟踪目录树的变更。在 SVN 中，文件和目录都是版本化的。

（2）真实的版本历史。由于只能跟踪单个文件的变更，CVS 无法支持如文件复制、改名这些常见的操作——这些操作改变了目录的内容。同样，在 CVS 中，一个目录下的文件只要名称相同即拥有相同的历史，即使这些同名文件在历史上毫无关系。而在 SVN 中，可以对文件或目录进行增加、复制和改名操作，也解决了同名而无关的文件之间的历史联系问题。

（3）原子提交。一系列相关的更改，要么全部提交到版本库，要么一个也不提交。这样用户就可以将相关的更改组成一个逻辑整体，防止出现只有部分修改提交到版本库的情况。

（4）版本化的元数据。每一个文件和目录都有自己的一组属性——键和对应的值。可以根据需要建立并存储任何键/值对。和文件本身的内容一样，属性也在版本控制之下。

（5）可选的网络层。SVN 在版本库访问的实现上具有较高的抽象程度，有利于人们实现新的网络访问机制。SVN 可以作为一个扩展模块嵌入 Apache。这种方式在稳定性和交互性方面有很大的优势，可以直接使用服务器的成熟技术——认证、授权和传输压缩等。此外，SVN 自身也实现了一个轻型的、可独立运行的服务器软件，这个服务器软件使用了一个自定义协议，可以轻松使用 SSH 封装。

（6）一致的数据操作。SVN 用一个二进制差异算法描述文件的变化，对文本（可读）文件和二进制（不可读）文件的操作方式一致。这两种类型的文件压缩存储在版本库中，而差异信息则在网络上双向传递。

（7）高效的分支和标签操作。在 SVN 中，分支和标签操作的开销与工程的大小无关。SVN 的分支和标签操作只是使用一种类似于硬链接的机制复制整个工程。因而，这些操作通常只会花费很少且相对固定的时间。

（8）可修改性。SVN 没有历史负担，它以一系列优质的共享 C 程序库的方式实现，具有定义良好的 API。这使 SVN 非常容易维护，和其他语言的互操作性也很强。

9.3　SVN 的架构剖析

如图 9-1 所示，一端是保存所有版本数据的 SVN 版本库，另一端是 SVN 的客户端应用程序，管理所有版本数据的本地影射（称为"工作备份"），在这两极之间是各种各样的版本库访问（RA）层，某些层通过网络服务器访问版本库，某些层则绕过网络服务器直接访问版本库。

SVN 版本管理主要是以文件变更列表的方式存储信息，SVN 将保存的信息看作一组基本文件和每个文件随时间逐步累积的差异，如图 9-2 所示。

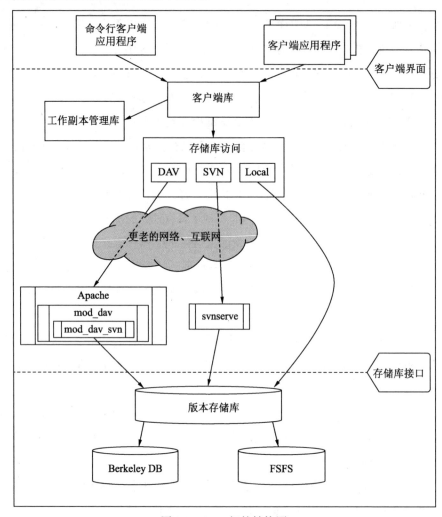

图 9-1　SVN 拓扑结构图

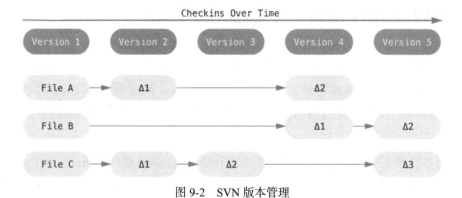

图 9-2　SVN 版本管理

9.4 SVN 的组件模块

SVN 由以下几部分组成。

（1）svn。

命令行客户端程序。

（2）svnversion。

用来显示工作备份的状态，即当前项目的修订版本。

（3）svnlook。

直接查看 SVN 版本库的工具。

（4）svnadmin。

建立、调整和修复 SVN 版本库的工具。

（5）svndumpfilter。

过滤 SVN 版本库转储数据流的工具。

（6）mod_dav_svn。

Apache HTTP 服务器的插件，使版本库可以通过网络访问。

（7）svnserve。

单独运行的服务器程序，或可以作为守护进程由 SSH 调用。这是另一种让版本库可以通过网络访问的方式。

（8）svnsync。

通过网络增量镜像版本库的程序。

9.5 SVN 分支概念剖析

假设你的工作是维护本公司一个部门的手册文档，有一天，另一个部门需要近乎相同的手册，但有一些地方会有区别，因为他们的部分需求不同。

这种情况下你会怎样做？显而易见的方法是：做一个版本的备份，然后分别维护这两个版本，需要时在对应的版本进行更改。

也许你希望在两个版本同时作修改。举个例子，你在第一个版本发现了一个拼写错误，很显然这个错误也会出现在第二个版本里。因为两份文档几乎相同，只有特定的微小区别。

分支的基本概念正如它的名字——开发的一条线独立于另一条线，如果回顾历史，可以发现两条线分享共同的历史，一个分支总是从一个备份开始，然后发展自己独有的历史。

SVN 允许并行维护文件和目录的分支，允许通过复制数据建立分支。注意，分支互相联系，它从一个分支复制修改到另一个分支。最终，它可以让工作备份反映到不同的分支上，所以日常工作中可以"混合和比较"不同的开发线。

9.6 基于 YUM 构建 SVN 服务器

（1）基于 CentOS 7.x 操作系统，通过 YUM 安装 SVN，其操作指令如下。

```
yum install subversion -y
```

（2）建立版本库目录的操作指令如下。

```
mkdir -p /data/svn/
```

（3）建立 SVN 版本库的操作指令如下。

```
svnadmin create /data/svn
```

（4）修改版本库配置文件，在/data/svn/conf/svnserve.conf 文件中加入如下代码。

```
cat>/data/svn/conf/svnserve.conf<<EOF
[general]
#使非授权用户无法访问
anon-access = none
#使授权用户有写权限
auth-access = write
#指明密码文件路径
password-db = passwd
#访问控制文件
authz-db = authz
#认证命名空间，SVN 会在认证提示里显示,并作为凭证缓存的关键字
realm = /data/svn
EOF
```

（5）配置用户及权限，修改/data/svn/conf/passwd 文件，添加两个用户访问 SVN 服务端，操作方法和指令如下。

```
cat>/data/svn/conf/passwd<<EOF
[users]
jfedu1 = 123456
jfedu2 = 123456
EOF
```

（6）配置用户及权限，修改/data/svn/conf/authz 文件，添加两个用户访问 SVN 服务端，操作方法和指令如下。

```
cat>/data/svn/conf/authz<<EOF
[/]
jfedu1 = rw
jfedu2 = rw
EOF
```

（7）启动 SVN 服务并设置系统服务的操作指令如下。

```
svnserve -d -r /data/svn/ --listen-port=8001
```

（8）查看本地 8001 端口是否启动的操作指令如下。

```
netstat -ntl|grep 8001
```

（9）SVN 平台部署成功，可以在客户端使用 SVN 指令，所有操作如图 9-3 所示。

```
svn co svn://101.34.116.235:8001/
```

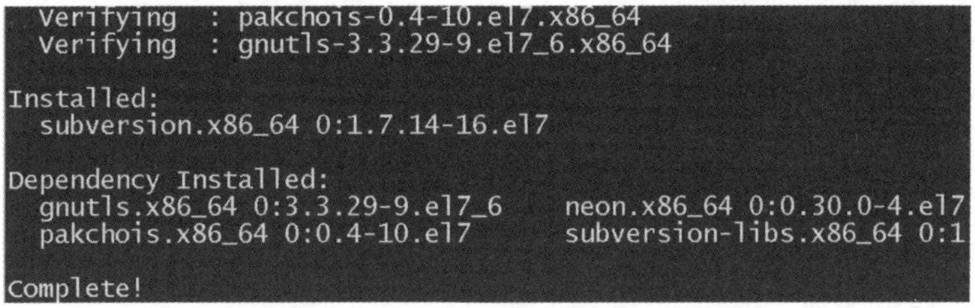

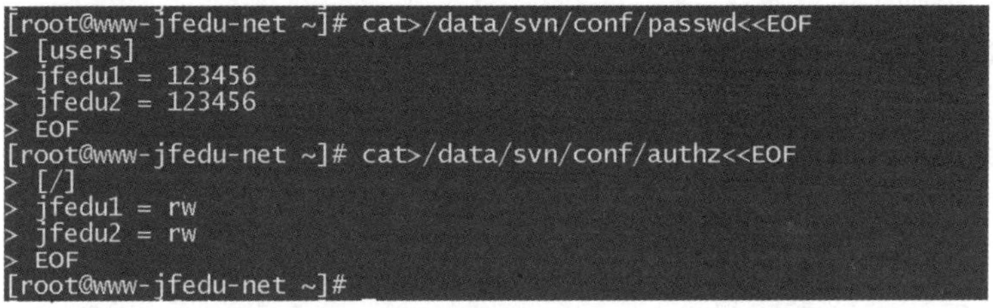

图 9-3　在客户端使用 SVN 命令

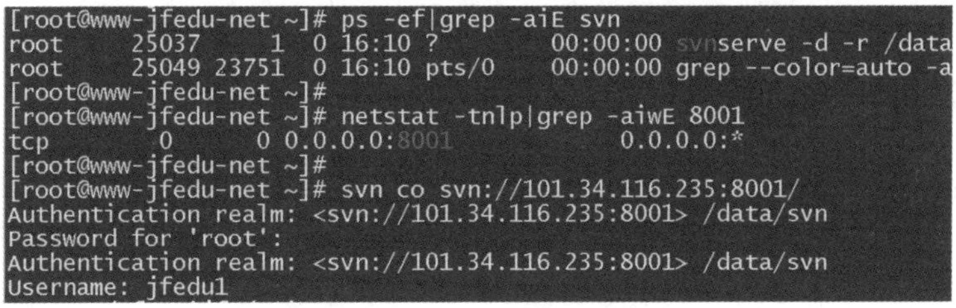

（a）安装 subversion 服务；（b）创建 SVN 用户名和密码；（c）查看 SVN 进程状态

9.7　SVN 二进制+Apache 整合实战

（1）安装 httpd 软件包的操作指令如下。

```
yum install httpd httpd-devel mod_dav_svn -y
```

（2）检测 SVN 和 Apache 整合模块是否安装成功的操作指令如下。

```
ls -l/etc/httpd/modules/ | grep svn
```

（3）修改 httpd 主配置文件/etc/httpd/conf/httpd.conf，加入以下代码。

```
LoadModule dav_module modules/mod_dav.so
LoadModule dav_svn_module modules/mod_dav_svn.so
```

（4）修改 httpd 主配置文件，末行添加如下代码。

```
<Location /svn>
      DAV svn
      SVNPath /data/svn
      AuthType Basic
      AuthName "svn for project"
      AuthUserFile  /etc/httpd/conf/passwd
      AuthzSVNAccessFile /data/svn/conf/authz
      Satisfy all
      Require valid-user
</Location>
```

（5）生成 HTTP 访问密钥的操作指令如下。

```
htpasswd -c /etc/httpd/conf/passwd jfedu1
htpasswd /etc/httpd/conf/passwd jfedu2
```

（6）重启 Apache httpd 服务的操作指令如下。

```
service httpd restart
```

（7）如果没有权限，需要授权给/data/svn 目录，其操作指令如下。

```
chown -R  apache /data/svn
```

（8）SVN 和 Apache 整合成功，通过浏览器访问 SVN，输入用户名和密码后，结果如图 9-4 所示。

图 9-4 浏览器访问 SVN

9.8 基于 MAKE 构建 SVN 服务器

（1）基于 CentOS 7.x 操作系统，通过 MAKE 源码编译安装 SVN 的操作指令如下。

```
#下载 SVN 相关软件包
wget https://archive.apache.org/dist/subversion/subversion-1.8.9.tar.bz2
wget http://www.sqlite.org/sqlite-amalgamation-3071502.zip
wget https://mirrors.cnnic.cn/apache/httpd/httpd-2.4.37.tar.gz
#解压 SVN 软件包
tar -xzvf subversion-1.8.9.tar.bz2
unzip sqlite-amalgamation-3071502.zip
mv sqlite-amalgamation-3071502  subversion-1.8.9/sqlite-amalgamation
#切换至 SVN 源代码包
cd subversion-1.8.9
#SVN 预编译（提前编译安装 Apache）
./configure --prefix=/usr/local/subversion --with-apxs=/usr/local/apache/bin/apxs --enable-mod-activation
#Apache 预编译参数
#./configure --prefix=/usr/local/apache/ --enable--so --enable-dav --enable-maintainer-mode --enable-rewrite
#编译
make
#安装
make install
```

（2）在/etc/profile 文件中添加以下内容的操作指令。

```
cat>>/etc/profile<<EOF
export PATH = /usr/local/subversion/bin:\$PATH
EOF
source /etc/profile
```

（3）安装完成后，查看 SVN 版本信息是否已经安装的操作指令如下。

```
/usr/local/subversion/bin/svn --version
svn --version
```

（4）建立版本库目录的操作指令如下。

```
mkdir -p /data/svn/
```

（5）建立 SVN 版本库的操作指令如下。

```
svnadmin create /data/svn
```

（6）修改版本库配置文件，在/data/svn/conf/svnserve.conf文件中加入以下代码。

```
cat>/data/svn/conf/svnserve.conf<<EOF
[general]
#使非授权用户无法访问
anon-access = none
#使授权用户有写权限
auth-access = write
#指明密码文件路径
password-db = passwd
#访问控制文件
authz-db = authz
#认证命名空间，SVN会在认证提示里显示,并作为凭证缓存的关键字
realm = /data/svn
EOF
```

（7）配置用户及权限，修改/data/svn/conf/passwd文件，添加两个用户访问SVN服务端，其操作方法和指令如下。

```
cat>/data/svn/conf/passwd<<EOF
[users]
jfedu1 = 123456
jfedu2 = 123456
EOF
```

（8）配置用户及权限，修改/data/svn/conf/authz文件，添加两个用户访问SVN服务端，其操作方法和指令如下。

```
cat>/data/svn/conf/authz<<EOF
[/]
jfedu1 = rw
jfedu2 = rw
EOF
```

（9）启动SVN服务并设置系统服务的操作指令如下。

```
svnserve -d -r /data/svn/ --listen-port=8001
```

（10）查看本地8001端口有没有启动的操作指令如下。

```
netstat -ntl|grep 8001
```

（11）SVN平台部署成功，可以在客户端使用SVN的以下指令，所有操作如图9-5所示。

```
svn co svn://101.34.116.235:8001/
```

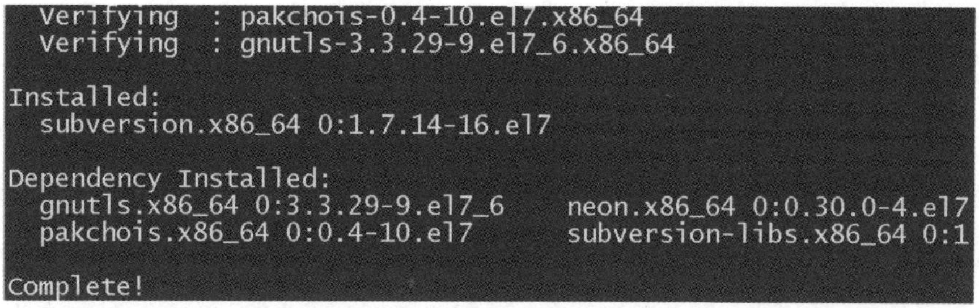

图 9-5 在客户端使用 SVN 命令

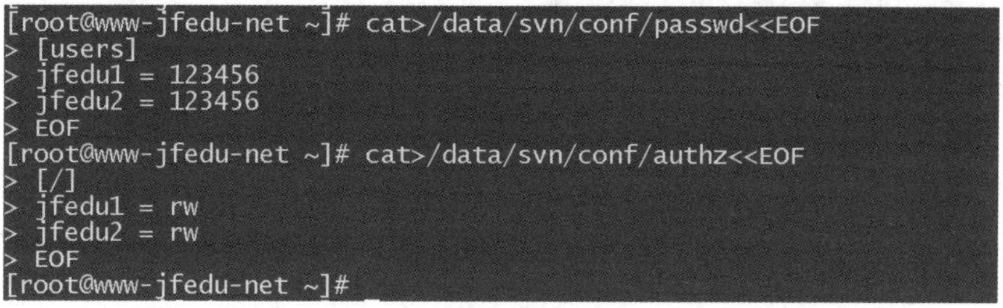

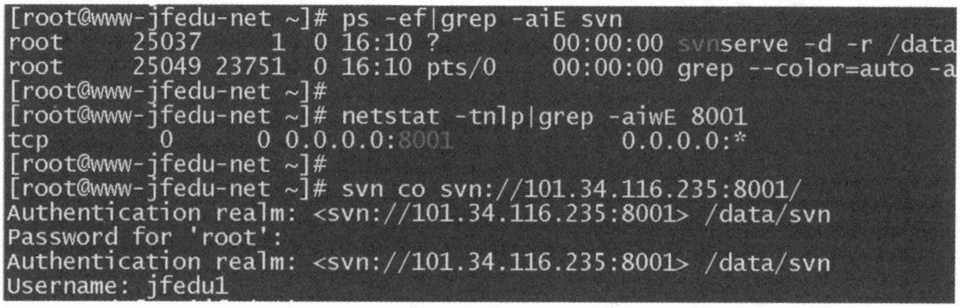

（a）安装 subversion 服务；（b）创建 SVN 用户名和密码；（c）查看 SVN 进程状态

9.9 SVN 源码+Apache 整合实战

（1）复制 SVN 模块至 Apache 模块目录（如果 httpd.conf 配置文件已存在，无须添加以下代码）。

```
cp /usr/local/subversion/libexec/mod_dav_svn.so /usr/local/apache/modules/
cp /usr/local/subversion/libexec/mod_authz_svn.so /usr/local/apache/modules/
```

（2）修改配置文件/usr/local/apache/conf/httpd.conf 需添加以下代码。

```
LoadModule dav_module modules/mod_dav.so
```

```
LoadModule dav_svn_module modules/mod_dav_svn.so
```

（3）在httpd.conf配置文件末行添加如下代码。

```
<Location /svn>
        DAV svn
        SVNPath /data/svn
        AuthType Basic
        AuthName "svn for project"
        AuthUserFile  /data/svn/conf/.passwd
        AuthzSVNAccessFile /data/svn/conf/authz
        Satisfy all
        Require valid-user
</Location>
```

（4）生成HTTP访问密钥的操作指令如下。

```
/usr/local/apache/bin/htpasswd -c /data/svn/conf/.passwd jfedu1
/usr/local/apache/bin/htpasswd  /data/svn/conf/.passwd jfedu2
```

（5）重启Apache httpd服务的操作指令如下。

```
/usr/local/apache/bin/apachectl restart
```

（6）如果没有权限，需要授权给/data/svn目录，其操作指令如下。

```
chown -R apache /data/svn
```

（7）SVN和Apache整合成功，通过浏览器访问SVN，输入用户名和密码后，结果如图9-6所示。

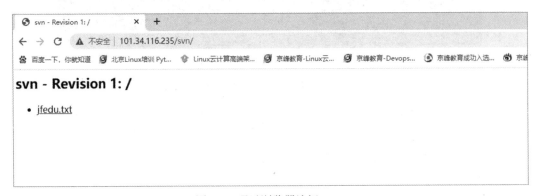

图9-6　通过浏览器访问SVN

9.10　SVN客户端命令实战

作为运维人员，要想维护好SVN服务器，必须掌握常见的SVN操作指令，以下为常见的SVN操作指令及其含义详解。

(1) 迁出 SVN 代码至 www 目录的操作指令如下。

```
svn co svn://101.34.116.235:8001/ www
svn co http://106.12.133.186/svn/ www
```

(2) 创建文件并添加和提交，如图 9-7 所示。

```
svn add jfedu1.txt jfedu2.txt
svn commit -m v1
```

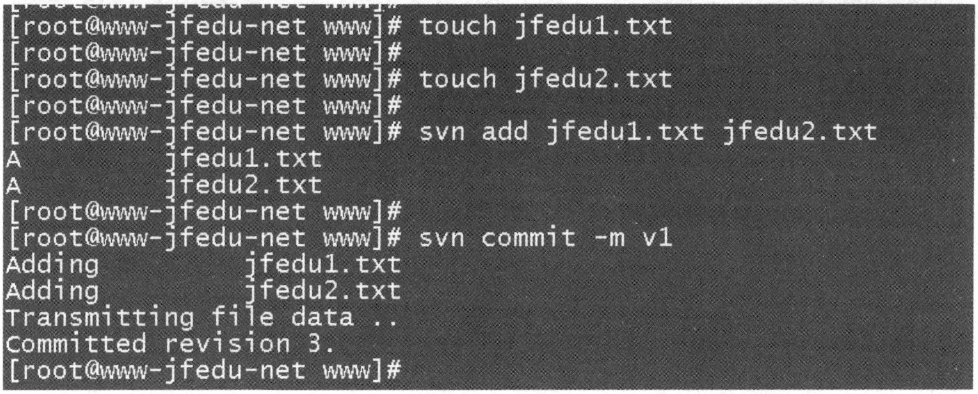

```
[root@www-jfedu-net www]# touch jfedu1.txt
[root@www-jfedu-net www]#
[root@www-jfedu-net www]# touch jfedu2.txt
[root@www-jfedu-net www]#
[root@www-jfedu-net www]# svn add jfedu1.txt jfedu2.txt
A         jfedu1.txt
A         jfedu2.txt
[root@www-jfedu-net www]#
[root@www-jfedu-net www]# svn commit -m v1
Adding          jfedu1.txt
Adding          jfedu2.txt
Transmitting file data ..
Committed revision 3.
[root@www-jfedu-net www]#
```

图 9-7　SVN 服务器提交文件

(3) 从 SVN 服务器更新软件至本地目录，如图 9-8 所示。

```
svn up
```

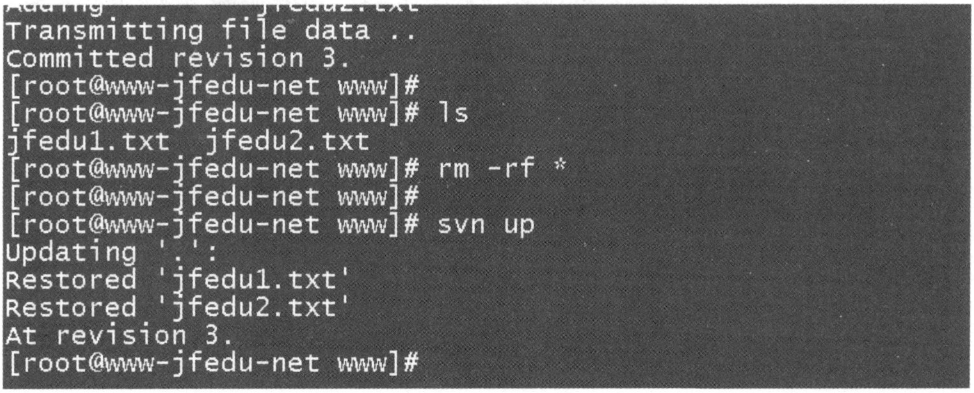

```
Adding          jfedu2.txt
Transmitting file data ..
Committed revision 3.
[root@www-jfedu-net www]#
[root@www-jfedu-net www]# ls
jfedu1.txt  jfedu2.txt
[root@www-jfedu-net www]# rm -rf *
[root@www-jfedu-net www]#
[root@www-jfedu-net www]# svn up
Updating '.':
Restored 'jfedu1.txt'
Restored 'jfedu2.txt'
At revision 3.
[root@www-jfedu-net www]#
```

图 9-8　SVN 更新代码文件

(4) 添加本地所有修改的操作指令如下。

```
svn add . --no-ignore --force
```

(5) 提交本地所有修改的操作指令如下，如图 9-9 所示。

```
svn commit . -m vv1 --force-log
```

```
[root@www-jfedu-net www]# svn commit . -m vv1
svn: E205005: The log message is a pathname (was -F intend
[root@www-jfedu-net www]#
[root@www-jfedu-net www]# svn commit . -m vv1 -F
svn: missing argument: F
Type 'svn help' for usage.
[root@www-jfedu-net www]# svn commit . -m vv1 --force-log
Adding          vv1
Adding          vv1/jfedu1.txt
Adding          vv1/jfedu2.txt
Transmitting file data ..
Committed revision 6.
[root@www-jfedu-net www]#
```

图 9-9 SVN 提交代码文件

（6）更多 SVN 常见操作指令如下。

```
#将文件 checkout 到本地目录
svn checkout path                       #path 是服务器上的目录
svn checkout svn://101.34.116.235:8001 www
svn co svn://101.34.116.235:8001/ www
#往版本库中添加新的文件
svn add file
svn add jfedu.txt                       #添加 jfedu.txt 文件
svn add *.txt                           #添加当前目录下所有的.txt 文件
#将改动的文件提交到版本库
svn commit -m "LogMessage" [-N] [--no-unlock] PATH #如果选择了保持锁，就使用
                                                   #--no-unlock 开关
svn commit -m "add test file for my test" jfedu.txt
svn ci -m "add test file for my test" jfedu.txt
#SVN 加锁/解锁
svn lock -m "LockMessage" [--force] PATH
svn lock -m "lock test file" jfedu.txt
svn unlock PATH
#更新到某个版本
svn update -r m path
#svn update 如果后面没有目录，默认将当前目录以及子目录下的所有文件都更新到最新版本
svn update -r 200 jfedu.txt             #将版本库中的文件 jfedu.txt 还原到版本 200
svn update jfedu.txt                    #更新，与版本库同步。如果在提交的时候提示过期，是因为冲突，
                                        #需要先更新，然后修改文件，再清除 svn resolved,最后再提交
                                        #查看文件或者目录状态
#（1）
svn status path                         #显示目录下的文件及其子目录的状态，正常状态不显示
#【?：不在 svn 的控制中;M：内容被修改;C：发生冲突;A：预定加入版本库;K：被锁定】
#（2）
```

```
svn status -v path                    #显示文件和子目录状态
#第一列保持相同,第二列显示工作版本号,第三和第四列显示最后一次修改的版本号和修改人
#注: svn status、svn diff 和 svn revert 这三条命令在没有网络的情况下也可以执行,原
#因是 SVN 在本地的 .svn 文件中保留了本地版本的原始拷贝
#简写为 svn st
#SVN 删除文件
svn delete path -m "delete test fle"
#例如,svn delete svn://101.34.116.235:8001/pro/domain/jfedu.txt -m "delete
#test file"
#或者直接 svn delete jfedu.txt 然后再 svn ci -m 'delete test file',推荐使用
#简写 svn (del, remove, rm)
#SVN 查看日志
svn log path
#例如,svn log jfedu.txt 显示这个文件的所有修改记录及其版本号的变化
#查看文件详细信息
svn info path
svn info jfedu.txt
#比较差异
svn diff path                         #将修改的文件与基础版本比较
svn diff jfedu.txt
svn diff -r m:n path                  #对版本 m 和版本 n 比较差异
svn diff -r 200:201 jfedu.txt
svn di
#将两个版本之间的差异合并到当前文件
svn merge -r m:n path
svn merge -r 200:205 jfedu.txt  #将版本 200 与版本 205 之间的差异合并到当前文件,但
                                #是一般都会产生冲突,需要处理一下
#SVN 查看帮助
svn help
svn help ci
```

9.11 Svnserve.conf 文件配置参数剖析

Svnserve 是 SVN 服务器核心主配置文件,主要用于对 SVN 服务器进行访问权限、目录定义等的配置。该文件由一个[general]配置段组成,格式为<配置项>=<值>,常见参数详解如下。

(1) anon-access。

控制非鉴权用户访问版本库的权限。取值范围为 write、read 和 none,其中 write 为可读可写,read 为只读,none 为无访问权限。默认值为 read。

(2) auth-access。

控制鉴权用户访问版本库的权限。取值范围为 write、read 和 none,其中 write 为可读可写,

read 为只读，none 为无访问权限。默认值为 write。

（3）password-db。

指定用户名口令文件名。除非指定绝对路径，否则文件位置为相对 conf 文件目录的相对路径。默认值为 passwd。

（4）authz-db。

指定权限配置文件名，通过该文件可以实现以路径为基础的访问控制。除非指定绝对路径，否则文件位置为相对 conf 文件目录的相对路径。默认值为 authz。

（5）realm。

指定版本库的认证域，即在登录时提示的认证域名称。若两个版本库的认证域相同，建议使用相同的用户名口令数据文件。默认值为全局唯一标示（Universal Unique Identifier，UUID）。

在企业生产环境中，Svnserve.conf 配置文件的案例代码如下。

```
[general]
#使非授权用户无法访问
anon-access = none
#使授权用户有写权限
auth-access = write
#指明密码文件路径
password-db = passwd
#访问控制文件
authz-db = authz
#认证命名空间,SVN 会在认证提示里显示,并且作为凭证缓存的关键字
realm = /data/svn
```

9.12 Passwd 文件参数剖析

Passwd 是 SVN 用户名和密码配置文件，该文件由一个[users]配置段组成，格式为<用户名>=<口令>（注：口令为未经过任何处理的明文）。在企业生产环境中，Passwd 配置文件的案例代码如下。

```
[users]
jfedu1 = 123456
jfedu2 = 123456
```

9.13 Authz 文件参数剖析

Authz 是 SVN 用户名和密码配置文件，该文件由[groups]配置段和若干版本库路径权限段组成，常见参数详解如下。

（1）[groups]配置段格式：<用户组>=<用户列表>。

用户列表由若干用户组或用户名构成，用户组或用户名之间用逗号“,”分隔，引用用户组时要使用前缀“@”。

（2）版本库路径权限段格式：[<版本库名>:<路径>]。

如版本库 abc 路径/tmp 的版本库路径权限段的段名为"[abc:/tmp]"，可省略段名中的版本库名。若省略版本库名，则该版本库路径权限段对所有版本库中相同路径的访问控制都有效，如[/tmp]。

版本库路径权限段中配置行格式有以下 3 种。

① <用户名> = <权限>。

② <用户组> = <权限>。

③ * = <权限>。

其中，“*”表示任何用户；权限的取值范围为空值、r 和 rw，空值表示对该版本库路径无任何权限，r 表示具有只读权限，rw 表示具有读写权限。

在企业生产环境中，Authz 配置文件的案例代码如图 9-10 所示。

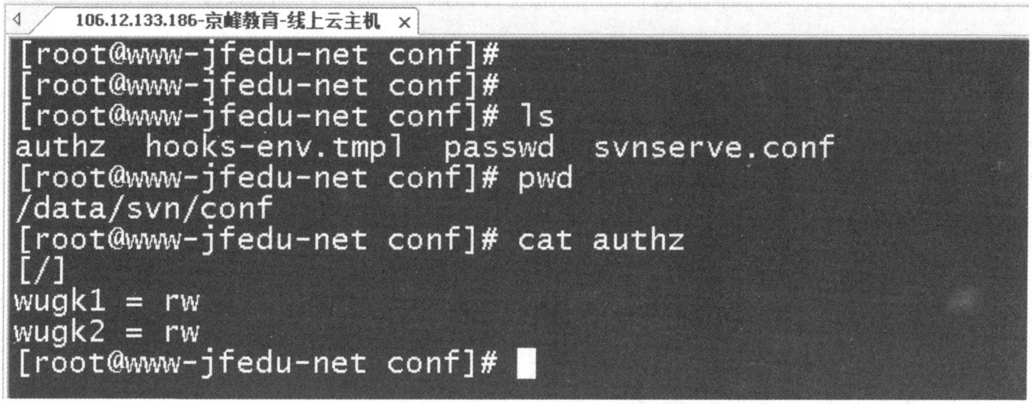
106.12.133.186-京峰教育-线上云主机
[root@www-jfedu-net conf]#
[root@www-jfedu-net conf]#
[root@www-jfedu-net conf]# ls
authz hooks-env.tmpl passwd svnserve.conf
[root@www-jfedu-net conf]# pwd
/data/svn/conf
[root@www-jfedu-net conf]# cat authz
[/]
wugk1 = rw
wugk2 = rw
[root@www-jfedu-net conf]#

图 9-10　Authz 配置文件的案例代码

第 10 章 Git 版本管理企业实战

10.1 版本控制的概念

什么是版本控制？版本控制是一种记录一个或若干文件内容变化，以便将来查阅特定版本修订情况的系统。一般对保存着软件源代码的文件做版本控制，也可以对任何类型的文件进行版本控制。

如果需要保存某一幅图片或页面布局文件的所有修订版本，采用版本控制系统（Version Control Systems，VCS）是个明智的选择。有了它就可以将某个文件回溯到之前的状态，甚至将整个项目都回退到过去某个时间点的状态；可以比较文件的变化细节，查出最后是谁修改了哪个地方，从而找出导致异常的原因，又是谁在何时报告了某个功能缺陷；等等。

使用版本控制系统通常还意味着，就算对整个项目中的文件胡乱改动，还是可以轻松将其恢复到原先的样子，但额外增加的工作量却微乎其微。

10.2 本地版本控制系统

许多人习惯用复制整个项目目录的方式来保存不同的版本，或许还会改名加上备份时间以示区别。这样做唯一的好处就是简单，但是特别容易出错。有时候会混淆所在的工作目录，一不小心会写错文件或者覆盖意外的文件。

为了解决这个问题，人们开发了许多种本地版本控制系统，大多采用某种简单的数据库记录文件的历次更新差异，如图 10-1 所示。

其中最流行的一种叫作修订控制系统（Revision Control System，RCS），现今许多计算机系统上都还看得到它的踪影，甚至在流行的 macOS X 系统上安装了开发者工具包之后，也可以使用 RCS 命令。它的工作原理是在硬盘上保存补丁集（补丁是指文件修订前后的变化），通过应用所有的补丁，可以重新计算出各个版本的文件内容。

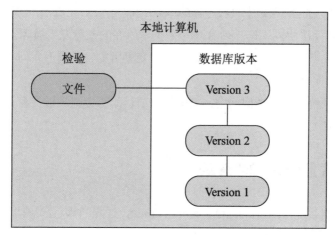

图 10-1　本地版本控制系统

10.3　集中化版本控制系统

接下来人们又遇到一个问题，即如何让不同系统上的开发者协同工作？于是，集中化版本控制系统（Centralized Version Control Systems，CVCS）应运而生。这类系统，如 CVS、SVN、Perforce 等，都有一个单一的集中管理的服务器，保存所有文件的修订版本，而协同工作的人们都通过客户端连接这台服务器，读取最新的文件或者提交更新。多年以来，这已成为版本控制系统的标准做法，如图 10-2 所示。

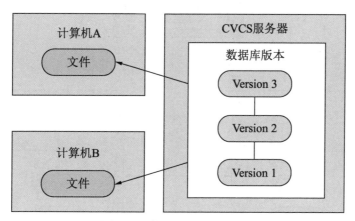

图 10-2　集中化版本控制系统

这种做法带来了许多好处，特别是相较于老式的本地版本控制系统来说。现在，每个人都可以在一定程度上看到项目中的其他人正在做些什么，而管理员也可以轻松掌控每个开发者的权限，且管理一个集中化版本控制系统要远比在各个客户端上维护本地数据库轻松。

这样做最显而易见的缺点是中央服务器的单点故障。如果死机 1h，那么在这 1h 内，谁都无法提交更新，也就无法协同工作。如果中心数据库所在的磁盘发生损坏，又没有做恰当备份，那么毫无疑问将丢失所有数据——包括项目的整个变更历史，只剩下人们在各自计算机上保留的单独快照。

本地版本控制系统也存在类似问题，只要整个项目的历史记录被保存在单一位置，就有丢失所有历史更新记录的风险。

10.4 分布式版本控制系统

分布式版本控制系统（Distributed Version Control System，DVCS）面世后，在这类系统中，像 Git、Mercurial、Bazaar、Darcs 等，客户端不只提取最新版本的文件快照，还把代码仓库完整地镜像保存下来。

协同工作用的服务器发生故障后，可以通过任何一个保存有镜像文件的本地仓库恢复。因为每一次的克隆操作，实际上都是一次对代码仓库的完整备份，如图 10-3 所示。

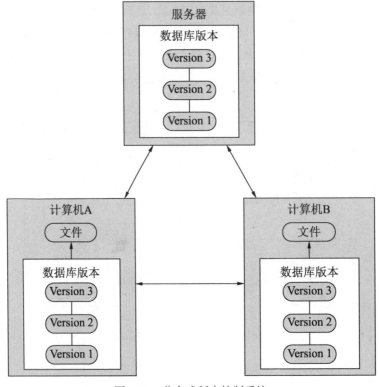

图 10-3　分布式版本控制系统

10.5 Git 版本控制系统简介

同生活中的许多伟大事物一样，Git 诞生于一个极富纷争、大举创新的年代。

Linux 内核开源项目有着为数众多的参与者。绝大多数的 Linux 内核维护工作都花在了提交补丁和保存归档的烦琐事务上（1991 年～2002 年）。到了 2002 年，整个项目组开始启用一个专有的分布式版本控制系统 BitKeeper 管理和维护代码。

到了 2005 年，开发 BitKeeper 的商业公司同 Linux 内核开源社区的合作关系结束，他们收回了 Linux 内核社区免费使用 BitKeeper 的权力。这就迫使 Linux 开源社区（特别是 Linux 操作系统的缔造者 Linus Torvalds）基于使用 BitKeeper 时的经验教训，开发出自己的版本系统。对新的系统也制订了若干目标。

（1）速度快。

（2）设计简单。

（3）对非线性开发模式强力支持（允许成千上万个并行开发的分支）。

（4）完全分布式。

（5）有能力高效管理类似于 Linux 内核一样的超大规模项目（速度和数据量）。

自 2005 年诞生以来，Git 日臻成熟完善，在高度易用的同时，仍然保留着初期设定的目标。它的速度飞快，极其适合管理大项目，有着令人难以置信的非线性分支管理系统。

10.6 Git 和 SVN 的区别

开始学习 Git 的时候，请努力分清对其他版本控制系统的已有认识，如 SVN 和 Perforce 等，这样做有助于避免使用工具时发生混淆。Git 在保存和对待各种信息的时候与其他版本控制系统有很大差异，尽管命令形式非常相近。而理解这些差异将有助于避免使用中的困惑。

（1）直接记录快照，而非差异比较。

Git 和其他版本控制系统（包括 SVN 和近似工具）的主要差别在于对待数据的方法。其他大部分系统（CVS、SVN、Perforce、Bazaar 等）以文件变更列表的方式存储信息，将其保存的信息看作一组基本文件和每个文件随时间逐步累积的差异，如图 10-4 所示。

（2）存储每个文件与初始版本的差异。

Git 不按照以上方式对待或保存数据。Git 更像是把数据看作对小型文件系统的一组快照。每次提交更新，或在 Git 中保存项目状态时，它主要对当时的全部文件制作一个快照并保存这个快照的索引。为了高效，如果文件没有修改，Git 不再重新存储该文件，而是只保留一个链接指向之前存储的文件。Git 对待数据更像是一个快照流，如图 10-5 所示。

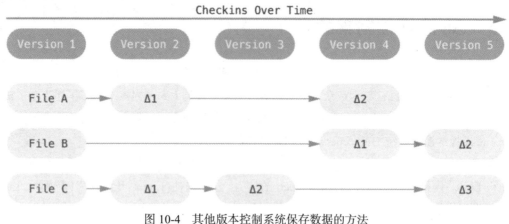

图 10-4　其他版本控制系统保存数据的方法

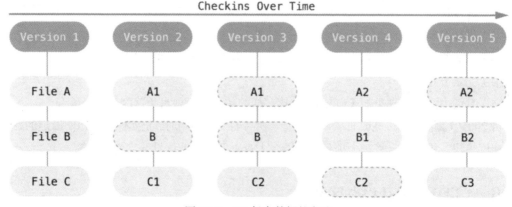

图 10-5　Git 保存数据的方法

（3）存储项目随时间改变的快照。

这是 Git 与几乎所有其他版本控制系统的重要区别。Git 重新考虑了以前每一代版本控制系统延续下来的诸多方面。Git 更像是一个小型的文件系统，提供了许多以此为基础构建的超强工具，而不只是一个简单的版本控制系统。接下来在讨论 Git 分支管理时，将探究用这种方式对待数据所能获得的益处。

（4）所有操作近乎都是本地执行。

在 Git 中的绝大多数操作都只需要访问本地文件和资源，一般不需要来自网络上其他计算机的信息。如果用户习惯于所有操作都有网络延时开销的集中化版本控制系统，那么 Git 在这方面会让用户感到速度之神赐给了 Git 超凡的能量。因为在本地磁盘上就有项目的完整历史，所以大部分操作看起来瞬间完成。

举个例子，要浏览项目的历史，Git 无须外连到服务器获取历史，然后再显示出来——它只需要直接从本地数据库中读取，就能立即看到项目历史。如果想查看当前版本与一个月前的版本之间引入的修改，Git 会查找一个月前的文件作一次本地的差异计算，而不是由远程服务器处理或从远程服务器拉回旧版本文件再来本地处理。

这也意味着离线或者没有 VPN 时，还可以进行几乎任何操作。如在飞机或火车上想做些工作，你能愉快地提交，直到有网络连接时再上传；如回家后 VPN 客户端不正常，你仍能工作。而使用其他系统，做到如此是不可能或很费力的。比如，用 Perforce，没有连接服务器时几乎不能做什么事；用 SVN 和 CVS，能修改文件，但不能向数据库提交修改（因为本地数据库离线了）。这看起来不是大问题，但是你可能会惊喜地发现它带来的巨大不同。

（5）Git 保证完整性。

Git 中所有数据在存储前都计算校验和，然后用以校验和引用。这意味着不可能在 Git 不知情时更改任何文件内容或目录内容。这个功能建构在 Git 底层，是构成 Git 哲学不可或缺的部分。在传送过程中丢失信息或损坏文件，Git 都能发现。

Git 用于计算校验和的机制叫作 SHA-1 散列（Hash）。这是一个由 40 个十六进制字符（0~9 和 a~f）组成的字符串，基于 Git 中文件的内容或目录结构计算出来。SHA-1 哈希值看起来是这样的。

24b9da6552252987aa493b52f8696cd6d3b00373

Git 中使用这种哈希值的情况很多，你将经常看到这种哈希值。实际上，Git 数据库中保存的信息都是以文件内容的哈希值为索引，而不是文件名。

（6）Git 只添加数据。

用户执行的 Git 操作，几乎只往 Git 数据库中增加数据。很难让 Git 执行任何不可逆操作，或者让它以任何方式清除数据。同别的版本控制系统，未提交更新时有可能丢失或弄乱修改的内容；但是一旦提交快照到 Git 中，就难以再丢失数据，特别是当定期地推送数据库到其他仓库时。

这使使用 Git 成为一个安心愉悦的过程，因为深知可以尽情地做各种尝试，而没有把事情弄糟的危险。更深度探讨 Git 如何保存数据及恢复丢失数据的话题，请参考撤销操作。

Git 有三种状态：已提交（committed）、已修改（modified）和已暂存（staged）。已提交表示数据已经安全地保存在本地数据库中；已修改表示修改了文件，但尚未保存到数据库中；已暂存表示对一个已修改文件的当前版本做了标记，使之包含在下次提交的快照中。

由此引入 Git 项目的三个工作区域的概念：Git 仓库、工作目录以及暂存区域，如图 10-6 所示。

Git 仓库是 Git 用来保存项目的元数据和对象数据库的地方。这是 Git 中最重要的部分，从其他计算机克隆仓库时，复制的就是这里的数据。

工作目录是对项目的某个版本独立提取出来的内容。这些从 Git 仓库的压缩数据库中提取出来的文件，放在磁盘上供使用或修改。

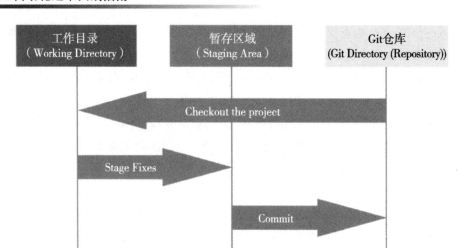

图 10-6　Git 三个工作区域结构图

暂存区域是一个文件，保存了下次提交的文件列表信息，一般在 Git 仓库中。有时候也被称作"索引"。

基本的 Git 工作流程如下。

（1）在工作目录中修改文件。

（2）暂存文件，将文件快照放入暂存区域。

（3）提交更新，找到暂存区域的文件，将快照永久性存储到 Git 仓库。

如果 Git 仓库中保存着特定版本的文件，就属于已提交状态。如果做了修改并已放入暂存区域，就属于已暂存状态。如果自上次取出后，作了修改但还没有放到暂存区域，就是已修改状态。

可以使用原生的命令行模式，也可以使用 GUI 模式，这些 GUI 软件也能提供多种功能。本书中将使用命令行模式。这是因为只有在命令行模式下才能执行 Git 的所有命令，而大多数的 GUI 软件只实现了 Git 所有功能的一个子集以降低操作难度。

如果学会了在命令行模式下如何操作，那么在操作 GUI 软件时应该也不会遇到什么困难，反之则不成立。此外，由于每个人的想法与侧重点不同，不同的人常常会安装不同的 GUI 软件，但一定都会有命令行工具。

Git 和 SVN 版本控制系统的区别总结如下。

（1）Git 是分布式，SVN 是集中式。

（2）Git 的每个历史版本存储的是完整的文件，而 SVN 只是存储文件的差异。

（3）Git 可以离线完成大部分操作，SVN 不可以。

（4）Git 有着更优雅的分支和合并实现。

（5）Git 有更强的撤销修改和修改版本历史的能力。

（6）Git 速度更快，效率更高。

10.7 Git 版本控制系统实战

在开始使用 Git 前,需要将它安装在计算机上。即便已经安装,也最好将它升级到最新的版本。可以通过软件包或者其他安装程序安装,或者下载源码编译安装。

(1) YUM 二进制方式部署 Git。

基于二进制安装程序安装 Git,可以使用发行版包含的基础软件包管理工具安装。以 CentOS 7.x 操作系统为例,可以使用 YUM 二进制方式部署 Git 服务。其操作指令如下。

```
yum install git -y
```

(2) MAKE 源码方式部署 Git。

基于 MAKE 源码方式编译安装 Git,需要安装 Git 依赖的库:Curl、zlib、OpenSSL、Expat,还有 libiconv。如果操作系统中有 YUM(如 Fedora)或者 apt-get(如基于 Debian 的系统),其操作方法和指令如下。

```
#安装 Git 编译所需依赖环境、库文件
yum install curl-devel expat-devel gettext-devel openssl-devel zlib-devel
libcurl4-gnutls-dev libexpat1-dev gettext libz-dev libssl-dev -y
yum install asciidoc xmlto docbook2x -y
#下载 Git 软件包
wget http://mirrors.edge.kernel.org/pub/software/scm/git/git-2.33.1.tar.gz
#解压 Git 软件包
tar -xzvf git-2.33.1.tar.gz
#切换至 Git 源代码目录
cd git-2.33.1/
#预编译 Git
make prefix=/usr/local/git/ all
#安装 Git
make prefix=/usr/local/git/ install
#查看 Git 是否部署成功
ls -l /usr/local/git/
#将 Git 程序 bin 目录下内容软链接至/usr/bin/目录下
ln -s /usr/local/git/bin/* /usr/bin/
#将 Git 程序 bin 目录加入 PATH 环境变量中
cat>>/etc/profile<<EOF
export PATH=\$PATH:/usr/local/git/bin/
EOF
source /etc/profile
#查看 Git 版本信息
git --version
```

10.8 配置 Git 版本仓库

（1）初次运行 Git 配置。

在 Linux 操作系统上安装 Git，需要做几件事定制 Git 环境。每台计算机上只需要配置一次，程序升级时会保留配置信息。可以在任何时候通过再次运行命令修改它们。

Git 自带一个 git config 工具帮助设置控制 Git 外观和行为的配置变量。这些变量存储在三个不同的位置。

① /etc/gitconfig 文件：包含系统上每一个用户及其仓库的通用配置。如果使用带有 --system 选项的 git config 文件，它会从此文件读写配置变量。

② ~/.gitconfig 文件或 ~/.config/git/config 文件：只针对当前用户。可以传递 --global 选项让 Git 读写此文件。

③ 当前使用仓库的 Git 目录中的 config 文件即 .git/config 文件，针对该仓库，每一个级别覆盖上一级别的配置，所以 .git/config 文件中的配置变量会覆盖 /etc/gitconfig 文件中的配置变量。

（2）用户信息。

安装完 Git 后应该做的第一件事就是设置用户名称与邮件地址。这样做很重要，因为每一个 Git 的提交都会使用这些信息，它会写入每一次提交的文件，并且不可更改。

```
git config --global user.name "support"
git config --global user.email support@jfedu.net
```

如果使用了 --global 选项，那么该命令只需要运行一次，因为之后无论在该系统上做任何事情，Git 都会使用那些信息。当想针对特定项目使用不同的用户名称与邮件地址时，可以在那个项目目录下运行没有 --global 选项的命令配置。

（3）创建一个 Git 用户，用来运行 Git 服务。

```
useradd git
```

（4）创建证书登录。

收集所有需要登录的用户的公钥，就是他们自己的 id_rsa.pub 文件，把所有公钥导入 /home/git/.ssh/authorized_keys 文件中，一行一个。

（5）初始化 Git 仓库。

创建目录作为 Git 仓库：/data/jfedu.git/，在 /data/ 目录下输入命令，如图 10-7 所示。

```
mkdir -p /data/
git init --bare /data/jfedu.git
```

Git 会创建一个裸仓库，裸仓库没有工作区，因为服务器上的 Git 仓库纯粹是为了共享，所以不让用户直接登录到服务器上改工作区。服务器上的 Git 仓库通常都以 .git 结尾，把 owner 改为 git.git 即可。

```
[root@www-jfedu-net ~]# cd /data/
[root@www-jfedu-net data]#
[root@www-jfedu-net data]# git init --bare jfedu.git
Initialized empty Git repository in /data/jfedu.git/
[root@www-jfedu-net data]#
[root@www-jfedu-net data]# ll jfedu.git/
total 32
drwxr-xr-x 2 root root 4096 Dec  9 18:11
-rw-r--r-- 1 root root   66 Dec  9 18:11 config
-rw-r--r-- 1 root root   73 Dec  9 18:11 description
-rw-r--r-- 1 root root   23 Dec  9 18:11 HEAD
drwxr-xr-x 2 root root 4096 Dec  9 18:11
drwxr-xr-x 2 root root 4096 Dec  9 18:11
```

图 10-7　创建目录作为 Git 仓库

（6）禁用 Shell 登录。

```
chown -R git.git /data/jfedu.git/
```

为了安全考虑，第二步创建的 Git 用户不允许登录 Shell，这可以通过编辑/etc/passwd 文件实现。找到类似下面的一行：

```
git:x:1001:1001:,,,:/home/git:/bin/bash
```

改为：

```
git:x:1001:1001:,,,:/home/git:/usr/bin/git-shell
sed -i '/git/s#/bin/bash#/usr/bin/git-shell#g' /etc/passwd
echo 123456|passwd --stdin git
```

这样，Git 用户可以正常通过 ssh 使用 Git，但无法登录 Shell，因为为 Git 用户指定的 git-shell 每次登录就会自动退出。

（7）克隆远程仓库。

接下来可以通过 git clone 命令克隆远程仓库，在客户端执行以下指令，如图 10-8 所示。

```
ssh-keygen
ssh-copy-id -i /root/.ssh/id_rsa.pub git@101.34.116.235
git clone git@101.34.116.235:/data/jfedu.git
```

```
[root@www-jfedu-net ~]# git clone git@127.0.0.1:/data/jfedu.git
Cloning into 'jfedu'...
git@127.0.0.1's password:
warning: You appear to have cloned an empty repository.
[root@www-jfedu-net ~]#
[root@www-jfedu-net ~]# cd jfedu/
[root@www-jfedu-net jfedu]#
[root@www-jfedu-net jfedu]# ls
[root@www-jfedu-net jfedu]# ls -al
total 12
drwxr-xr-x  3 root root 4096 Dec  9 18:49
dr-xr-x--- 19 root root 4096 Dec  9 18:49
drwxr-xr-x  7 root root 4096 Dec  9 18:49
```

图 10-8　Git 客户端访问服务端

（8）使用 touch 命令创建 1.txt、2.txt…10.txt 文件，如图 10-9 所示。

```
[root@www-jfedu-net jfedu]# touch {1..10}.txt
[root@www-jfedu-net jfedu]#
[root@www-jfedu-net jfedu]# ll
total 0
-rw-r--r-- 1 root root 0 Dec  9 18:50 10.txt
-rw-r--r-- 1 root root 0 Dec  9 18:50 1.txt
-rw-r--r-- 1 root root 0 Dec  9 18:50 2.txt
-rw-r--r-- 1 root root 0 Dec  9 18:50 3.txt
-rw-r--r-- 1 root root 0 Dec  9 18:50 4.txt
-rw-r--r-- 1 root root 0 Dec  9 18:50 5.txt
-rw-r--r-- 1 root root 0 Dec  9 18:50 6.txt
```

图 10-9　Git 创建多个文件命令

（9）添加新增文件，执行如下命令，如图 10-10 所示。

```
git add *
git status
```

```
[root@www-jfedu-net jfedu]# git add *
[root@www-jfedu-net jfedu]#
[root@www-jfedu-net jfedu]# git status
# On branch master
#
# Initial commit
#
# Changes to be committed:
#   (use "git rm --cached <file>..." to unstage)
#
#       new file:   1.txt
#       new file:   10.txt
#       new file:   2.txt
```

图 10-10　Git 添加新增文件

（10）将代码提交至本地仓库，如图 10-11 所示。

```
git commit -m 1
```

```
[root@www-jfedu-net jfedu]# git commit -m 1
[master (root-commit) 9901bf0] 1
 10 files changed, 0 insertions(+), 0 deletions(-)
 create mode 100644 1.txt
 create mode 100644 10.txt
 create mode 100644 2.txt
 create mode 100644 3.txt
 create mode 100644 4.txt
 create mode 100644 5.txt
 create mode 100644 6.txt
 create mode 100644 7.txt
```

图 10-11　Git 提交代码文件

（11）将本地库的代码提交至远程仓库。

```
git push origin master
```

10.9 Git 获取帮助

（1）有以下三种方法可以找到 Git 命令的使用手册。

```
git help <verb>
git <verb> --help
man git-<verb>
```

（2）要想获取 config 命令的手册，需执行以下指令。

```
git help config
```

第 11 章 ELK 日志平台企业实战

运维工程师需要每天对服务器进行故障排错,最先能帮助定位问题的就是查看服务器日志,通过日志可以快速定位问题。

目前所说的日志主要包括系统日志、应用程序日志和安全日志。系统运维和开发人员可以通过日志了解服务器软硬件信息、检查配置过程中的错误以及错误发生的原因。经常分析日志可以了解服务器的负荷和性能安全性,从而及时采取措施纠正错误。日志都被分散地存储在不同的设备上。

每台服务器创建开发普通用户权限,只运行查看日志、查看进程,运维、开发通过命令 tail、head、cat、more、find、awk、grep、sed 统计分析。

如果管理数百台服务器,通过登录每台机器的传统方法查阅日志,将很烦琐,且效率低下。当务之急是使用集中化的日志管理系统,例如,开源的 Syslog,将所有服务器上的日志收集汇总。

集中化管理日志后,日志的统计和检索又成为了一件比较麻烦的事情。一般使用 find、grep、awk 及 wc 等 Linux 命令实现检索和统计,但是对于要求更高的查询、排序和统计等需求及庞大的机器数量依然使用这样的方法,难免有点力不从心。

开源实时日志分析 ELK 平台能够完美解决上述问题。ELK 原来是由 ElasticSearch、Logstash 和 Kibana 三个开源工具组成,现在还新增了一个 Beats。它是一个轻量级的日志收集处理工具(Agent),适合于在各个服务器上搜集日志后传输给 Logstash,官方也推荐此工具。由于原本的 ELK Stack 成员中加入了 Beats 工具,所以现已改名为 Elastic Stack。

(1) ElasticSearch 是基于 Lucene 的全文检索引擎架构,用 Java 语言编写的,对外开源、免费,它的特点有分布式、零配置、自动发现、索引自动分片、索引副本机制、舒适的风格接口、多数据源、自动搜索负载等。ELK 官网为 https://www.elastic.co/。

(2) Logstash 主要用于日志的搜集、分析和过滤,支持大量的数据获取方式。工作方式为 C/S 架构,客户端安装在需要收集日志的主机上,服务器负责对收到的各节点日志进行过滤、修改等操作，再一并发往 ElasticSearch 服务器。

（3）Kibana 也是一个开源和免费的工具，可以为 Logstash 和 ElasticSearch 提供日志分析友好的 Web 界面，可以帮助汇总、分析和搜索重要数据日志。

（4）FileBeat 是一个轻量级日志采集器，属于 Beats 家族的 6 个成员之一。早期的 ELK 架构中使用 Logstash 收集、解析并过滤日志，但是 Logstash 对 CPU、内存、I/O 等资源消耗比较高，相比 Logstash，Beats 所占系统的 CPU 和内存几乎可以忽略不计。

（5）Logstash 和 ElasticSearch 是用 Java 语言编写的，而 Kibana 使用的是 node.js 框架，所以在配置 ELK 环境时要保证系统有 Java JDK 开发库。

11.1 ELK 架构原理深入剖析

图 11-1 为 ELK 企业分布式实时日志平台结构图，如果没有使用 Filebeat，Logstash 将直接收集日志，进行过滤处理，并将数据发往 ElasticSearch。

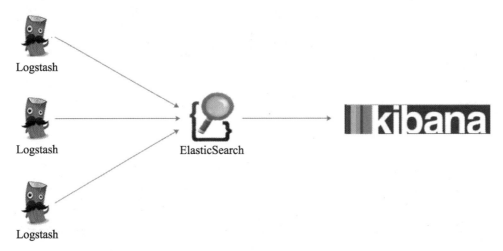

图 11-1　ELK 企业分布式实时日志平台结构图

如果使用了 Filebeat，Logstash 则从 FileBeat 获取日志文件，称为 ELFK。Filebeat 作为 Logstash 的输入（input）将获取到的日志进行处理，将处理好的日志文件输出到 ElasticSearch 进行处理，如图 11-2 所示。

（1）ELK 工作流程。

客户端安装 Logstash 日志收集工具，通过 Logstash 收集客户端应用程序的日志数据，将所有的日志过滤出来，存入 ElasticSearch 搜索引擎，然后通过 Kibana GUI 在 Web 前端展示给用户，用户可以查看指定的日志内容。同时也可以加入 Redis 通信队列，如图 11-3 所示。

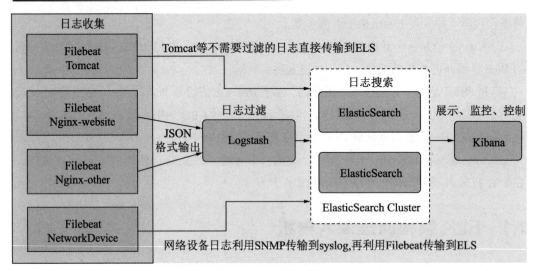

图 11-2　ELFK 日志平台结构图

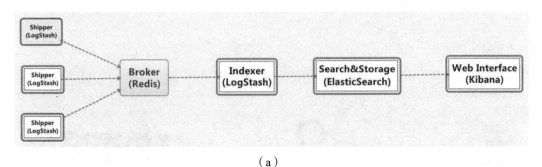

（a）

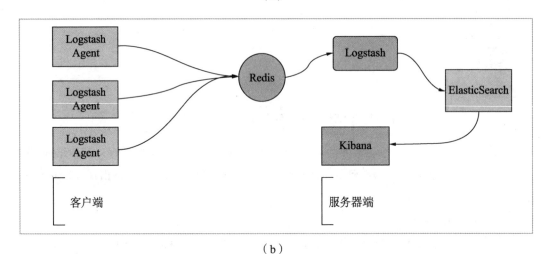

（b）

图 11-3　ELK+Redis 日志平台结构图

（a）ELK+Redis 工作流程；（b）ELK 客户端和服务端流程

（2）加入 Redis 队列后的工作流程。

Logstash 包含 Index 和 Agent（shipper），Agent 负责客户端监控和过滤日志，而 Index 负责收集日志并将日志交给 ElasticSearch，ElasticSearch 将日志存储在本地，建立并提供搜索，Kibana 可以从 ElasticSearch 集群中获取想要的日志信息。

（3）ELFK 工作流程。

① 使用 Filebeat 获取 Linux 服务器上的日志。当启动 Filebeat 时，将启动一个或多个检测者（prospectors），查找服务器上指定的日志文件，作为日志的源头等待输出到 Logstash。

② Logstash 从 Filebeat 获取日志文件。Filebeat 对获取到的日志进行处理后作为 Logstash 的输入，Logstash 将处理好的日志文件输出到 ElasticSearch 进行处理。

③ ElasticSearch 得到 Logstash 的数据之后进行相应的搜索存储操作，使写入的数据可以被检索和聚合等，以便于搜索，最后 Kibana 通过 ElasticSearch 提供的 API 将日志信息可视化。

11.2 ElasticSearch 配置实战

部署配置 ElasticSearch，需要配置 JDK 环境。JDK（Java Development Kit）是 Java 语言的软件开发工具包（SDK），此处采用 JDK 11 版本，配置 Java 环境变量，在 vi/etc/profile 文件中加入以下代码。

```
export JAVA_HOME=/usr/java/jdk-11.0.10/
export CLASSPATH=$CLASSPATH:$JAVA_HOME/lib:$JAVA_HOME/jre/lib
export PATH=$JAVA_HOME/bin:$JAVA_HOME/jre/bin:$PATH:$HOME/bin
```

使环境变量立刻生效，同时查看 Java 版本，如显示版本信息，则证明安装成功。

```
source /etc/profile
java -version
```

分别下载 ELK 软件包。

（1）ELK 安装信息如下。

```
192.168.111.128    ElasticSearch
192.168.111.129    Kibana
192.168.111.130    Logstash
```

（2）在 IP 地址 192.168.111.128 上安装 ElasticSearch。

```
#安装 JDK 版本信息
tar xzf jdk-11.0.10_linux-x64_bin.tar.gz
mkdir -p /usr/java/ ;mv jdk-11.0.10/  /usr/java/
#同时在/etc/profile 文件末尾加入如下三行
export JAVA_HOME=/usr/java/jdk-11.0.10
```

```
export CLASSPATH=$CLASSPATH:$JAVA_HOME/lib:$JAVA_HOME/jre/lib
export PATH=$JAVA_HOME/bin:$JAVA_HOME/jre/bin:$PATH:$HOME/bin
```

（3）基于二进制 Tar 包方式，安装 ElasticSearch 程序，其操作步骤如下。

```
#官网下载 ElasticSearch 软件包
ls -l elasticsearch-7.2.0-linux-x86_64.tar.gz
#通过 Tar 工具对其解压缩
tar xzf elasticsearch-7.2.0-linux-x86_64.tar.gz
#将解压后的 ElasticSearch 程序部署至 /usr/local/ 目录下
mv elasticsearch-7.2.0 /usr/local/elasticsearch
#查看 ElasticSearch 是否部署成功
ls -l /usr/local/elasticsearch/
#切换至 ElasticSearch 程序目录
cd /usr/local/elasticsearch/
```

（4）修改 /usr/local/elasticsearch/config/elasticsearch.yml 文件，设置监听地址为 network.hosts：0.0.0.0，同时设置 node 名称和集群 node 名称，其指令如下。

```
cd /usr/local/elasticsearch/config/
sed -i -e '/network\.host/s/#//g' -e'/network\.host/s/192.168.0.1/0.0.0.0/g' elasticsearch.yml
sed -i '/node.name/s/#//g' elasticsearch.yml
sed -i '/cluster\.initial/s/\, \"node-2\"//g' elasticsearch.yml
sed -i -e '/cluster\.initial/s/\, \"node-2\"//g' -e '/cluster\.initial/s/#//g' elasticsearch.yml
```

（5）创建 ElasticSearch 用户和组，同时授权访问，启动 ElasticSearch 服务。

```
useradd elk
chown -R elk:elk /usr/local/elasticsearch/
su - elk
/usr/local/elasticsearch/bin/elasticsearch -d
```

（6）查看 ElasticSearch 服务日志，如图 11-4 所示。

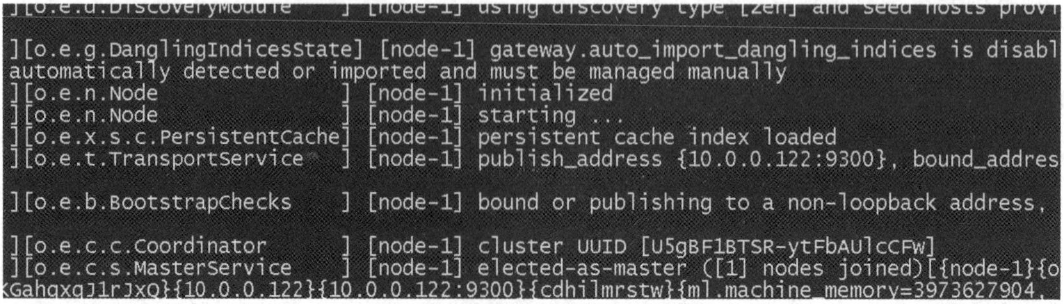

图 11-4 ElasticSearch 服务日志

11.3　ElasticSearch 配置故障演练

启动后可能会报错，报错内容及需要修改的内核参数如下。

（1）SecComp 功能不支持。

报错内容如下。

```
ERROR: bootstrap checks failed
system call filters failed to install; check the logs and fix your
configuration or disable system call filters at your own risk;
```

因为 CentOS 6 操作系统不支持 SecComp，而 ES 5.3.0 默认 bootstrap.system_call_filter 为 true 进行检测，所以导致检测失败，失败后直接导致 ElasticSearch 不能启动。

Seccomp（securecomputing mode）是 Linux Kernel 从 2.6.23 版本开始支持的一种安全机制。

在 Linux 操作系统里，大量地系统调用（systemcall）直接暴露给用户态程序。但并不是所有的系统调用都被需要，不安全的代码滥用系统调用会对系统造成安全威胁。通过 Seccomp 限制程序使用某些系统调用，这样可以减少系统的暴露面，同时使程序进入一种"安全"的状态。

解决方法：在 elasticsearch.yml 文件中配置 bootstrap.system_call_filter 为 false，注意要在 Memory 下面配置。

```
bootstrap.memory_lock: false
bootstrap.system_call_filter: false
```

（2）内核参数设置问题。

报错内容如下。

```
max file descriptors [4096] for elasticsearch process is too low, increase
to at least [65536]
max number of threads [1024] for user [hadoop] is too low, increase to at
least [2048]
max virtual memory areas vm.max_map_count [65530] is too low, increase to
at least [262144]
```

解决方法如下。

```
vim /etc/security/limits.conf
* soft nofile 65536
* hard nofile 65536
vim /etc/security/limits.d/90-nproc.conf
soft  nproc  2048
vi /etc/sysctl.conf
vm.max_map_count=655360
```

Max_map_count 文件中包含一个值，用于限制一个进程可以拥有的虚拟内存区域（VMA）

的数量。虚拟内存区域是一个连续的虚拟地址空间区域。在进程的生命周期中，每当程序尝试在内存中映射文件，链接到共享内存段，或者分配堆空间的时候，这些区域都将被创建。

调整优化这个值将限制进程可拥有 VMA 的数量。限制一个进程拥有 VMA 的总数可能会导致应用程序出错，因为当进程达到了 VMA 上限但又只能释放少量的内存给其他内核进程使用时，操作系统会抛出内存不足的错误。

如果操作系统在 Normal 区域仅占用少量的内存，那么调低这个值可以帮助释放内存给内核用。

至此，ElasticSearch 配置完毕，如果想配置 ElasticSearch 集群模式，只需要复制 ElasticSearch 副本集，然后修改相应的参数即可。

11.4 ElasticSearch 插件部署实战

ElasticSearch 老版本（5.x 以下）部署 ElasticSearch-head 插件方法如下。

```
cd /usr/local/elasticsearch;
./bin/plugin install mobz/elasticsearch-head
```

访问 ElasticSearch 插件，地址为 http://192.168.111.128:9200/_plugin/head/，如图 11-5 所示。

图 11-5　ElasticSearch-head 插件界面

ElasticSearch-head 是 ElasticSearch 的集群管理工具，是完全由 HTML5 编写的独立网页程序，通过插件安装到 ElasticSearch，然后重启 ElasticSearch，通过界面来实现访问和管理。

ElasticSearch 新版本（5.x 以上）部署 ElasticSearch-head 插件方法如下。

（1）安装 nodejs 和 npm。

```
yum -y install nodejs npm
```

（2）下载源码并安装。

```
git clone https://github.com/mobz/elasticsearch-head.git
cd elasticsearch-head/
#基于国内淘宝网站镜像安装 grunt
npm install -g grunt --registry=https://registry.npm.taobao.org
#安装 elasticsearch-head 插件
npm install
#npm install --registry=https://registry.npm.taobao.org
#ElasticSearch 的配置修改和 elasticsearch-head 的插件源码修改
```

（3）修改 elasticsearch.yml 文件，增加跨域的配置。

```
http.cors.enabled: true
http.cors.allow-origin: "*"
```

（4）编辑 elasticsearch-head/Gruntfile.js 文件，修改服务器监听地址。

增加 hostname 属性，将其值设置为*，以下两种配置任选一种。

```
# Type1
connect: {
    hostname: '*',
    server: {
        options: {
            port: 9100,
            base: '.',
            keepalive: true
        }
    }
}
# Type 2
connect: {
    server: {
        options: {
            hostname: '*',
            port: 9100,
            base: '.',
            keepalive: true
        }
    }
}
```

（5）编辑 head/_site/app.js 文件，修改 ElasticSearch-head 连接 ElasticSearch 的地址。

将以下 app-base_uri 中的 localhost 修改为 ElasticSearch 的 IP 地址，其代码如下。

```
this.base_uri = this.config.base_uri || this.prefs.get("app-base_uri") ||
"http://localhost:9200";
```

修改完成后，最终显示代码如下。

```
this.base_uri = this.config.base_uri || this.prefs.get("app-base_uri") ||
"http://192.168.1.161:9200";
```

（6）启动 ElasticSearch-head 独立服务。

```
cd elasticsearch-head/
nohup ./node_modules/grunt/bin/grunt server &
```

访问 ElasticSearch-head 插件，如图 11-6 所示。

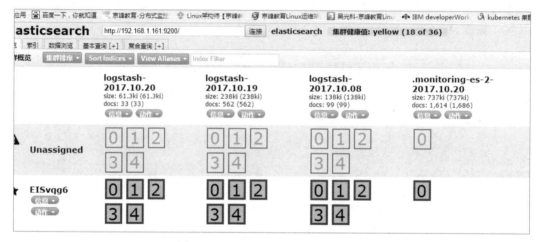

图 11-6　ElasticSearch-head Web 界面

11.5　Kibana Web 安装配置

（1）基于二进制 Tar 压缩包方式安装 Kibana 程序的操作步骤如下。

```
#官网下载 Kibana 软件包
ls -l kibana-7.2.0-linux-x86_64.tar.gz
#通过 Tar 工具对其解压缩
tar xzf kibana-7.2.0-linux-x86_64.tar.gz
#将解压后的 Kibana 程序部署至 /usr/local/ 目录下
mv kibana-7.2.0-linux-x86_64 /usr/local/kibana
#查看 Kibana 是否部署成功
ls -l /usr/local/kibana/
#切换至 Kibana 程序目录
cd /usr/local/kibana/
```

（2）修改 Kibana 配置文件信息，设置 ElasticSearch 地址，其代码如下，结果如图 11-7 所示。

```
vim /usr/local/kibana/config/kibana.yml
```

```
# Kibana is served by a back end server. This controls which
server.port: 5601

# The host to bind the server to.
server.host: "0.0.0.0"

# If you are running kibana behind a proxy, and want to mount
# specify that path here. The basePath can't end in a slash.
# server.basePath: ""

# The maximum payload size in bytes on incoming server reques
# server.maxPayloadBytes: 1048576

# The Elasticsearch instance to use for all your queries.
elasticsearch.url: "http://192.168.111.128:9200"

# preserve_elasticsearch_host true will send the hostname spe
# then the host you use to connect to *this* Kibana instance
```

图 11-7 Kibana ElasticSearch 地址设置结果

（3）将 Kibana 设置为中文。

```
cat>>/usr/local/kibana/config/kibana.yml <<EOF
i18n.locale: "zh-CN"
EOF
```

（4）启动 Kibana 服务。

```
nohup sh kibana --allow-root &
```

（5）通过浏览器访问 Kibana 5601 端口，如图 11-8 所示。

(a)

图 11-8 Kibana Web 界面

(a) 将数据添加到 Kibana

（b）

图 11-8　Kibana Web 界面（续）

（b）创建索引模式

11.6　Logstash 客户端配置实战

（1）配置 JDK 环境，在 etc/profile 文件中添加以下代码。

```
export JAVA_HOME=/usr/java/jdk11.0_10
export CLASSPATH=$CLASSPATH:$JAVA_HOME/lib:$JAVA_HOME/jre/lib
export PATH=$JAVA_HOME/bin:$JAVA_HOME/jre/bin:$PATH:$HOME/bin
```

（2）基于二进制 Tar 压缩包方式安装 Kibana 程序的操作步骤如下。

```
#官网下载 Logstash 软件包
ls -l logstash-7.2.0.tar.gz
#通过 Tar 工具对其解压缩
tar xzf logstash-7.2.0.tar.gz
#将解压后的 Logstash 程序部署至/usr/local/目录下
mv logstash-7.2.0 /usr/local/logstash/
#查看 Logstash 是否部署成功
ls -l /usr/local/logstash/
#切换至 Logstash 程序目录
cd /usr/local/logstash/
```

11.7　ELK 收集系统标准日志

创建收集日志配置目录。

```
mkdir -p /usr/local/logstash/config/etc/
cd /usr/local/logstash/config/etc/
```

创建 ELK 整合配置文件 vim logstash.conf，文件内容如下。

```
input {
 stdin { }
}
output {
 stdout {
  codec => rubydebug {}
 }
 elasticsearch {
  hosts => "192.168.111.128" }
}
```

启动 Logstash 服务，如图 11-9 所示。

```
/usr/local/logstash/bin/logstash -f logstash.conf
```

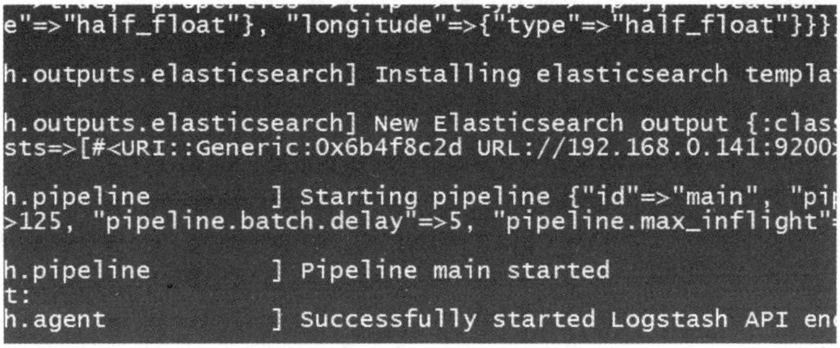

图 11-9　Logstash 客户端启动日志

11.8　ELK-Web 日志数据图表

在 Logstash 启动窗口中输入任意信息，会自动输出相应格式的日志信息，如图 11-10 所示。

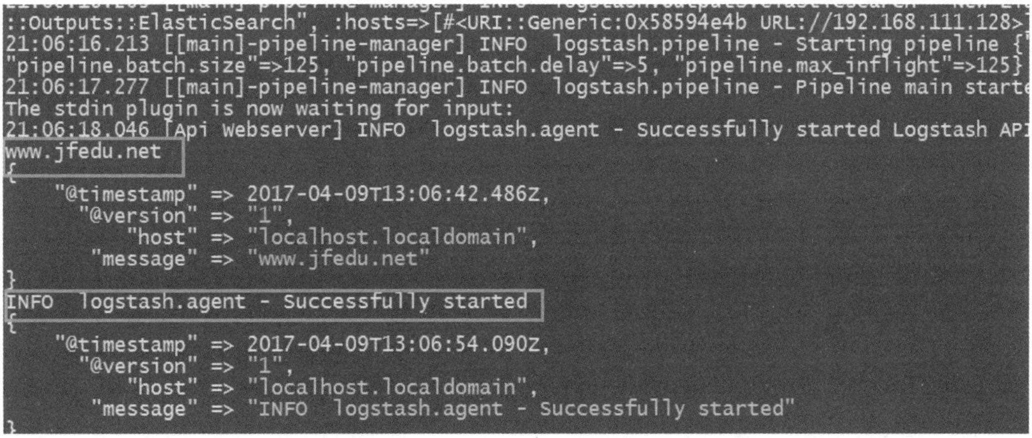

图 11-10　Logstash 客户输入日志信息

通过浏览器访问 Kibana，地址为 http://192.168.111.129:5601/，如图 11-11 所示。

（a）

（b）

图 11-11 Kibana Web 界面

（a）创建索引模式；（b）查看 logstash 所有信息

为了使用 Kibana，必须配置至少一个索引模式，用于确认 ElasticSearch 索引，也用于运行搜索和分析，还可以用于配置字段。

```
Index contains time-based events                        #索引基于时间的事件
Use event times to create index names [DEPRECATED]      #使用事件时间创建索引名字
                                                        #【过时】
Index name or pattern                                   #索引名字或者模式
#模式允许定义动态的索引名字,使用*作为通配符,例如默认:
logstash-*
#选择
Time field name;
```

单击 Discover，可以搜索和浏览 ElasticSearch 中的数据，默认搜索的是最近 15 min 的数据。

也可以自定义选择时间，如图 11-12 所示。

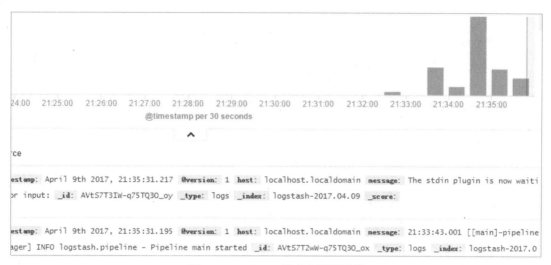

图 11-12　Kibana 搜索和浏览数据

11.9　ELK-Web 中文汉化支持

Kibana Web 平台所有的字段均显示为英文。ELK 7.x 自带中文，但是需要手动开启（在 Kibana 配置文件末尾加入 i18n.locale: "zh-CN"）。ELK 5.x 版本默认没有中文汉化插件或者汉化包，感谢 Github 开源贡献者开发了汉化包，汉化包插件地址为 https://github.com/anbai-inc/Kibana_Hanization，如图 11-13 所示。

Kibana 汉化包适用于 Kibana 5.x ~ Kibana 6.x 的任意版本，汉化过程不可逆，汉化前请注意备份。汉化资源会慢慢更新完善，已汉化过的 Kibana 可以重复使用此汉化包更新的资源。

除一小部分资源外，大部分资源无须重启 Kibana，刷新页面即可看到效果。Kibana 汉化方法和步骤如下。

（1）从 Github 仓库下载 Kibana 中文汉化包，下载指令如下。

```
git clone https://github.com/anbai-inc/Kibana_Hanization.git
#wget http://bbs.jingfengjiaoyu.com/download/Kibana_Hanization_2021.tar.gz
```

（2）切换至 Kibana_Hanization 目录，并执行汉化过程。

```
cd Kibana_Hanization/
python main.py /usr/local/kibana/        #此处为系统 Kibana 安装路径
```

（3）重启 Kibana 服务。通过浏览器访问，如图 11-14 所示。

(a)

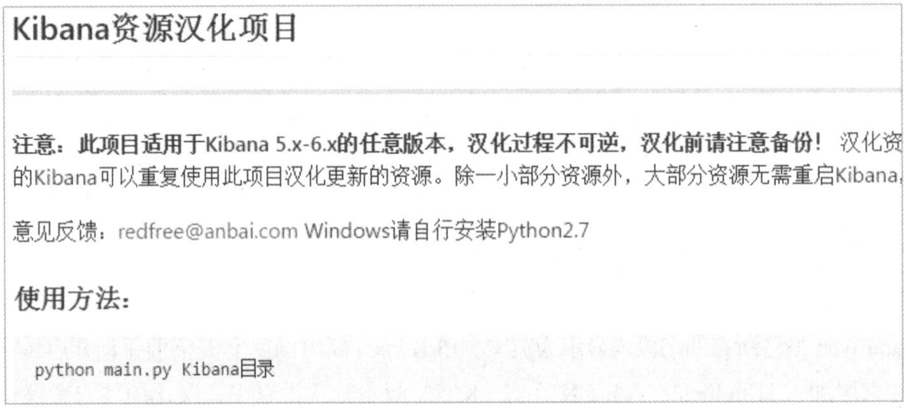

Kibana资源汉化项目

注意：此项目适用于Kibana 5.x-6.x的任意版本，汉化过程不可逆，汉化前请注意备份！汉化资的Kibana可以重复使用此项目汉化更新的资源。除一小部分资源外，大部分资源无需重启Kibana。

意见反馈：redfree@anbai.com Windows请自行安装Python2.7

使用方法：

```
python main.py Kibana目录
```

(b)

图 11-13　Kibana 汉化插件

（a）Kibana 汉化；（b）汉化项目简介

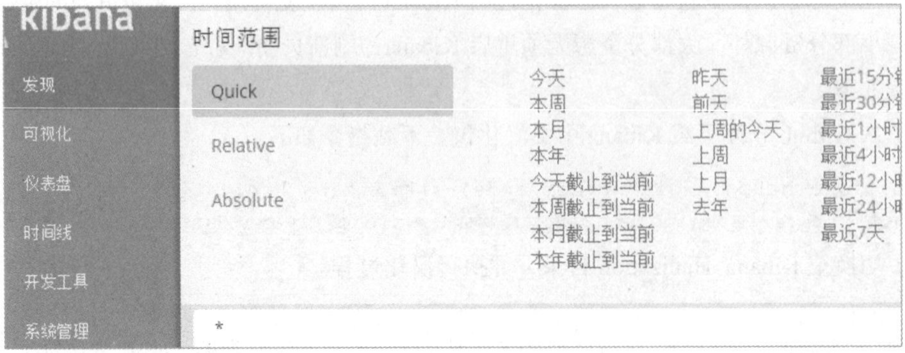

(a)

图 11-14　Kibana 汉化界面

（a）Kibana 汉化效果

（b）

图 11-14　Kibana 汉化界面（续）

（b）查看 Kibana 汉化索引

11.10　Logstash 配置详解

Logstash 是一个开源的数据收集引擎，具有传输实时数据的能力。它可以统一过滤来自不同源的数据，并按照开发者制定的规范输出到目的地。

Logstash 通过管道进行运作，管道有两个必需的元素——输入和输出，还有一个可选的元素——过滤器。输入插件从数据源获取数据，过滤器插件根据用户指定的数据格式修改数据，输出插件则将数据写到目的地，如图 11-15 所示。

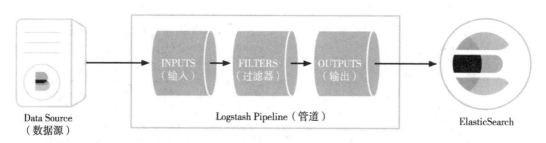

图 11-15　Logstash 管道结构图

使用 Logstash 之前，需要了解一个概念：事件。Logstash 每读取一次数据的行为叫作一个事件。在 Logstach 目录中创建一个配置文件，命名为 logstash.conf 或其他。

Logstash 过滤器插件位于 Logstash 管道的中间位置，对事件执行过滤处理，配置在 filter {} 中，且可以配置多个。如使用 Grok 插件演示，Grok 插件用于过滤杂乱的内容，将其结构化，增加可读性。输入以下代码，效果如图 11-16 所示。

```
input {
    stdin {}
}
filter {
    grok {
    match => { "message" => "%{IP:client} %{WORD:method} %{URIPATHPARAM:request} %{NUMBER:bytes} %{NUMBER:duration}" }
    }
}
output {
    stdout {
       codec => "rubydebug"
    }
}
```

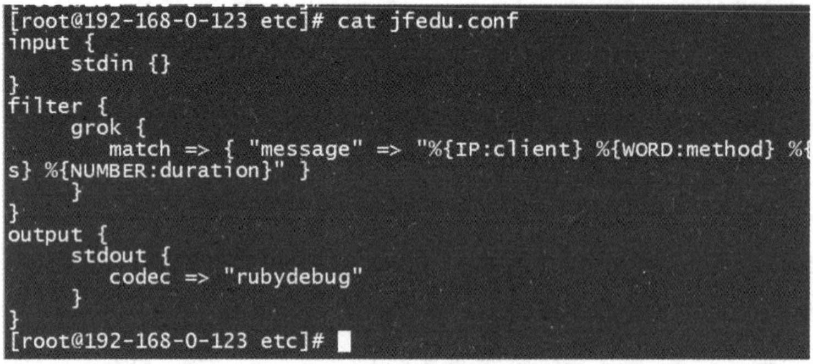

图 11-16 配置 Grok

输入以下代码,效果如图 11-17 所示。

```
192.168.0.111 GET /index.html 13843 0.059
```

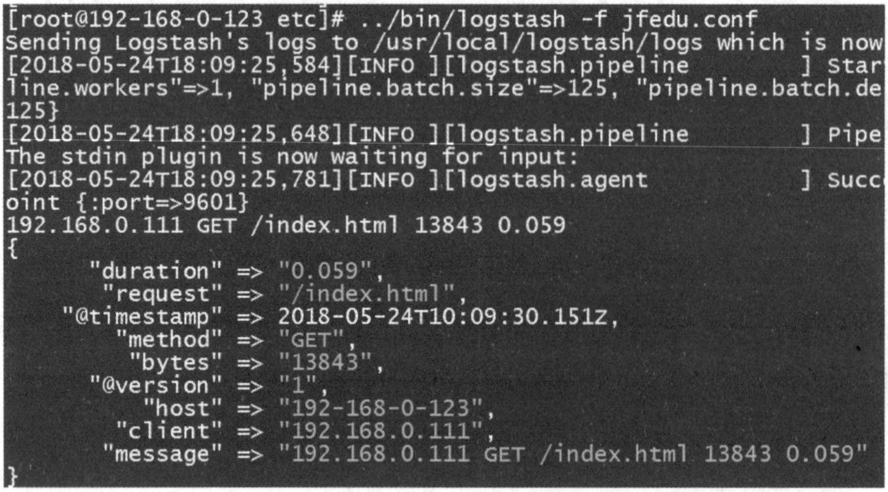

图 11-17 Logstash 日志过滤效果

Grok 内置的默认类型有很多种，可以通过官网查看支持的类型。

Logstash Grok 内置正则如下。

```
https://github.com/logstash-plugins/logstash-patterns-core/blob/master/patterns/grok-patterns
USERNAME [a-zA-Z0-9._-]+
USER %{USERNAME}
EMAILLOCALPART [a-zA-Z][a-zA-Z0-9_.+-=:]+
EMAILADDRESS %{EMAILLOCALPART}@%{HOSTNAME}
INT (?:[+-]?(?:[0-9]+))
BASE10NUM (?<![0-9.+-])(?>[+-]?(?:(?:[0-9]+(?:\.[0-9]+)?)|(?:\.[0-9]+)))
NUMBER (?:%{BASE10NUM})
BASE16NUM (?<![0-9A-Fa-f])(?:[+-]?(?:0x)?(?:[0-9A-Fa-f]+))
BASE16FLOAT \b(?<![0-9A-Fa-f.])(?:[+-]?(?:0x)?(?:(?:[0-9A-Fa-f]+(?:\.[0-9A-Fa-f]*)?)|(?:\.[0-9A-Fa-f]+)))\b
POSINT \b(?:[1-9][0-9]*)\b
NONNEGINT \b(?:[0-9]+)\b
WORD \b\w+\b
NOTSPACE \S+
SPACE \s*
DATA .*?
GREEDYDATA .*
QUOTEDSTRING (?>(?<!\\)(?>"(?>\\.|[^\\"]+)+"|""|(?>'(?>\\.|[^\\']+)+')|''|(?>`(?>\\.|[^\\`]+)+`)|``))
UUID [A-Fa-f0-9]{8}-(?:[A-Fa-f0-9]{4}-){3}[A-Fa-f0-9]{12}
# URN, allowing use of RFC 2141 section 2.3 reserved characters
URN urn:[0-9A-Za-z][0-9A-Za-z-]{0,31}:(?:%[0-9a-fA-F]{2}|[0-9A-Za-z()+,.:=@;$_!*'/?#-])+
```

使用自定义的 Grok 类型。更多时候 Logstash Grok 没办法提供需要的匹配类型，这个时候就可以使用自定义。

（1）直接使用 oniguruma 语法匹配文本片段的语法如下。

```
(?<field_name>the pattern here)
```

假设需要匹配的文本片段为一个长度为 10 或 11 的十六进制的值，使用下列语法可以获取该片段，并把值赋于 queue_id，其配置如下。

```
(?<queue_id>[0-9A-F]{10,11})
192.168.0.111 GET /index.html 15824 0.043 1
```

（2）创建自定义 pattern 文件。

创建文件夹 patterns，在此文件夹下创建一个文件，文件名随意，如：

```
postfix;
mkdir patterns
POSTFIX_QUEUEID [0-9A-F]{10,11}
input {
    stdin {}
}
filter {
```

```
grok {
 patterns_dir => ["./patterns"]
 match => { "message" => "%{IP:client_id_address} %{WORD:method}
%{URIPATHPARAM:request} %{NUMBER:bytes} %{NUMBER:http_response_time}
%{POSTFIX_QUEUEID:queue_id}" }
 }
}
output {
    stdout {
       codec => "rubydebug"
    }
}
```

然后在命令行终端输入以下代码，如图 11-18 所示。

```
192.168.0.111 GET /index.html 15824 0.043 ABC24C98567
```

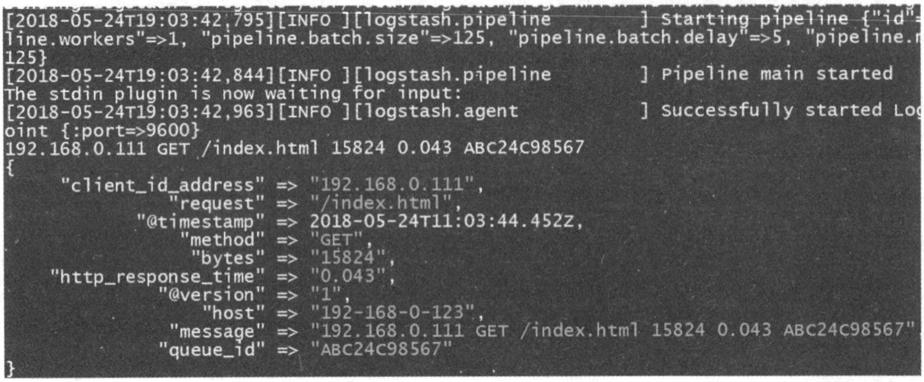

图 11-18 Logstash 日志正则

11.11 Logstash 自定义索引实战

默认情况下，Logstash 客户端采集的日志，在 ElasticSearch 存储中索引名称为 logstash-*，也可以根据需求修改索引的名称，例如，根据不同的日志类型命名，以 Nginx 为例，可命名为 nginx-*。

如何将默认 Logstash 索引名称修改为以 nginx 开头呢？

（1）修改 Logstash 日志收集配置文件 jfedu.conf，添加以下代码。

```
input {
 stdin { }
}
output {
  elasticsearch {
    hosts => ["localhost:9200"]
    index => ["nginx-%{+YYYY-MM-dd}"]
  }
}
```

(2)重启 Logstash 服务。

```
cd /usr/local/logstash/config/
../bin/logstash -f jfedu.conf
```

(3)查看 Logstash 服务启动日志的代码如下,结果如图 11-19 所示。

```
tail -fn 30 nohup.out
```

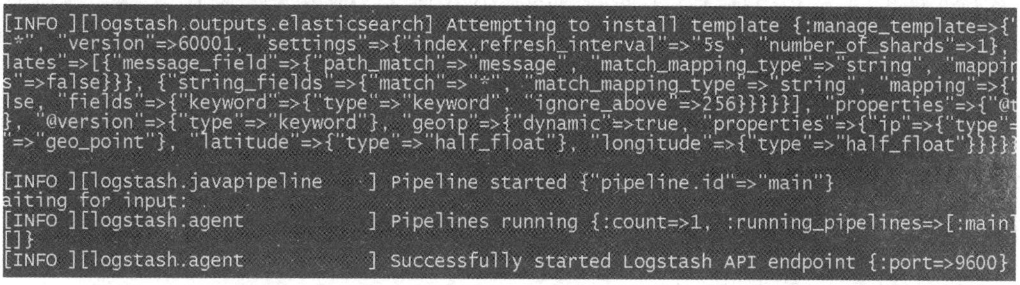

图 11-19 Logstash 启动日志结果

(4)登录 Kibana Web 界面,并创建索引,如图 11-20 所示。

(a)

(b)

图 11-20 Kibana Web 界面

(a)创建 nginx 索引;(b)选择索引排序类型

11.12 Grok 语法格式剖析

Grok 支持以正则表达式的方式提取所需要的信息，正则表达式又分两种：一种是内置的正则表达式，另一种是自定义的正则表达式。

（1）内置的正则表达式式。下面的写法表示从输入的消息中提取 IP 字段，并命名为 sip。

```
%{IP:sip}
```

（2）自定义的正则表达式，开始符与终止符分别为 "(?" 与 "?)"。下面的写法表示获取除 "," 以外的字符，并命名为 log_type。

```
(?<log_type>[^,]+?)
```

Grok 具体的语法参数如下。

```
%{SYNTAX:SEMANTIC}
#SYNTAX 代表匹配值的类型,例如,0.11 可以用 NUMBER 类型所匹配,#10.222.22.25 可以用 IP
#地址匹配
#SEMANTIC 表示存储该值的一个变量声明,它会存储在 ElasticSearch
#中方便 Kibana 做字段搜索和统计,可以将一个 IP 地址定义为客户端 IP 地址 client_
#ip_address
#例如,%{IP:client_ip_address},匹配值就会存储到 client_ip_address
#这个字段里边,类似数据库的列名,也可以把 event log 中的数字当成数字类型存储在一个指定
#的变量当中,响应时间为 http_response_time;
```

Grok 文本片段切分的案例如下。

```
input {
    stdin {}
}
filter {
    grok {
    match => { "message" => "%{IP:client} %{WORD:method} %{URIPATHPARAM:request} %{NUMBER:bytes} %{NUMBER:http_response_time}" }
    }
}
output {
    stdout {
       codec => "rubydebug"
    }
}
192.168.0.111 GET /index.html 13843 0.059
```

11.13 Redis 高性能加速实战

相关代码如下。

```
wget    http://download.redis.io/releases/redis-2.8.13.tar.gz
tar     zxf     redis-2.8.13.tar.gz
cd      redis-2.8.13
make    PREFIX=/usr/local/redis  install
cp      redis.conf     /usr/local/redis/
```

将目录/usr/local/redis/bin/加入环境变量配置文件/etc/profile 的末尾,然后在 Shell 终端执行 source /etc/profile 让环境变量生效。

```
export PATH=/usr/local/redis/bin:$PATH
```

Nohup 后台启动及停止 Redis 服务命令。

```
nohup /usr/local/redis/bin/redis-server  /usr/local/redis/redis.conf  &
/usr/local/redis/bin/redis-cli  -p  6379 shutdown
```

11.14 ELK 收集 MySQL 日志实战

切换到目录 usr/local/logstash/config/etc/,创建以下配置文件。

(1) 日志采集:存入 Redis 缓存数据库。

创建 agent.conf 文件,内容如下。

```
input {
  file {
    type => "mysql-access"
    path => "/var/log/mysqld.log"
  }
}
output {
  redis {
    host => "localhost"
    port => 6379
    data_type => "list"
    key => "logstash"
  }
}
```

启动 Agent 的代码如下。

```
../bin/logstash -f agent.conf
```

(2) Redis 数据:存入 ElasticSearch。

创建 index.conf 文件的内容如下。

```
input {
  redis {
```

```
        host => "localhost"
        port => "6379"
        data_type => "list"
        key => "logstash"
        type => "redis-input"
        batch_count => 1
    }
}
output {
    elasticsearch {
        hosts => "192.168.111.128"
    }
}
```

启动 index 的代码如下。

```
../bin/logstash -f index.conf
```

查看启动进程，如图 11-21 所示。

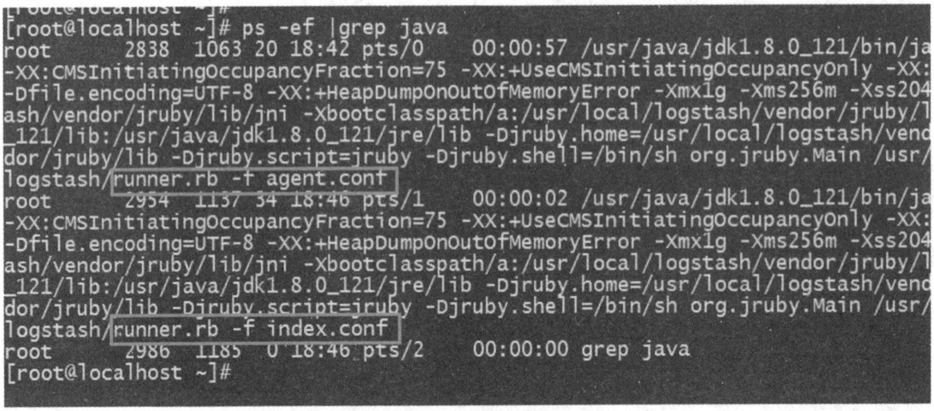

图 11-21　查看启动进程

11.15　ELK 收集 Kernel 日志实战

切换到目录 usr/local/logstash/config/etc/，创建以下配置文件。

（1）日志采集：存入 Redis 缓存数据库。

创建 agent.conf 文件的内容如下。

```
input {
    file {
        type => "kernel-message"
        path => "/var/log/messages"
    }
}
output {
```

```
    redis {
        host => "localhost"
        port => 6379
        data_type => "list"
        key => "logstash"
    }
}
```

启动 Agent 的代码如下。

```
../bin/logstash -f agent.conf
```

(2) Redis 数据：存入 ElasticSearch。

创建 index.conf 文件的内容如下。

```
input {
  redis {
    host => "localhost"
    port => "6379"
    data_type => "list"
    key => "logstash"
    type => "redis-input"
    batch_count => 1
  }
}
output {
  elasticsearch {
    hosts => "192.168.111.128"
  }
}
```

启动 index 的代码如下。

```
../bin/logstash -f index.conf
```

查看启动进程，如图 11-22 所示。

```
[root@localhost ~]# ps -ef |grep java
root      2838  1063 20 18:42 pts/0    00:00:57 /usr/java/jdk1.8.0_121/bin/ja
-XX:CMSInitiatingOccupancyFraction=75 -XX:+UseCMSInitiatingOccupancyOnly -XX:
-Dfile.encoding=UTF-8 -XX:+HeapDumpOnOutOfMemoryError -Xmx1g -Xms256m -Xss204
ash/vendor/jruby/lib/jni -Xbootclasspath/a:/usr/local/logstash/vendor/jruby/l
_121/lib:/usr/java/jdk1.8.0_121/jre/lib -Djruby.home=/usr/local/logstash/vend
dor/jruby/lib -Djruby.script=jruby -Djruby.shell=/bin/sh org.jruby.Main /usr/
logstash/runner.rb -f agent.conf
root      2954  1137 34 18:46 pts/1    00:00:02 /usr/java/jdk1.8.0_121/bin/ja
-XX:CMSInitiatingOccupancyFraction=75 -XX:+UseCMSInitiatingOccupancyOnly -XX:
-Dfile.encoding=UTF-8 -XX:+HeapDumpOnOutOfMemoryError -Xmx1g -Xms256m -Xss204
ash/vendor/jruby/lib/jni -Xbootclasspath/a:/usr/local/logstash/vendor/jruby/l
_121/lib:/usr/java/jdk1.8.0_121/jre/lib -Djruby.home=/usr/local/logstash/vend
dor/jruby/lib -Djruby.script=jruby -Djruby.shell=/bin/sh org.jruby.Main /usr/
logstash/runner.rb -f index.conf
root      2986  1185  0 18:46 pts/2    00:00:00 grep java
[root@localhost ~]#
```

图 11-22 ELK 收集 Kernel 日志

11.16　ELK 收集 Nginx 日志实战

切换到目录 usr/local/logstash/config/etc/，创建以下配置文件。

（1）日志采集：存入 Redis 缓存数据库。

创建 agent.conf 文件的内容如下。

```
input {
  file {
      type => "nginx-access"
      path => "/usr/local/nginx/logs/access.log"
  }
}
output {
    redis {
        host => "localhost"
        port => 6379
        data_type => "list"
        key => "logstash"
    }
}
```

启动 Agent，代码如下：

```
../bin/logstash -f agent.conf
```

（2）Redis 数据：存入 ElasticSearch。

创建 index.conf 文件的内容如下。

```
input {
  redis {
    host => "localhost"
    port => "6379"
    data_type => "list"
    key => "logstash"
    type => "redis-input"
    batch_count => 1
  }
}
output {
  elasticsearch {
    hosts => "192.168.111.128"
  }
}
```

启动 index 的代码如下。

```
../bin/logstash -f index.conf
```

查看启动进程，如图 11-23 所示。

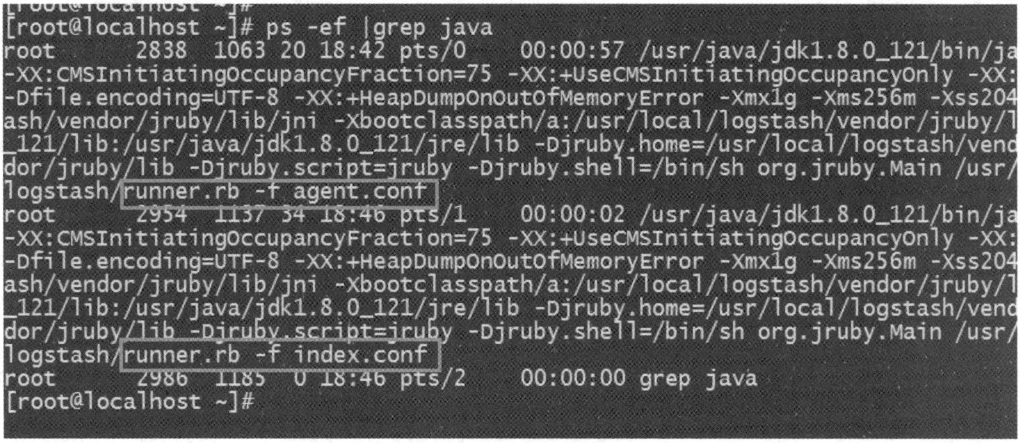

图 11-23 ELK 收集 Nginx 日志

浏览器访问 Kibana Web，地址为 http://192.168.149.129:5601，如图 11-24 所示。

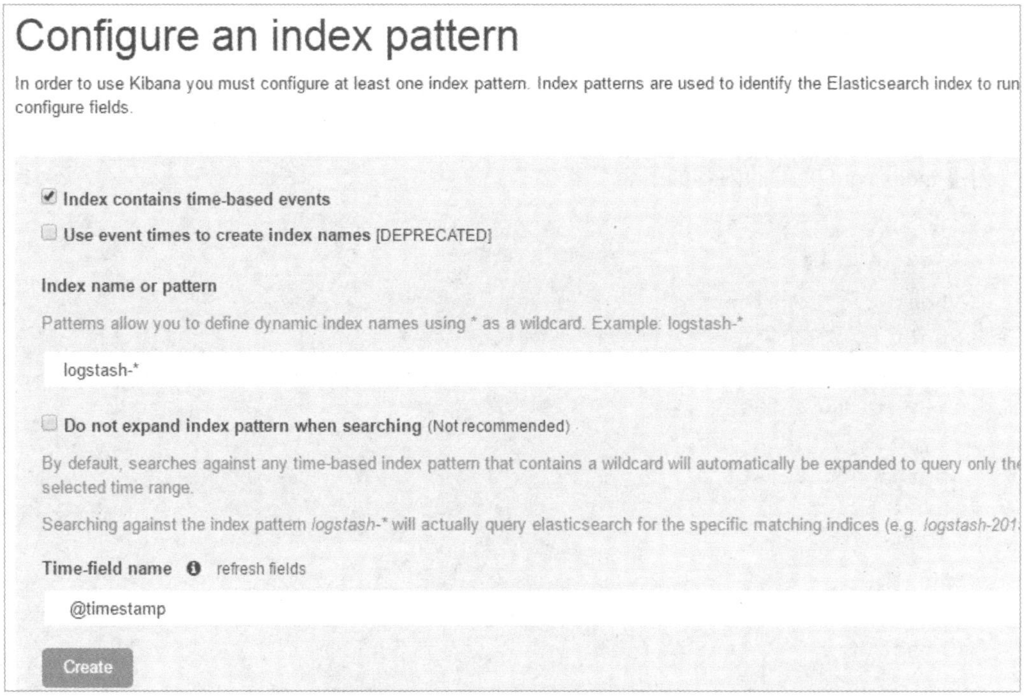

图 11-24 Kibana Web 界面

11.17　ELK 收集 Tomcat 日志实战

切换到目录 usr/local/logstash/config/etc/，创建以下配置文件。

（1）日志采集：存入 Redis 缓存数据库。

创建 agent.conf 文件的内容如下。

```
input {
  file {
    type => "nginx-access"
    path => "/usr/local/tomcat/logs/catalina.out"
  }
}
output {
  redis {
    host => "localhost"
    port => 6379
    data_type => "list"
    key => "logstash"
  }
}
```

启动 Agent 的代码如下。

```
../bin/logstash -f agent.conf
```

（2）Redis 数据：存入 ElasticSearch。

创建 index.conf 文件的内容如下。

```
input {
  redis {
    host => "localhost"
    port => "6379"
    data_type => "list"
    key => "logstash"
    type => "redis-input"
    batch_count => 1
  }
}
output {
  elasticsearch {
    hosts => "192.168.111.128"
  }
}
```

启动 index 的代码如下。

```
../bin/logstash -f index.conf
```

查看启动进程，如图 11-25 所示。

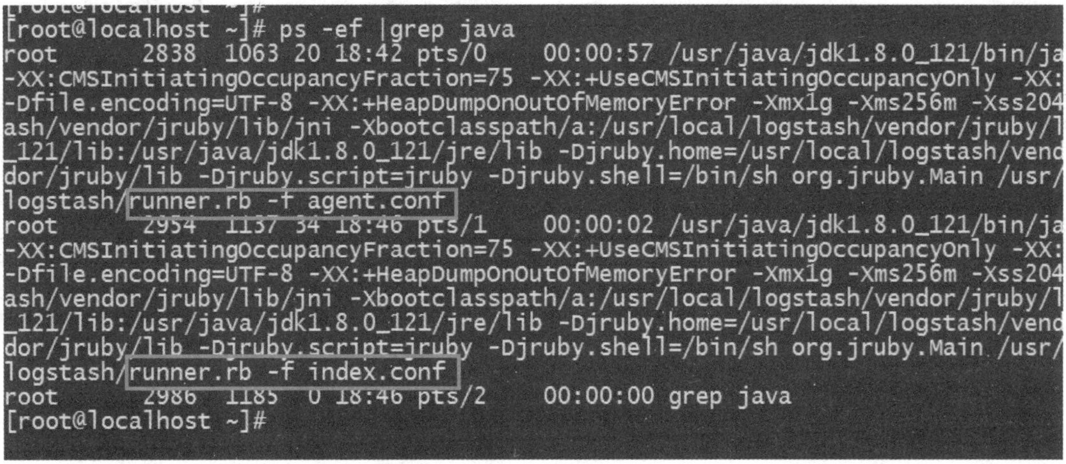

图 11-25　Logstash 客户端日志

11.18　ELK 批量日志集群实战

如上配置完毕，可以正常收集单台服务器的日志。如何批量收集其他服务器的日志信息呢？可以基于 Shell 脚本将配置完毕的 Logstash 文件夹同步至其他服务器，也可以通过 Ansible 和 Saltstack 服务器同步。例如，收集 Nginx 日志，index.conf 文件和 agent.conf 文件的内容保持不变，配置文件目录/usr/local/logstash/config/etc/，修改配置文件原 Redis 服务器的 localhost 为 Redis 服务器的 IP 地址。

（1）日志采集：存入 Redis 缓存数据库。

创建 agent.conf 文件的内容如下。

```
input {
  file {
    type => "nginx-access"
    path => "/usr/local/nginx/logs/access.log"
  }
}
output {
  redis {
    host => "localhost"
    port => 6379
    data_type => "list"
```

```
      key => "logstash"
  }
}
```

启动 Agent 的代码如下。

```
../bin/logstash -f agent.conf
```

（2）Redis 数据：存入 ElasticSearch。

创建 index.conf 文件的内容如下。

```
input {
  redis {
    host => "localhost"
    port => "6379"
    data_type => "list"
    key => "logstash"
    type => "redis-input"
    batch_count => 1
  }
}
output {
  elasticsearch {
    hosts => "192.168.111.128"
  }
}
```

启动 index 的代码如下。

```
../bin/logstash -f index.conf
```

查看启动进程，如图 11-26 所示。

```
[root@localhost ~]# ps -ef |grep java
root      2838  1063 20 18:42 pts/0    00:00:57 /usr/java/jdk1.8.0_121/bin/ja
-XX:CMSInitiatingOccupancyFraction=75 -XX:+UseCMSInitiatingOccupancyOnly -XX:
-Dfile.encoding=UTF-8 -XX:+HeapDumpOnOutOfMemoryError -Xmx1g -Xms256m -Xss204
ash/vendor/jruby/lib/jni -Xbootclasspath/a:/usr/local/logstash/vendor/jruby/l
_121/lib:/usr/java/jdk1.8.0_121/jre/lib -Djruby.home=/usr/local/logstash/vend
dor/jruby/lib -Djruby.script=jruby -Djruby.shell=/bin/sh org.jruby.Main /usr/
logstash/runner.rb -f agent.conf
root      2954  1137 34 18:46 pts/1    00:00:02 /usr/java/jdk1.8.0_121/bin/ja
-XX:CMSInitiatingOccupancyFraction=75 -XX:+UseCMSInitiatingOccupancyOnly -XX:
-Dfile.encoding=UTF-8 -XX:+HeapDumpOnOutOfMemoryError -Xmx1g -Xms256m -Xss204
ash/vendor/jruby/lib/jni -Xbootclasspath/a:/usr/local/logstash/vendor/jruby/l
_121/lib:/usr/java/jdk1.8.0_121/jre/lib -Djruby.home=/usr/local/logstash/vend
dor/jruby/lib -Djruby.script=jruby -Djruby.shell=/bin/sh org.jruby.Main /usr/
logstash/runner.rb -f index.conf
root      2986  1185  0 18:46 pts/2    00:00:00 grep java
[root@localhost ~]#
```

图 11-26　Logstash 服务进程

11.19 ELK 报表统计 IP 地域访问量

（1）配置 Nginx 日志格式（采用默认）。

```
log_format  main  '$remote_addr - $remote_user [$time_local] "$request" '
                  '$status $body_bytes_sent "$http_referer" '
                  '"$http_user_agent" "$http_x_forwarded_for"';
```

（2）客户端部署 IP 库工具。

```
cd /usr/local/logstash/config/
wget http://geolite.maxmind.com/download/geoip/database/GeoLite2-City.mmdb.gz
gunzip GeoLite2-City.mmdb.gz
```

（3）Logstash 客户端配置。

```
vim /usr/local/logstash/config/nginx.conf
input {
        file {
            path => "/usr/local/nginx/logs/access.log"
            type => "nginx"
            start_position => "beginning"
            }
}
filter {
        grok {
            match => { "message" => "%{IPORHOST:remote_addr} - - \[%{HTTPDATE:time_local}\] \"%{WORD:method} %{URIPATHPARAM:request} HTTP/%{NUMBER:httpversion}\" %{INT:status} %{INT:body_bytes_sent} %{QS:http_referer} %{QS:http_user_agent}"
            }
}
geoip {
    source => "remote_addr"
    target => "geoip"
    database => "/usr/local/logstash/config/GeoLite2-City.mmdb"
    add_field => ["[geoip][coordinates]","%{[geoip][longitude]}"]
    add_field => ["[geoip][coordinates]","%{[geoip][latitude]}"]
    }
}
output {
  elasticsearch {
        hosts => ["47.98.151.187:9200"]
        manage_template => true
        index => "logstash-%{+YYYY-MM}"
    }
}
```

(4) Logstash 配置文件详解如下。

```
geoip        #IP 地址定位查询插件
source       #需要通过 geoip 插件处理的 field,remote_addr
Target       #解析后的 geoip 地址数据,应该存放在哪一个字段中,默认是 geoip 这个字段
database     #指定下载的数据库文件
add_field    #添加经纬度,地图中地区显示根据经纬度识别
```

(5) Kibana 使用高德地图。

在文件 vim/usr/local/kibana/config/kibana.yml 末尾添加如下代码。

```
tilemap.url: 'http://webrd02.is.autonavi.com/appmaptile?lang=zh_cn&size=1&scale=1&style=7&x={x}&y={y}&z={z}'
```

重启 Kibana,刷新 Kibana 页面,可以看到 geoip 相关字段,如图 11-27 所示。

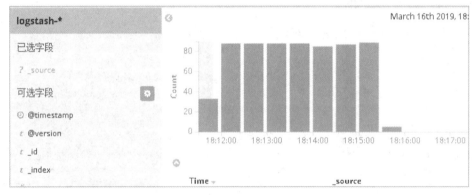

(a)

(b)

图 11-27　Kibana Web 界面

(a) 查看 logstash 索引列表;(b) 查看 logstash 索引 geoip 字段

(6) 刷新索引信息,如图 11-28 所示。

图 11-28　Kibana Web 界面刷新索引信息

（7）按图 11-29 选择 Tile map。

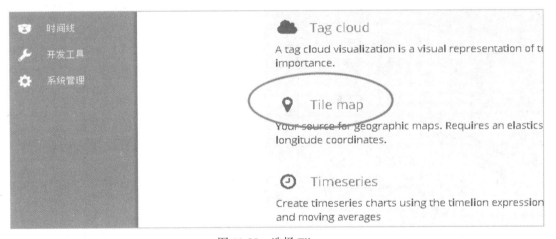

图 11-29　选择 Tile map

11.20　ELK 报表统计 Nginx 访问量

ELK 平台构建完成之后，基于 Kibana 平台可以实现日志的统一管理，包括收集、统计、查看、搜索、分析报表等。

（1）基于 ELK 日志平台可以分析 Nginx 日志，分析指标如下。

① Nginx 总访问量。

② Nginx 某个时间的访问量。

③ Nginx 来源客户的 IP 地址所在地。

④ 请求方法占比统计。

⑤ HTTP Referer 来源统计。

⑥ 客户端 user_agent 统计。

⑦ URL 慢响应时间统计。

（2）基于 ELK 平台统计 Nginx 总访问量，创建报表。

配置步骤：选择 Visualize 视图，然后创建视图，选择视图类型 Pie chart（饼图），如图 11-30 所示。

图 11-30　统计 Nginx 总访问量

（a）选择 Pie；（b）添加 Nginx 访问量

（3）基于 ELK 平台统计 Nginx 访问高峰期，创建报表。

配置步骤：选择 Visualize 视图，然后创建视图，选择视图类型 Vertical Bar chart（直方图），如图 11-31 所示。

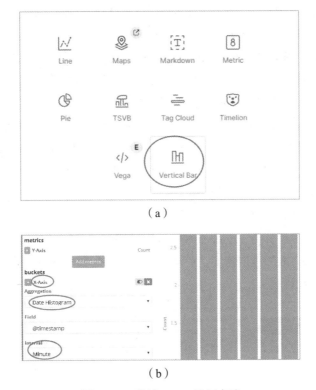

（a）

（b）

图 11-31　统计 Nginx 访问高峰

（a）选择 Vertical Bar chart；（b）添加字段

11.21　Filebeat 日志收集实战

Filebeat 是一款日志文件托运工具，基于 Go 语言开发，一般 Filebeat 安装在客户端上，监控日志目录或指定的日志文件，追踪读取这些文件的变化，并转发这些信息到 ElasticSearch 或者 Logstash 中存放。

Filebeat 工作原理：开启 Filebeat 程序的时候，它会启动一个或多个探测器（prospectors）检测指定的日志目录或文件，对于探测器找出的每一个日志文件，Filebeat 启动收割进程（harvester），每一个收割进程读取一个日志文件的新内容，并发送这些新的日志数据到处理程序（spooler），处理程序会集合这些事件，最后 Filebeat 会发送集合的数据到指定的地点。

可以将 Filebeat 简单地理解为一个轻量级的 Logstash，当需要收集信息的机器配置或资源并不是特别多时，使用 Filebeat 收集日志。日常使用中，Filebeat 十分稳定，而且占用内存、CPU 资源相对于 Logstash 更少。

Filebeat 可以通过提供一种轻量级的方式转发和集中日志和文件，帮助用户把事情简单化。

在任何环境中，应用程序总是时不时地停机。在读取和转发日志的过程中，如果被中断，Filebeat 会记录中断的位置。当重新联机时，Filebeat 会从中断的位置开始。

Filebeat 附带了内部模块（auditd、Apache、Nginx、System 和 MySQL），这些模块简化了普通日志格式的收集、解析和可视化。结合使用基于操作系统的自动默认设置，使用 ElasticSearch Ingest Node 的管道定义，以及 Kibana 仪表盘来实现这一点。

当发送数据到 Logstash 或 ElasticSearch 时，Filebeat 使用一个反压力敏感（backpressure-sensitive）的协议解释高负荷的数据量。当 Logstash 处理数据繁忙时，Filebeat 会放慢它的读取速度。一旦压力解除，Filebeat 将恢复到原来的速度，继续传输数据。

在任何环境下，应用程序都有停机的可能。Filebeat 读取并转发日志行，如果中断，则会记住所有事件恢复联机状态时的所在位置。

Filebeat 不会让管道超负荷。Filebeat 如果是向 Logstash 传输数据，当 Logstash 忙于处理数据时，会通知 Filebeat 放慢读取速度。一旦拥塞得到解决，Filebeat 将恢复到原来的速度并继续传播，如图 11-32 所示。

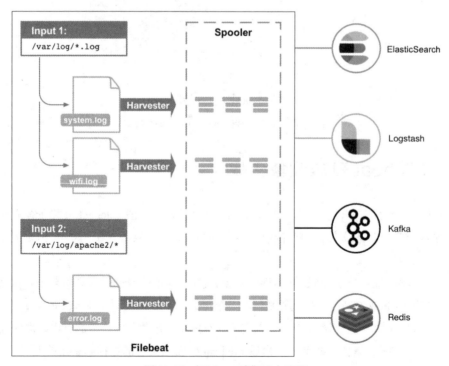

图 11-32　Filebeat 采集日志流程

11.22 Filebeat 案例实战

基于二进制 Tar 压缩包方式安装 Filebeat 程序的操作步骤如下。

```
#官网下载 Filebeat 软件包
https://artifacts.elastic.co/downloads/beats/filebeat/filebeat-7.2.0-linux-x86_64.tar.gz
#通过 Tar 工具对其解压缩
tar xzf filebeat-7.2.0-linux-x86_64.tar.gz
#将解压后的 Filebeat 程序部署到/usr/local/目录下
mv filebeat-7.2.0-linux-x86_64 /usr/local/filebeat/
#查看 Filebeat 是否部署成功
ls -l /usr/local/filebeat/
#切换至 Filebeat 程序目录
cd /usr/local/filebeat/
grep -aivE "#|^$" filebeat.yml
```

11.23 Filebeat 收集 Nginx 日志

（1）修改/usr/local/filebeat/filebeat.yml 配置文件，将 enable：false 改为 enable：true，同时将 paths 路径设置为 Nginx 日志路径，Filebeat 配置文件的代码如下。

```
filebeat.inputs:
- type: log
  enabled: true
  paths:
    - /var/log/nginx/access.log
filebeat.config.modules:
  path: ${path.config}/modules.d/*.yml
  reload.enabled: false
setup.template.settings:
  index.number_of_shards: 1
setup.kibana:
output.elasticsearch:
  hosts: ["192.168.111.128:9200"]
processors:
  - add_host_metadata: ~
  - add_cloud_metadata: ~
```

（2）默认 Filebeat 配置文件收集客户端自身/var/log/下的所有日志，可以根据自身的需求修改，同时将 ES 9200 的地址修改为自己的 ES IP，然后启动 Filebeat，采集日志。

（3）启动 Filebeat 服务的命令如下。

```
cd /usr/local/filebeat/
```

```
nohup ./filebeat -e -c filebeat.yml &
-c                          #配置文件位置
-path.logs                  #日志位置
-path.data                  #数据位置
-path.home                  #根目录位置
-e                          #关闭日志输出
-d                          #启用对指定选择器的调试
```

（4）查看 Filebeat 启动日志，如图 11-33 所示。

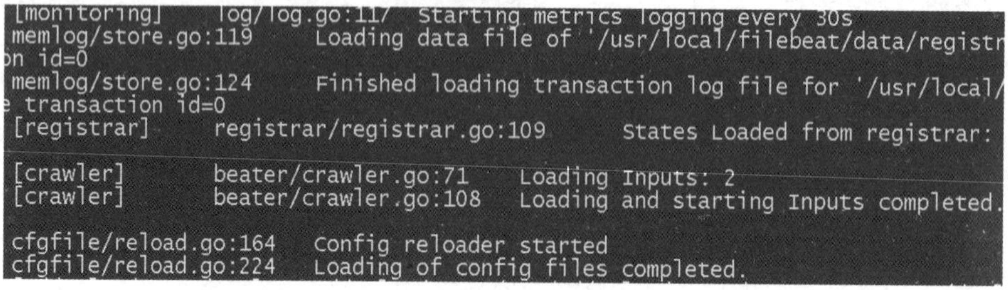

图 11-33　Filebeat 启动服务日志

（5）登录 Kibana，添加索引数据，如图 11-34 所示。

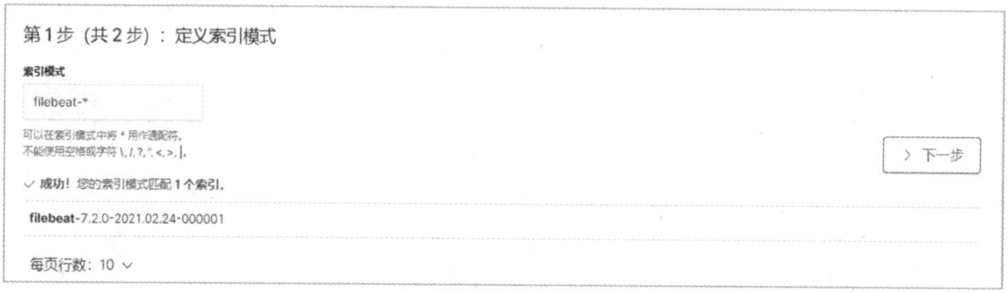

图 11-34　Filebeat 日志采集展示

（a）添加 Filebeat 索引；（b）查看 Filebeat 索引

（c）

图 11-34　Filebeat 日志采集展示（续）

（c）Filebeat 日志展示

11.24　Filebeat 自定义索引

在默认情况下，Filebeat 客户端采集的日志，在 ElasticSearch 存储中索引名称为 filebeat-*，可以根据需求修改索引的名称，例如，根据不同的日志类型命名。以 Nginx 为例，命名为 nginx-*。

如何将默认 Filebeat 索引名称修改为 Nginx 开头呢？

（1）修改 Filebeat 默认配置文件，添加以下代码。

```
#---------------------- ElasticSearch output ---------------------------
setup.ilm.enabled: false
output.elasticsearch:
  hosts: ["localhost:9200"]
  index: "nginx-%{+yyyy.MM.dd}"
setup.template.name: "nginx"
setup.template.pattern: "nginx-*"
```

（2）重启 Filebeat 服务。

```
cd /usr/local/filebeat/
nohup ./filebeat -e -c filebeat.yml &
```

（3）查看启动日志的代码如下，Nginx 索引已经生效，如图 11-35 所示。

```
tail -fn 30 nohup.out
```

图 11-35　Filebeat 自定义索引

(4)登录 Kibana Web 界面,创建索引,如图 11-36 所示。

(a)

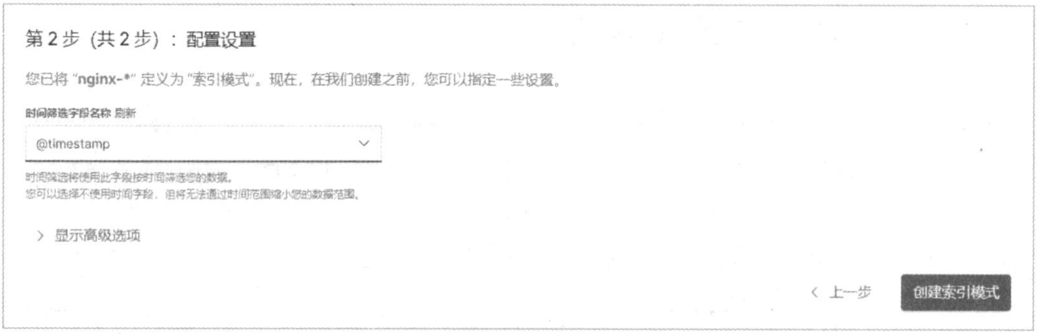

(b)

图 11-36　Filebeat 索引选择

(a)添加 nginx 索引;(b)选择索引字段名称

(5)创建索引时,Kibana 添加 index pattern 时出现问题,提示 403 Forbidden,解决方法如下。

```
#curl -XPUT -H 'Content-Type: application/json' http://localhost:9200/_settings -d '
{
    "index": {
        "blocks": {
            "read_only_allow_delete": "false"
        }
    }
}'
#curl -XPUT -H 'Content-Type: application/json' http://localhost:9200/$index_name/_settings -d '
{
    "index": {
        "blocks": {
```

```
            "read_only_allow_delete": "false"
        }
    }
}'
```

（6）根据步骤（5）方法解决完毕，如果再次报错如下。

```
ClusterBlockException[blocked by: [FORBIDDEN/12/index read-only / allow delete (api)];
```

则采用以下代码解决，最终如图 11-37 所示。

```
curl -XPUT -H "Content-Type: application/json" http://localhost:9200/_all/_settings -d '{"index.blocks.read_only_allow_delete": null}'
```

图 11-37　Filebeat 日志展示

11.25　Filebeat 收集多个日志

默认情况下，Filebeat 只收集/var/log/*.log 日志，如何让 Filebeat 既可以收集日志内核日志和 Nginx，又可以收集 MySQL 日志呢？

（1）修改 Filebeat 默认配置文件，添加多个 Type-log，其代码如下。

```
filebeat.inputs:
- type: log
  enabled: true
  paths:
    - /var/log/nginx/access.log
  tags: ["nginx-log"]
- type: log
  enabled: true
  paths:
    - /var/log/mysql/mysqld.log
  tags: ["mysql-log"]
filebeat.config.modules:
```

```
  path: ${path.config}/modules.d/*.yml
  reload.enabled: false
setup.template.settings:
  index.number_of_shards: 1
setup.kibana:
setup.ilm.enabled: false
output.elasticsearch:
  hosts: ["localhost:9200"]
  index: "nginx-%{+yyyy.MM.dd}"
setup.template.name: "nginx"
setup.template.pattern: "nginx-*"
processors:
  - add_host_metadata: ~
  - add_cloud_metadata: ~
```

(2）修改 Filebeat 默认配置文件，在一个 Type log 下添加多个日志路径，其代码如下。

```
filebeat.inputs:
- type: log
  enabled: true
  paths:
  - /var/log/nginx/access.log
  - /var/log/mysql/mysqld.log
  - /var/log/tomcat/catalina.out
filebeat.config.modules:
  path: ${path.config}/modules.d/*.yml
  reload.enabled: false
setup.template.settings:
  index.number_of_shards: 1
setup.kibana:
setup.ilm.enabled: false
output.elasticsearch:
  hosts: ["localhost:9200"]
  index: "nginx-%{+yyyy.MM.dd}"
setup.template.name: "nginx"
setup.template.pattern: "nginx-*"
processors:
  - add_host_metadata: ~
  - add_cloud_metadata: ~
```

(3）重启 Filebeat 服务。

```
cd /usr/local/filebeat/
nohup ./filebeat -e -c filebeat.yml &
```

(4）查看启动日志的代码如下。Filebeat 收集多个日志已经生效，如图 11-38 所示。

```
tail -fn 30 nohup.out
```

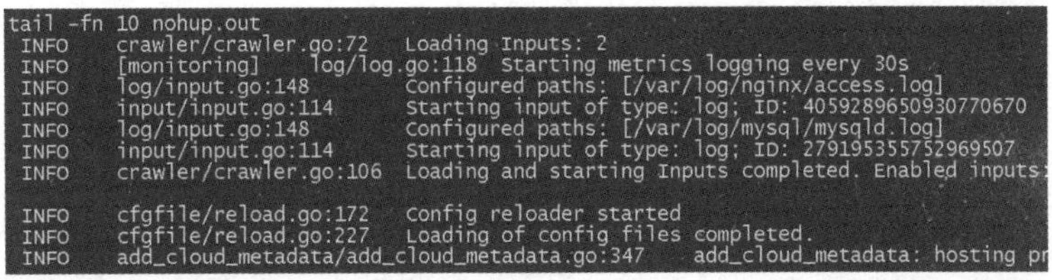

（a）

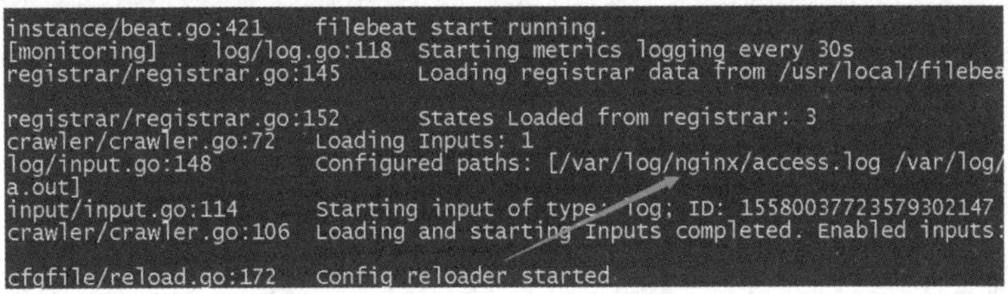

（b）

图 11-38　Filebeat 收集多日志生效

（a）查看 filebeat 日志；（b）查看 filebeat 启动信息

（5）登录 Kibana Web 界面，查看日志收集情况，如图 11-39 所示。

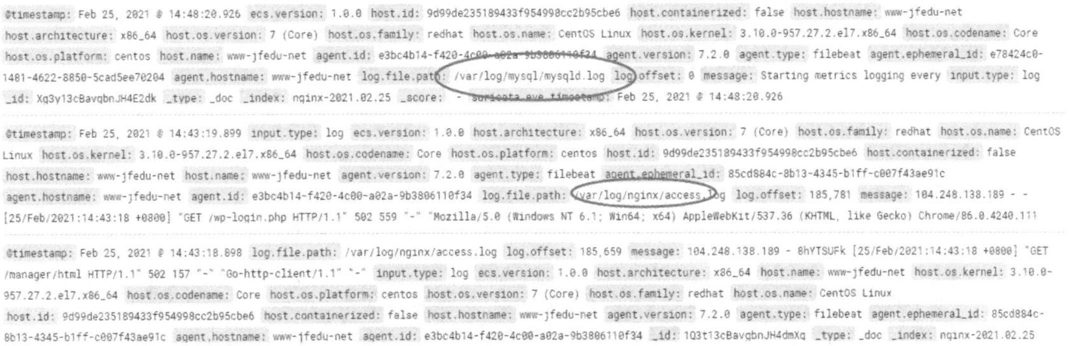

图 11-39　Filebeat 收集多日志

11.26　Kibana Web 安全认证

安装完 ElasticSearch 和 Kibana 的启动进程，如果没有开启 X-pack 插件，用户可以直接通过浏览器访问，这样不利于数据安全。接下来利用 Apache 的密码认证进行安全配置。通过访问 Nginx 转发至 ElasticSearch 和 Kibana 服务器，来安装 Nginx。

```
yum install pcre-devel pcre -y
wget -c http://nginx.org/download/nginx-1.16.0.tar.gz
tar -xzf nginx-1.16.0.tar.gz
useradd www ;./configure --user=www --group=www --prefix=/usr/local/nginx
```

```
--with-http_stub_status_module
--with-http_ssl_module
make
make install
```

修改 Nginx.conf 配置文件的代码如下。

```
worker_processes  1;
events {
    worker_connections  1024;
}
http {
    include       mime.types;
    default_type  application/octet-stream;
    sendfile        on;
    keepalive_timeout  65;
    upstream  jvm_web1 {
    server    127.0.0.1:5601 weight=1 max_fails=2 fail_timeout=30s;
}
    server {
       listen 80;
       server_name  localhost;
       location / {
        proxy_set_header  Host  $host;
        proxy_set_header  X-Real-IP $remote_addr;
        proxy_set_header X-Forwarded-For $proxy_add_x_forwarded_for;
        proxy_pass http://jvm_web1;
       }
    }
}
```

修改 Kibana 配置文件的监听 IP 地址为 127.0.0.1，如图 11-40 所示。

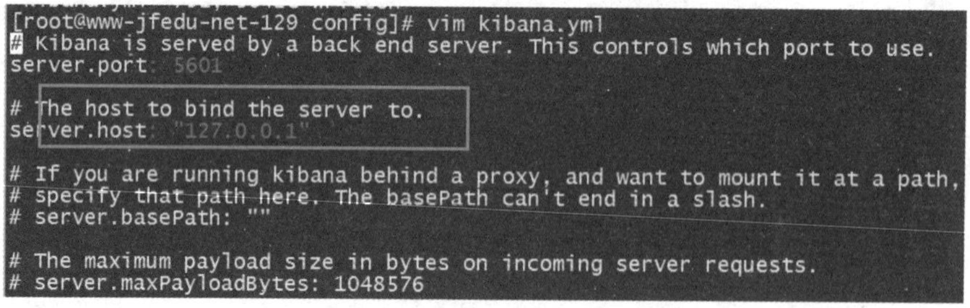

图 11-40 Kibana 日志展示

添加 Nginx 权限认证，在 Nginx.conf 配置文件 location /中加入以下代码。

```
auth_basic "ELK Kibana Monitor Center";
auth_basic_user_file /usr/local/nginx/html/.htpasswd;
```

通过 Apache 加密工具 htpasswd 生成用户名和密码，其代码如下，如图 11-41 所示。

```
htpasswd -c /usr/local/nginx/html/.htpasswd admin
```

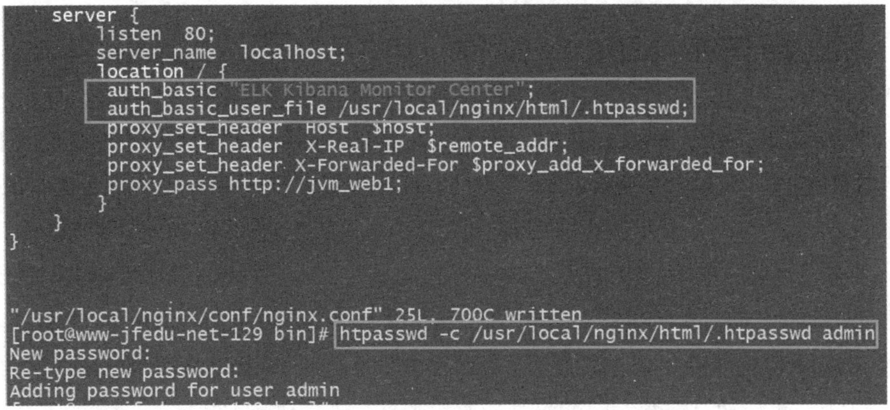

图 11-41 ELK Web 认证配置

重启 Nginx Web 服务并访问，如图 11-42 所示。

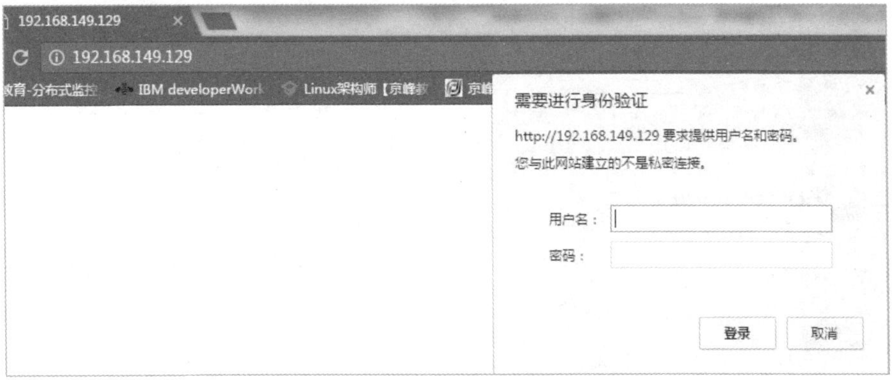

图 11-42 ELK Web 认证配置后需要进行身份验证

用户名和密码正确，即可成功登录，如图 11-43 所示。

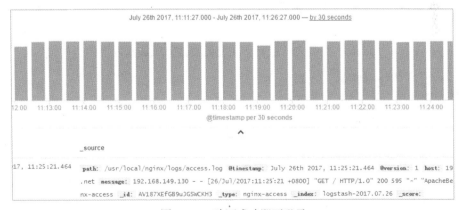

图 11-43 验证成功即可登录

11.27 ELK 增加 X-pack 插件

X-pack 是一个 Elastic Stack 的扩展，将安全、警报、监视、报告和图形功能包含在一个易于安装的软件包中。在 ElasticSearch 5.0.0 之前，必须安装单独的 Shield 插件、Watcher 插件和 Marvel 插件才能获得 X-pack 中所有的功能。

X-pack 监控组件让用户可以通过 Kibana 轻松地监控 ElasticSearch，还可以实时查看集群的健康和性能，以及分析过去的集群、索引和节点度量，同时还可以监视 Kibana 本身性能。

将 X-pack 安装在群集上，监控代理运行在每个节点上从 ElasticSearch 收集和指数指标；安装在 Kibana 上，可以通过一套专门的仪表板监控数据，如图 11-44 所示。

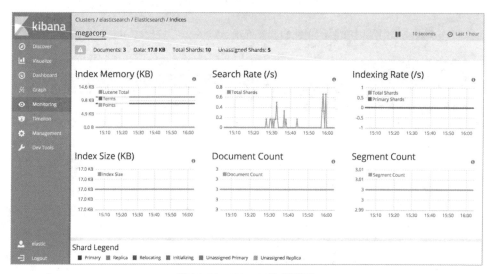

图 11-44　X-pack 监控界面

ELK 7.x 默认已经集成了 X-pack 插件，所以无须安装，直接启用即可。操作的步骤和方法如下。

（1）修改 ElasticSearch 主配置文件代码，在配置文件末尾加入以下两行代码。

```
xpack.security.enabled: true
xpack.security.transport.ssl.enabled: true
```

（2）重新启动 ElasticSearch 服务。

```
su - elk
/usr/local/elasticsearch/bin/elasticsearch -d
```

（3）给 ELK 集群设置密码。为了统一管理 ElasticSearch、Kibana 和 Logstash，密码统一使用 123456，根据提示一直按 Enter 键即可。安装代码如下。

```
cd /usr/local/elasticsearch/bin/
./elasticsearch-setup-passwords interactive
```

（4）如果忘记密码，也可以手动修改密码，其命令如下。

```
curl -H "Content-Type:application/json" -XPOST -u elastic 'http://localhost:
9200/_xpack/security/user/elastic/_password' -d '{ "password" : "123456" }'
```

（5）配置 Kibana X-pack，修改 Kibana 配置文件，其代码如下。

```
elasticsearch.username: "elastic"
elasticsearch.password: "123456"
```

（6）重启 Kibana 服务，并通过浏览器访问，如图 11-45 所示。

图 11-45　Kibana 访问界面